Logic in Asia: Studia Logica Library

Logic in Asia: Studia Logica Library

This book series promotes the advance of scientific research within the field of logic in Asian countries. It strengthens the collaboration between researchers based in Asia with researchers across the international scientific community and offers a platform for presenting the results of their collaborations. One of the most prominent features of contemporary logic is its interdisciplinary character, combining mathematics, philosophy, modern computer science, and even the cognitive and social sciences. The aim of this book series is to provide a forum for current logic research, reflecting this trend in the field's development.

The series accepts books on any topic concerning logic in the broadest sense, i.e., books on contemporary formal logic, its applications and its relations to other disciplines. It accepts monographs and thematically coherent volumes addressing important developments in logic and presenting significant contributions to logical research. In addition, research works on the history of logical ideas, especially on the traditions in China and India, are welcome contributions.

The scope of the book series includes but is not limited to the following:

- Monographs written by researchers in Asian countries.
- Proceedings of conferences held in Asia, or edited by Asian researchers.
- Anthologies edited by researchers in Asia.
- Research works by scholars from other regions of the world, which fit the goal of "Logic in Asia".

The series discourages the submission of manuscripts that contain reprints of previously published material and/or manuscripts that are less than 165 pages/ 90,000 words in length.

Please also visit our webpage: http://tsinghualogic.net/logic-in-asia/background/

Relation with Studia Logica Library

This series is part of the Studia Logica Library, and is also connected to the journal Studia Logica. This connection does not imply any dependence on the Editorial Office of Studia Logica in terms of editorial operations, though the series maintains cooperative ties to the journal.

This book series is also a sister series to Trends in Logic and Outstanding Contributions to Logic.

For inquiries and to submit proposals, authors can contact the editors-in-chief Fenrong Liu at fenrong@tsinghua.edu.cn or Hiroakira Ono at ono@jaist.ac.jp.

More information about this series at http://www.springer.com/series/13080

Syraya Chin-Mu Yang · Duen-Min Deng
Hanti Lin
Editors

Structural Analysis of Non-Classical Logics

The Proceedings of the Second Taiwan Philosophical Logic Colloquium

Springer

Editors
Syraya Chin-Mu Yang
Department of Philosophy
National Taiwan University
Taipei
Taiwan

Hanti Lin
Department of Philosophy
University of California
Davis, CA
USA

Duen-Min Deng
Department of Philosophy
National Taiwan University
Taipei
Taiwan

ISSN 2364-4613 ISSN 2364-4621 (electronic)
Logic in Asia: Studia Logica Library
ISBN 978-3-662-48356-5 ISBN 978-3-662-48357-2 (eBook)
DOI 10.1007/978-3-662-48357-2

Library of Congress Control Number: 2015948710

Springer Heidelberg New York Dordrecht London

Printed on acid-free paper

Springer-Verlag GmbH Berlin Heidelberg is part of Springer Science+Business Media
(www.springer.com)

To Wendy Huang

Preface

The flourishing of non-classical logics since the 1950s has had a tremendous impact on a wide scope of subjects not only in philosophy (including metaphysics, epistemology, ethics, and so on), but also in many related disciplines such as economics (including decision theory and game theory), cognitive science, computer science, and linguistics, to mention a few. Ever since then, a movement known as 'philosophical logic' has emerged, with a Russellian motto at its core: 'Logic is fundamental to philosophy'. On the other hand, a majority of philosophers believe that without philosophical import, logic is merely a collection of vacuous intelligence games. In the last few decades, more and more logicians and philosophers have devoted their research to a closer and stronger connection between logic and philosophy. In particular, more attention has been paid to the philosophical perspective of logic, and to the construction and application of logical frameworks for analyzing philosophical concepts and theorizing philosophical doctrines.

Following this tendency, many researchers in the Asian area have already been engaged in this movement. To promote mutual understanding and collaboration for future researchers in Asia on logic, a series of biennial conferences was established and held in Asian countries since 2012, known as the Asian Workshop on Philosophical Logic (AWPL).

Almost at the same time, we were awarded a funding from personal annual donation to establish a second series of biennial conferences, entitled the 'Taiwan Philosophical Logic Colloquium' (TPLC), based at the Department of Philosophy, National Taiwan University. The TPLC-series aims to provide a solid and accessible forum for dialogs amongst logic-minded philosophers and philosophically orientated logicians in the Asian and Australasian regions on a variety of significant issues from philosophical and/or logical perspectives. We hope that the establishment of TPLC and AWPL will promote the development of logic and analytic philosophy in the Asian area, especially philosophical logic.

The scope of the TPLC-series covers philosophical logic (in a broad sense), non-classical logics, algebraic logic, all kinds of semantics/logics relating to philosophical concepts (in metaphysics, epistemology, and philosophy of language), philosophy of logic/mathematics, and their applications in computer science and cognitive science. It is dedicated to promoting both theoretical and empirical studies of logic (typically non-classical logics), with a close connection to some related disciplines, drawing on diverse methods and approaches from philosophy, computer science, mathematics, psychology, and linguistics.

This volume collects papers from the participants of the Second Taiwan Philosophical Colloquium (TPLC-2014) held during October 24–25, 2014. Though the topics are diverse, a majority of papers share two noticeable features in common: (i) the fundamental setting falls within the category of non-classical logics—modal logic, epistemic logic, logic of public announcement, logic of games, logic of truth-making, dynamic logics of speech acts, etc.; (ii) almost every paper involves, one way or the other, models of some sorts—ultraproducts, (causal) structural models, Kripke models, models for channel theory, and so on.

The title 'Structural Analysis of Non-Classical Logics' was suggested by Robert Goldblatt. It indicates implicitly that all authors have been working on the construction of various types of structures for non-classical logic of some sort. In doing so they provide analysis for the construction of various models as required in the framework they are working on. With an emphasis on the philosophical perspective, it therefore shows a somewhat dynamic aspect of constructing appropriate models for some desired non-classical logics.

In the opening chapter 'Semantical Approach to Cut Elimination and Subformula Property in Modal Logic', Hiroakira Ono discusses semantical study of cut elimination and subformula property in modal logics. A unified exposition is given for model-theoretic approach to finite model property, subformula property and cut elimination. At the same time, an attempt is made to clarify connections between model-theoretic and algebraic approaches to cut elimination.

Robert Goldblatt's 'Ultraproducts of Admissible Models for Quantified Modal Logic' (Chap. 2) continues work on models for quantified modal logic which have a restriction on which sets of worlds are admissible as propositions. In his 2011 book 'Quantifiers, Propositions and Identity', he showed that the problem of incompleteness of some such logics under their Kripkean possible-worlds semantics could be overcome, by showing that for any propositional modal logic S there is a quantificational proof system QS that is complete for validity in models whose algebra of admissible propositions validates S. In the present article he constructs ultraproducts of admissible models and uses them to derive compactness theorems that then combine with completeness to yield strong completeness: any QS-consistent set of formulas is satisfiable in a model whose admissible propositions validate S. The Barcan Formula is analyzed separately and shown to axiomatize certain logics that are strongly complete over admissible models in which the quantifiers are given their Kripkean actualist interpretation.

In 'Logic and/of Truthmaking' (Chap. 3), Jamin Asay addresses some basic questions about how truthmaker theory relates to various concerns in the philosophy of logic. He first defends truthmaker theory from Timothy Williamson's attack on it, showing how Williamson's logic-driven objections to truthmaker theory are unsuccessful. Then he explores some issues in the logic of the truthmaking relation itself, arguing that theorists, when trying to understand the nature of the relation, have been attempting to reconcile what may be inconsistent desiderata.

Duen-Min Deng's chapter 'Structural Models for Williamson's Modal Epistemology' (Chap. 4) examines Williamson's (2007) counterfactual-based account of modal epistemology. Deng argues that Williamson's account faces two serious problems—the cotenability problem and the gap problem. As Deng diagnoses it, these problems somehow indicate that our standard way of understanding counterfactuals under the received possible-worlds semantics may have insufficient 'structures' to distinguish various constraints on our counterfactual thinking. The remedy, Deng suggests, is to invoke the 'structural semantics' as developed by Pearl (2009) and Halpern (2000). Based on this semantics, Deng offers some philosophical elucidation for various kinds of modality, and provides his own account of how our modal knowledge can be grounded in our knowledge of counterfactuals.

In 'Motivating the Causal Modeling Semantics of Counterfactuals, or, Why We Should Favor the Causal Modeling Semantics over the Possible-Worlds Semantics' (Chap. 5), Kok Yong Lee argues that, from the perspective of philosophical semantics, one should favor the causal modeling semantics of counterfactuals over the orthodox possible-worlds semantics. Lee offers two reasons for this thesis. First, the possible-worlds semantics suffers from a specific kind of counterexamples which the causal modeling semantics can handle with ease. Secondly, the causal modeling semantics, but not the possible-worlds one, has the theoretical resources enough for accounting for backtracking counterfactuals. Lee's own causal modeling semantics differs from the standard causal modeling semantics in that, while both accounts feature a kind of causal manipulation known as 'intervention', Lee's semantics also specifies a distinct causal manipulation that he calls 'extrapolation'.

Hanti Lin's paper, 'The Meaning of Epistemic Modality and the Absence of Truth' (Chap. 6), proposes a new approach to natural language semantics, with a focus on epistemic modals. Instead of evaluating sentences at possible worlds, the new approach evaluates sentence at possible information states; instead of evaluating sentences to be true or not, the new approach evaluates sentences to be acceptable or not.

In 'Revising a Labelled Sequent Calculus for Public Announcement Logic' (Chap. 7), Shoshin Nomura, Katsuhiko Sano, and Satoshi Tojo provide a cut-free labeled sequent calculus GPAL for Public Announcement Logic (PAL) based on Maffezioli and Negri's (2011) system G3PAL. The authors show that G3PAL lacks rules of accessibility relation in updated models so an axiom in Hilbert-style axiomatization of PAL cannot be derived. GPAL will be free of this deficiency. The soundness of GPAL with regard to Kripke semantics with certain specified constraints on possible worlds involved is proved, and a direct proof of the semantic completeness of GPAL for the link-cutting semantics of PAL is provided.

Joshua Sack's chapter 'Logics for Dynamic Epistemic Behavioral Strategies' (Chap. 8) is devoted to reasoning about epistemic behavioral strategies in extensive form games with incomplete or imperfect information with chance moves. Sack shows how the probabilistic logic of communication and change can capture not just behavioral strategies that depend on what players believe about the game structure, but also epistemic behavioral strategies that depend on beliefs players have of each other. An extension of this logic is also considered to compare one strategy with infinitely many alternatives and to express various game theoretic notions such as best response, Nash equilibrium, and rationality.

The ninth chapter 'Measurement-Theoretic Foundations of Observational-Predicate Logic' is devoted to an analysis of the Phenomenal Sorites Paradox. The Phenomenal Sorites Paradox is a version of the Sorites Paradox, where observational predicates occur. Satoru Suzuki proposes a new version of logic for observational predicates—Observational-Predicate Logic (OPL)—that makes it possible to reason about observational predicates without inviting the Phenomenal Sorites Paradox on perceptual indiscriminability in the statistical sense. To accomplish this aim, he provides the language of OPL with a statistical model in terms of measurement theory.

In 'Channel Theoretic Reflections on Dynamic Logics of Speech Acts' (Chap. 10), Tomoyuki Yamada examines how it is possible to capture the regularities that enable agents to perform illocutionary acts of commanding and the background conditions that support them in logical terms. For this purpose, Yamada models the relevant kind of regularities in the form of constraints of local logics introduced in Barwise and Seligman's channel theory by building information channels with the language and the models of 'dynamified' deontic logic he developed. In doing so, it is shown that the language of the dynamified deontic logic needs to be substantially extended in order to talk about the relation between acts of saying things and acts of commanding. The chapter concludes by hinting at how this can be done.

Sakiko Yamasaki and Katsuhiko Sano's chapter 'Constructive Embedding from Extensions of Logics of Strict Implication into Modal Logic' (Chap. 11) is concerned with a proof-theoretic approach to Gödel-Mckinsey-Tarski embedding, i.e., the embedding from intuitionistic logic to modal logic S4. Dyckhoff and Negri employed labeled sequent calculi to provide a constructive proof of Gödel-Mckinsey-Tarski embedding from intermediate logics to extensions of modal logic S4. The authors generalize Dyckhoff and Negri's result to *sub*-intuitionistic logics, i.e., extensions of logic of strict implication. For this purpose, the authors provide a cut-free, sound and complete labeled sequent calculus for Corsi's logic F of strict implication, and employ a variant of Gödel-Mckinsey-Tarski translation sending an atom P to $P\&\Box P$ to establish a constructive embedding result.

The final chapter 'Common Knowledge and the Knowledge Account of Assertion' is devoted to the assertion account of common knowledge, to be compared with the iteration account and fixed-point account. This chapter continues Syraya C.-M. Yang's recent work on models for epistemic logics, which justifies a majority of Williamson's theses in his knowledge-first epistemology. Yang extends the constructed models to a multi-agent system for epistemic logic of common knowledge with the knowledge account of assertion. Adhering to the communication-oriented notion of common knowledge—common knowledge rising from communication, he highlights the substantial role assertion plays in the acquisition and transition of knowledge in a group of agents, and proposes that the propositional content of a sentence *s* is common knowledge to a group of agents if and only if everyone knows that *s* holds and also that everyone knows that *s* is asserted. Details of the semantic rules and some fundamental semantic properties of common knowledge are studied in due course.

We owe thanks to the contributors, the anonymous referees of the manuscripts, all speakers, discussants, attendees, and the staff of the Department. In particular, we would like to express our gratitude to Chen Bo, Shi-Chung Chang, Jui-Lin Lee, Churn Jung Liau, Dan Marshall, Hsing-Chien Tsai, Yanjing Wang, Kai-Yee Wong, and Jiji Zhang for their contribution and assistance to TPLC-2014 and this volume. We are deeply indebted to Hiroakira Ono and Rob Goldblatt for their long-term support of the TPLC-series and the preparation of this volume. We are most grateful to Fenrong Liu and Hiroakira Ono, the editors-in-chiefs of the book series 'Logic in Asia' (LIAA) for their supportive recommendation of this volume to LIAA. Thanks also go to Leana Li, Team Leader of Editor Human Sciences & Mathematics, and Li Nina, Editorial Assistant in Springer, for their help. Finally and above all, we owe special thanks to Ms. Wendy Huang. Without her exclusively financial support for the TPLC-series, this collection could only be materialized in some merely, perhaps even inaccessible, possible worlds. This volume is thereby dedicated to her.

June 2015

Syraya Chin-Mu Yang
Duen-Min Deng
Hanti Lin

Contents

1 Semantical Approach to Cut Elimination and Subformula Property in Modal Logic 1
Hiroakira Ono

2 Ultraproducts of Admissible Models for Quantified Modal Logic 17
Robert Goldblatt

3 Logic and/of Truthmaking 37
Jamin Asay

4 Structural Models for Williamson's Modal Epistemology 57
Duen-Min Deng

5 Motivating the Causal Modeling Semantics of Counterfactuals, or, Why We Should Favor the Causal Modeling Semantics over the Possible-Worlds Semantics 83
Kok Yong Lee

6 The Meaning of Epistemic Modality and the Absence of Truth 111
Hanti Lin

7 Revising a Labelled Sequent Calculus for Public Announcement Logic 131
Shoshin Nomura, Katsuhiko Sano and Satoshi Tojo

8 Logics for Dynamic Epistemic Behavioral Strategies 159
Joshua Sack

9 Measurement-Theoretic Foundations of Observational-Predicate Logic 183
Satoru Suzuki

10 Channel Theoretic Reflections on Dynamic Logics of Speech Acts 201
Tomoyuki Yamada

11 Constructive Embedding from Extensions of Logics of Strict Implication into Modal Logics 223
Sakiko Yamasaki and Katsuhiko Sano

12 Common Knowledge and the Knowledge Account of Assertion 253
Syraya Chin-Mu Yang

Contributors

Jamin Asay The University of Hong Kong, Pokfulam, Hong Kong

Duen-Min Deng National Taiwan University, Taipei, Taiwan

Robert Goldblatt Victoria University of Wellington, Wellington, New Zealand

Kok Yong Lee Department of Philosophy, National Chung Cheng University, Min-hsiung, Taiwan

Hanti Lin Department of Philosophy, University of California, Davis, CA, USA

Shoshin Nomura School of Information Science, Japan Advanced Institute of Science and Technology, Nomi, Japan

Hiroakira Ono Japan Advanced Institute of Science and Technology, Nomi, Japan

Joshua Sack Department of Mathematics and Statistics, California State University Long Beach, Long Beach, USA

Katsuhiko Sano School of Information Science, Japan Advanced Institute of Science and Technology, Nomi, Japan

Satoru Suzuki Faculty of Arts and Sciences, Komazawa University, Setagaya-ku, Tokyo, Japan

Satoshi Tojo School of Information Science, Japan Advanced Institute of Science and Technology, Nomi, Japan

Tomoyuki Yamada Hokkaido University, Sapporo, Hokkaido, Japan

Sakiko Yamasaki Graduate School of Humanities, Tokyo Metropolitan University, Tokyo, Japan

Syraya Chin-Mu Yang National Taiwan University, Taipei, Taiwan

Chapter 1
Semantical Approach to Cut Elimination and Subformula Property in Modal Logic

Hiroakira Ono

Abstract This is a short survey of semantical study of cut elimination and subformula property in modal logics. Cut elimination is a basic proof-theoretic notion in sequent systems, and subformula property is the most important consequence of cut elimination. A special feature of our presentation is its unified semantical approach to them based on Kripke models. Along the same lines as Takano's works on subformula property, these properties, together with finite model property, will be discussed as modifications of standard construction of canonical Kripke models. These semantical approaches will be compared with algebraic approaches in modal logics, which often take the forms of various kinds of embedding theorems. In the last part of the paper, an attempt is made to clarify connections between semantical approach to cut elimination and algebraic one.

Keywords Cut elimination · Subformula property · Finite model property · Modal logics · Embedding theorems

1.1 Introduction

This is an exposition of semantical approach to cut elimination and subformula property, together with finite model property, in modal logics. The main aim of the present paper is to develop a unified semantical approach to them based on Kripke models, along the same lines as Takano's works [18–20]. We will touch also on algebraic approaches in modal logics, which often take the forms of various kinds of embedding theorems, in order to clarify connections between these two approaches. In the following, to denote the semantical approach based on Kripke models, we use the word *model-theoretic approach* in order to avoid confusions.

After describing standard construction of canonical models in Sect. 1.3, it is shown that the similar construction, but restricted to finite sets of formulas, will work well sometimes for showing the finite model property. The idea was stated first by Schütte

---

H. Ono (✉)
Japan Advanced Institute of Science and Technology, Nomi, Japan
e-mail: ono@jaist.ac.jp

S.C.-M. Yang et al. (eds.), *Structural Analysis of Non-Classical Logics*,
Logic in Asia: Studia Logica Library, DOI 10.1007/978-3-662-48357-2_1

[17] in showing the finite model property of intuitionistic logic. Then, the *finite embeddability property* of varieties of modal algebras will be discussed in connection with Schütte's method.

Sections 1.4 and 1.5 will be devoted mostly to Takano's results on *analytic cut property* and *subformula property* in [18, 20], and also on cut elimination in [19]. In fact, it is shown that subformula property and cut elimination can be proved along the same line as those in Sect. 1.3. Here, the analytic cut property of a given sequent system **GL** for a logic **L** says that if a sequent S is provable in **GL** then S has a proof in **GL** such that for each application of the cut rule in this proof the cut formula is a subformula of a formula in the lower sequent of the cut rule. Sometimes, cut elimination fails but still analytic cut property holds, for instance a standard sequent system for the modal logic **S5**. In most cases, the analytic cut property implies subformula property of **GL**, where subformula property of a system **GL** says that if a sequent S is provable in **GL** then S has a proof $\mathscr{P}$ in **GL** such that every formula in $\mathscr{P}$ is a subformula of a formula in S.

In addition to these model-theoretic approaches, certain developments have been made in algebraic approach to cut elimination (see [2, 9, 12]). The study have been recently developed further in [6]. On the other hand, most of algebraic works until now are concerned mainly with substructural logics, though techniques can be applied also to modal logics as pointed out in [2]). In the last section, some attempts are made to clarify connections of model-theoretic approaches to cut elimination in the present paper with these algebraic approaches.

The author would like to express a special thank to M. Takano for his approval for referring to his unpublished note [19] and for his helpful comments. He would like to express also many thanks to T. Kowalski for inspiring discussions and valuable comments on the initial draft of the present paper, and to C.-M. Yang for his constant encouragement.

1.2 Sequent Systems for Some Modal Logics

To make our discussions concrete, we will consider several sequent systems for basic modal logics, though results shown in the rest of the paper hold for a wider class of modal logics. For the non-modal part, we may take any standard sequent system for classical logic. For the simplicity's sake, we assume that each sequent is of the form $\Sigma \Rightarrow \Theta$, where both Σ and Θ are finite (possibly empty) sets of formulas. Thus, each system has neither exchange rules nor contraction rules. We follow usual convention. For instance, the set $\Gamma \cup \{\alpha, \beta\}$ will be expressed as Γ, α, β. We consider here the following four rules for the modality $\Box$.

$$\frac{\Gamma \Rightarrow \alpha}{\Box\Gamma \Rightarrow \Box\alpha}\ (\Box) \qquad\qquad \frac{\alpha, \Gamma \Rightarrow \Delta}{\Box\alpha, \Gamma \Rightarrow \Delta}\ (\Box\Rightarrow)$$

$$\frac{\Box\Gamma \Rightarrow \alpha}{\Box\Gamma \Rightarrow \Box\alpha}\ (\Rightarrow\Box 1) \qquad\qquad \frac{\Box\Gamma \Rightarrow \Box\Delta, \alpha}{\Box\Gamma \Rightarrow \Box\Delta, \Box\alpha}\ (\Rightarrow\Box 2)$$

As usual, $\Box\Gamma$ denotes the set of formulas $\{\Box\alpha_1, \ldots, \Box\alpha_m\}$ when Γ is a set of formulas $\{\alpha_1, \ldots, \alpha_m\}$. Also, $\Diamond\alpha$ is an abbreviation of $\neg\Box\neg\alpha$. Basic sequent systems **GK**, **GKT**, **GS4** and **GS5** for **K**, **KT**, **S4** and **S5** are given as follows.

GK: **LK** + ($\Box$),

GKT: **LK** + ($\Box$) + ($\Box \Rightarrow$),

GS4: **LK** + ($\Box \Rightarrow$) + ($\Rightarrow \Box$1),

GS5: **LK** + ($\Box \Rightarrow$) + ($\Rightarrow \Box$2).

Cut elimination is one of the most important property in sequent systems. Cut elimination in a sequent system **GL** means:

If a sequent $\Gamma \Rightarrow \Delta$ is provable in GL *it is provable in* GL *without using cut rule.*

Cut elimination implies the following subformula property:

If a sequent $\Gamma \Rightarrow \Delta$ is provable in GL *then there exists a proof $\mathscr{P}$ of $\Gamma \Rightarrow \Delta$ such that every formula appearing in $\mathscr{P}$ is a subformula of a formula either in Γ or in Δ. In fact, every cut-free proof satisfies this subformula property.*

From subformula property, many useful logical properties follow. See, e.g., [14]. For instance,

1. decidability, and often tractable proof search algorithms,
2. Maksimova's variable separation property,
3. Craig's interpolation property.

Gentzen gave a syntactic proof of cut elimination for **LK** by using double induction. For modal systems, the following is obtained by [7, 11]

Theorem 1 *Cut elimination holds for* **GK**, **GKT** *and* **GS4**.

On the other hand, cut elimination does not hold in **GS5** (see [11]). In fact, $p \Rightarrow \Box\neg\Box\neg p$, which is an instance of the axiom (B), is provable in **GS5**, but cannot be provable in **GS5** without using cut rule. Here is a proof of $p \Rightarrow \Box\neg\Box\neg p$ with cut.

$$\dfrac{\dfrac{\dfrac{\Box\neg p \Rightarrow \Box\neg p}{\Rightarrow \neg\Box\neg p, \Box\neg p}}{\Rightarrow \Box\neg\Box\neg p, \Box\neg p} \quad \dfrac{\dfrac{p \Rightarrow p}{\neg p, p \Rightarrow}}{\Box\neg p, p \Rightarrow}}{p \Rightarrow \Box\neg\Box\neg p}\ (cut)$$

Many attempts have been made to introduce a cut-free sequent system for **S5**. All such systems must be essentially different from GS5, and therefore, lack its intuitiveness and simplicity of formulation. Notice, however that the cut formula $\Box\neg p$ is a subformula of a formula in $p \Rightarrow \Box\neg\Box\neg p$, and hence, this proof satisfies the subformula property. This suggests that subformula property may hold for GS5 despite the lack of cut elimination. Indeed it is so, as we will see shortly

1.3 Kripke Completeness, Finite Model Property and Finite Embeddability Property

We give here a quick overview of Kripke completeness and finite model property. We assume standard notions and basic results on Kripke frames and models for modal logics. Thus, a Kripke frame $\mathscr{F}$ is a pair $\langle W, R\rangle$ of a nonempty set W and a binary relation R, and a valuation V on $\mathscr{F}$ is a function which associates with each propositional variable p, a subset of W. Then, each valuation can be extended to all formulas in a usual way. A pair consisting of a Kripke frame and a valuation on it is called a Kripke model. The *truth* of a formula α at a world x in a Kripke model $\langle \mathscr{F}, V\rangle$ can be defined inductively. A formula α is *valid* in a Kripke frame $\mathscr{F}$ iff it is true at every world in the Kripke model $\langle \mathscr{F}, V\rangle$ for every valuation V on $\mathscr{F}$. For more information, see [3, 5].

A modal logic **L** is complete with respect to a class $\mathscr{C}$ of Kripke frames, when for any formula α, if α is valid in all Kripke frames in $\mathscr{C}$ then it is provable in **L**. A standard way of showing completeness of **L** is obtained by using the canonical frame for **L**. To fix basic notions and notations in our paper, we will give an outline of such a proof for the modal logic **S4**, taken as an example. In the following, Ω denotes the set of all modal formulas.

- A pair (Σ, Θ) of subsets Σ and Θ of Ω is **S4**-*consistent* (in Ω) if for all $\alpha_1, \ldots, \alpha_m \in \Sigma$ and $\beta_1, \ldots, \beta_n \in \Theta$, the sequent $\alpha_1, \ldots, \alpha_m \Rightarrow \beta_1, \ldots, \beta_n$ is not provable in **GS4**.
- A pair (Σ, Θ) of subsets Σ and Θ of Ω is *maximal* **S4**-*consistent* (in Ω), if it is **S4**-consistent but neither $(\Sigma \cup \{\gamma\}, \Theta)$ nor $(\Sigma, \Theta \cup \{\gamma\})$ is **S4**-consistent for any $\gamma \in \Omega\backslash(\Sigma \cup \Theta)$,

We have the following lemma with the help of cut rule.

Lemma 1 (extension lemma) *If* (Σ, Θ) *is* **S4**-*consistent in* Ω, *then for any formula* γ *in* Ω *either* $(\Sigma \cup \{\gamma\}, \Theta)$ *or* $(\Sigma, \Theta \cup \{\gamma\})$ *is* **S4**-*consistent in* Ω.

We enumerate all formulas. Then, we take each formula one-by-one in this enumeration and put it either side of a given consistent pair while keeping its consistency. The above lemma ensures that this is possible. Eventually, we will get a maximal **S4**-consistent pair. Clearly, if (Σ, Θ) is maximal **S4**-consistent then Θ must be equal to $\Omega\backslash\Sigma$. In the following, we simply say that Σ is a *maximal* **S4**-*consistent set* (in Ω), when $(\Sigma, \Omega\backslash\Sigma)$ is a maximal **S4**-consistent pair.

Lemma 2 (Lindenbaum's lemma) *For every* **S4**-*consistent pair* (Σ, Θ), *there exists a maximal* **S4**-*consistent set* Σ^* *in* Ω *such that* $\Sigma \subseteq \Sigma^*$ *and* $\Theta \subseteq (\Omega\backslash\Sigma^*)$.

The canonical frame for **S4** is the structure $\mathscr{F}_{S4} = \langle W_{S4}, R_{S4} \rangle$, where

- W_{S4} is the set of all maximal **S4**-consistent sets in Ω,
- R_{S4} is a binary relation over W_{S4} such that the relation $\Pi R_{S4} \Lambda$ holds iff $\Pi_\Box \subseteq \Lambda$, for every $\Pi, \Lambda \in W_{S4}$, where $\Pi_\Box = \{\beta; \Box\beta \in \Pi\}$.

Similarly, we can introduce the canonical frame for other modal logics. It is easy to see that the condition $\Pi_\Box \subseteq \Lambda$ is equivalent to $\Lambda \subseteq \Pi_\Diamond$, where $\Pi_\Diamond = \{\beta; \Diamond\beta \in \Pi\}$. We can show also that the condition $\Pi_\Box \subseteq \Lambda$ is equivalent to $\Pi_\Box \subseteq \Lambda_\Box$ for **S4**, and is equivalent to $\Pi_\Box = \Lambda_\Box$ for **S5**. The *canonical valuation* V_{S4} is defined by $V_{S4}(p) = \{\Pi \in W_{S4}; p \in \Pi\}$ for each propositional variable p. The pair $\mathscr{M}_{S4}$ of $\mathscr{F}_{S4}$ and V_{S4} is called the *canonical model* of **S4**. We can show that

Lemma 3 (truth lemma)
(1) The canonical frame $\mathscr{F}_{S4}$ for **S4** *is in fact a Kripke frame for* **S4**.
(2) $V_{S4}(\alpha) = \{\Pi \in W_{S4}; \alpha \in \Pi\}$ for every formula α, i.e. $\mathscr{M}_{S4}, \Pi \models \alpha$ iff $\alpha \in \Pi$.

Suppose that a given sequent $\alpha_1, \ldots, \alpha_m \Rightarrow \beta_1, \ldots, \beta_n$ is not provable in **GS4**. Then $(\{\alpha_1, \ldots, \alpha_m\}, \{\beta_1, \ldots, \beta_n\})$ is **S4**-consistent and hence it can be extended to a maximal **S4**-consistent pair (Σ, Θ). Under the canonical valuation V_{S4} of the canonical frame, $\mathscr{M}_{S4}, \Sigma \models \alpha_i$ for each i and $\mathscr{M}_{S4}, \Sigma \not\models \beta_j$ for all j. Hence the above sequent is not true in the canonical model.

Theorem 2 (Kripke completeness) *If a sequent $\Gamma \Rightarrow \Delta$ is not provable in* **GS4** *it is false in the canonical model $\mathscr{M}_{S4}$ for* **S4**.

A normal modal logic is said to be *canonical* if the canonical frame $\mathscr{F}_L$ for **L** is a Kripke frame in which all formulas in **L** are valid. By the same argument as the above, the following well-known result can be obtained.

Theorem 3 *Every canonical modal logic is Kripke complete.*

A standard way of proving the finite model property is to use the filtration method combined with Kripke completeness. But, the finite model property can be shown in a way similar to the above proof of Kripke completeness using canonical frames, but by *localizing* it to a finite set of formulas. The idea was introduced first by K. Schütte in [17] and was applied to modal logics by M. Sato [15]. (See also [13] for an application to an intuitionistic modal logic.) We will explain below how it goes, by taking **S4** again as an example.

Suppose that a sequent $\Gamma \Rightarrow \Delta$ is not provable in **GS4**. Our goal is to find a finite Kripke frame for **S4** in which $\Gamma \Rightarrow \Delta$ is false. Let Ω_F be the set $Sub(\Gamma \cup \Delta)$ of all subformulas of formulas in $\Gamma \cup \Delta$, which is obviously finite. We say that a pair (Σ, Θ) is **S4**-consistent in Ω_F whenever it is **S4**-consistent in Ω and Σ and Θ are subsets of Ω_F. Also it is maximal **S4**-consistent in Ω_F, if it is **S4**-consistent in Ω_F but neither $(\Sigma \cup \{\gamma\}, \Theta)$ nor $(\Sigma, \Theta \cup \{\gamma\})$ is **S4**-consistent in Ω_F for any $\gamma \in \Omega_F \backslash (\Sigma \cup \Theta)$, Similarly as before, we have the following lemmas.

Lemma 4 (extension lemma) *If* (Σ, Θ) *is* **S4**-*consistent in* Ω_F *then for any formula* γ *in* Ω_F *either* $(\Sigma \cup \{\gamma\}, \Theta)$ *or* $(\Sigma, \Theta \cup \{\gamma\})$ *is* **S4**-*consistent.*

Lemma 5 (Lindenbaum's lemma restricted to Ω_F) *For every* **S4**-*consistent pair* (Σ, Θ) *in* Ω_F, *there exists a maximal* **S4**-*consistent pair* (Σ^*, Θ^*) *in* Ω_F *such that* $\Sigma \subseteq \Sigma^*$ *and* $\Theta \subseteq \Theta^*$.

Note that every maximal **S4**-consistent pair (Σ^*, Θ^*) in Ω_F consists of finite sets Σ^* and Θ^* of formulas, where $\Theta^* = \Omega_F \backslash \Sigma^*$. Similarly as before, we say that Σ^* is a maximal **S4**-consistent set when $(\Sigma^*, \Omega_F \backslash \Sigma^*)$ is a maximal **S4**-consistent pair. Now, for a given sequent $\Gamma \Rightarrow \Delta$ which is not provable in **GS4**, define a structure $\langle W^f, R^f \rangle$ as follows;

- W^f is the set of all maximal **S4**-consistent sets Π in Ω_F.
- For every $\Pi, \Lambda \in W^f$, the relation $\Pi R^f \Lambda$ holds iff $\Pi_\Box \subseteq \Lambda_\Box$.

The valuation V^f is defined by $V^f(p) = \{\Pi \in W^f; p \in \Pi\}$ for every propositional variable $p \in \Omega_F$. We can show the following.

Lemma 6 (truth lemma restricted to Ω_F)
(1) The structure $\langle W^f, R^f \rangle$ *is a finite Kripke frame for* **S4**,
(2) $V^f(\alpha) = \{\Pi \in W^f; \alpha \in \Pi\}$ *for every formula* $\alpha \in \Omega_F$.

Let Π be any maximal **S4**-consistent set such that $\Gamma \subseteq \Pi$ and $\Delta \subseteq (\Omega_F \backslash \Pi)$. Then in the present Kripke model, $\Pi \models \alpha$ for every $\alpha \in \Gamma$ and $\Pi \not\models \beta$ for every $\beta \in \Delta$. Thus, we have the following.

Theorem 4 (finite model property) *If a sequent* $\Gamma \Rightarrow \Delta$ *is not provable in* **GS4** *it is false in a finite Kripke frame for* **S4**.

The same method will work for, e.g., **K**, **KT** and **S5**. On the other hand, some modification of the definition of R^f is necessary, since we can deal only with formulas in $Sub(\Gamma \cup \Delta)$. While $\Pi R^f \Lambda$ can be defined by $\Pi_\Box \subseteq \Lambda$ as before for both **K** and **KT**, it must be defined by $\Pi_\Box = \Lambda_\Box$ for **S5**.

We consider algebraic aspect of these results. Let **L** be a normal modal logic, and $\mathcal{V}_{\mathbf{L}}$ be the class of all modal algebras in which all formulas in **L** are valid. Then, the class $\mathcal{V}_{\mathbf{L}}$ forms a variety. The following can be shown. (For more information, see, e.g., [3].)

Theorem 5 (canonical embedding)
(1) Each modal algebra **A** *can be embedded into its* canonical embedding algebra *(Jónsson-Tarski),*
(2) a modal logic **L** *is canonical iff the corresponding variety* $\mathcal{V}_{\mathbf{L}}$ *is closed under canonical embedding algebras.*

Theorem 5 is an algebraic counterpart of proving completeness by canonical frames, in the sense that whenever **L** can be proved complete by this method, then the corresponding variety is canonical. Next, we consider what an algebraic counterpart of the above Schütte's method, i.e., a *local form* of canonical frames, will be. A class $\mathcal{K}$ of modal algebras has the *finite embeddability property* when for any given finite *partial subalgebra* **B** of an algebra **A** in $\mathcal{K}$, there exists a finite algebra **D** in $\mathcal{K}$ in which **B** can be embedded. Here, we say that a subset B of A is a partial subalgebra, if $f^{\mathbf{A}}(b_1, \ldots, b_m) = c$ for $b_1, \ldots, b_m, c \in B$ then $f^{\mathbf{B}}(b_1, \ldots, b_m) = c$. See [9] for the details. In [1], S. Amano showed that the finite embeddability of the variety $\mathcal{V}_{\mathbf{L}}$ holds whenever Schütte's method mentioned above works well in showing the finite model property of a modal logic **L**. In fact in such a case, we can get a required finite algebra **D**, by mimicking the construction of the finite Kripke model but using algebraic terms and then by taking its dual algebra. In this way, we have the following for instance:

The variety $\mathcal{V}_{\mathbf{L}}$ *of* **L**-*modal algebras has the finite embeddability property, where* **L** *is anyone of* **K**, **KT**, **S4** *and* **S5**.

Clearly, when the variety $\mathcal{V}_{\mathbf{L}}$ is *locally finite*, its finite embeddability property is an obvious corollary. We remark also that it is known that for every normal modal logic **L**, the variety $\mathcal{V}_{\mathbf{L}}$ has the finite embeddability property iff **L** has the (strong) finite model property.

To conclude this section, we point out papers [4, 13] in which the finite model property of some *intuitionistic modal logics* was obtained by using the finite embeddability property of some varieties of modal Heyting algebras.

1.4 Takano's Approach to Subformula Property

As we mentioned before, the formula $p \Rightarrow \Box\neg\Box\neg p$ does not have any proof in **GS5** without using cut rule, while it has a proof in which a cut formula is restricted to a subformula of a formula in the lower sequent. Hence, the above formula has a proof satisfying the subformula property.

Since the non-modal fragment of the logics we consider is classical, without loss of generality we will always assume that every rule except cut and the rules for modality has the subformula property, that is, every formula in an upper sequent will appear as a subformula of a formula in the lower sequent. An application of a rule R which is either the cut rule or a rule for modality is *acceptable* if every formula in an upper sequent will appear as a subformula of a formula in the lower sequent in this application. Sometimes, an acceptable application of the cut rule in a given proof is said to be *analytic*. For a given sequent system **GL**, if every sequent $\Gamma \Rightarrow \Delta$ which is provable in **GL** has a proof $\mathcal{P}$ in which every application of the cut rule and rules for modality is acceptable, then all formulas in $\mathcal{P}$ are subformulas of a formula in Γ or Δ. In such a case, it is said that **GL** has *subformula property*. When cut elimination holds for **GL**, quite often it has subformula property. (But this is not always the case. In his personal communication to the author, Takano gave an example of a cut-free sequent system for **S4** without subformula property.) If every sequent $\Gamma \Rightarrow \Delta$ which is provable in **GL** has a proof $\mathscr{P}$ in which every application of the cut rule is acceptable (i.e., analytic), then **GL** is said to have *analytic cut property*. When all rules for modality are acceptable as well, then analytic cut property implies subformula property. The decidability of **GL** follows often from subformula property.

Subformula property of modal logics have been studied extensively by M. Takano in his papers [18–20], from both proof-theoretic and semantical approaches. In the following we will give a semantical proof of subformula property **GS5** due to M. Takano [19, 20]. As you will see, the proof goes quite similarly to the proof of finite model property given in the previous section. But, one should note that the proof here depends on the choice of a given sequent system, though the choice of a sequent system for **S4** in the previous section is irrelevant to its proof.

We take an arbitrary sequent $\Gamma \Rightarrow \Delta$ which is not provable in **GS5**. Again, let Ω_F be the set $Sub(\Gamma \cup \Delta)$ of all subformulas of formulas in $\Gamma \cup \Delta$. For all finite subsets Ψ and Π of Ω_F, we say that a sequent $\Psi \Rightarrow \Pi$ is **GS5**$[\Omega_F]$-provable if it has a proof $\mathscr{P}$ such that every formula appearing in $\mathscr{P}$ belongs to Ω_F. Otherwise, we say that $\Psi \Rightarrow \Pi$ is **GS5**$[\Omega_F]$*-consistent*.

Notice that the difference between **S5**-consistency localized to Ω_F and **GS5**$[\Omega_F]$-consistency is that in the former we allow all **S5** proofs, while in the latter we allow only *some* **S5** proofs: these that do not exceed the resources of Ω_F.

Now, to show our theorem, by taking the contraposition, we assume that the sequent $\Gamma \Rightarrow \Delta$ does not have any proof with the subformula property, that is, it is **GS5**$[\Omega_F]$-consistent. Our goal is to show that $\Gamma \Rightarrow \Delta$ is false in a Kripke frame for **S5** (and hence is not provable in **GS5**.) Similarly as before, we can show the following.

Lemma 7 (analytic extension lemma) *If* (Σ, Θ) *is* **GS5**$[\Omega_F]$*-consistent, then for any formula* γ *in* Ω_F *either* $(\Sigma \cup \{\gamma\}, \Theta)$ *or* $(\Sigma, \Theta \cup \{\gamma\})$ *is* **GS5**$[\Omega_F]$*-consistent.*

Proof Suppose that neither $(\Sigma, \Theta \cup \{\gamma\})$ nor $(\Sigma \cup \{\gamma\}, \Theta)$ is **GS5**$[\Omega_F]$-consistent. Then both sequents $\Sigma \Rightarrow \Theta, \gamma$ and $\gamma, \Sigma \Rightarrow \Theta$ and are **GS5**$[\Omega_F]$-provable. Since γ belongs to Ω_F, we can apply the cut rule to them. Hence $\Sigma \Rightarrow \Theta$ is **GS5**$[\Omega_F]$-provable, i.e. (Σ, Θ) is not **GS5**$[\Omega_F]$-consistent. By taking the contraposition, we have our lemma.

Lemma 8 (Lindenbaum's lemma relative to Ω_F-consistency) *For every* **GS5**$[\Omega_F]$*-consistent pair* (Σ, Θ)*, there exists a maximal* **GS5**$[\Omega_F]$*-consistent pair* (Σ^+, Θ^+) *such that* $\Sigma \subseteq \Sigma^+$ *and* $\Theta \subseteq \Theta^+$.

As before, we can easily see that if (Σ^+, Θ^+) is maximal **GS5**$[\Omega_F]$-consistent then $\Theta^+ = \Omega_F \backslash \Sigma^+$, and hence we can call Σ^+ a maximal **GS5**$[\Omega_F]$-consistent set. For a given sequent $\Gamma \Rightarrow \Delta$ which is **GS5**$[\Omega_F]$-consistent, we define a structure $\langle W^a, R^a \rangle$ as follows.

- W^a is the set of all maximal **GS5**[Ω_F]-consistent sets Σ.
- for every $\Sigma, \Lambda \in W^a$, the relation $\Sigma R^a \Lambda$ holds iff $\Sigma_\Box = \Lambda_\Box$.

The valuation V^a is defined by $V^a(p) = \{\Sigma \in W^a; p \in \Sigma\}$, for every propositional variable $p \in \Omega_F$. We can show the following.

Lemma 9 (truth lemma restricted to Ω_F)
(1) The structure $\langle W^a, R^a \rangle$ is a finite Kripke frame for **S5**,
(2) $V^a(\alpha) = \{\Sigma \in W^a; \alpha \in \Sigma\}$ for every formula $\alpha \in \Omega_F$.

Proof Item (2) can be proved by the induction. We will give a proof of it when α (in Ω_F) is of the form $\Box\beta$. Our goal is to show that $\Box\beta \in \Sigma$ iff $\Sigma R^a \Lambda$ implies $\beta \in \Lambda$ for all $\Lambda \in W^a$.

To show the only-if part, we assume that $\Box\beta \in \Sigma$. If $\Sigma R^a \Lambda$ then $\Box\beta \in \Lambda$. Since $\Box\beta \Rightarrow \beta$ is **GS5**[Ω_F]-provable, we have $\beta \in \Lambda$. Conversely, suppose that $\Box\beta \notin \Sigma$. Let $\Theta = \Omega_F \backslash \Sigma$. Since $\Box(\Sigma_\Box) \subseteq \Sigma$ and $\Box(\Theta_\Box) \subseteq \Theta$, the sequent $\Box(\Sigma_\Box) \Rightarrow \Box(\Theta_\Box), \Box\beta$ is not **GS5**[Ω_F]-provable because of **GS5**[Ω_F]-consistency of $(\Sigma, \Omega_F \backslash \Sigma)$. Due to the rule ($\Rightarrow \Box$2), neither the sequent $\Box(\Sigma_\Box) \Rightarrow \Box(\Theta_\Box), \beta$ is **GS5**[Ω_F]-provable. By Lemma 8, there exists a maximal **GS5**[Ω_F]-consistent set Λ such that $\Box(\Sigma_\Box) \subseteq \Lambda$ and $\Box(\Theta_\Box) \cup \{\beta\} \subseteq (\Omega_F \backslash \Lambda)$. Clearly, $\Sigma R^a \Lambda$ and $\beta \notin \Lambda$ holds.

Theorem 6 (subformula property) *If a sequent $\Gamma \Rightarrow \Delta$ is provable in* **GS5**, *there exists a proof $\mathscr{P}$ of $\Gamma \Rightarrow \Delta$ such that every formula appearing in $\mathscr{P}$ is a subformula of a formula either in Γ or in Δ.*

In fact, it is shown in [18], the following result is proved by using proof-theoretic method.[1]

Theorem 7 (analytic cut property) *If a sequent $\Gamma \Rightarrow \Delta$ is provable in* **GS5**, *there exists a proof of $\Gamma \Rightarrow \Delta$ in* **GS5** *in which every application of cut rule is analytic.*

It should be noticed that Takano [20] succeeded to extend the method by taking a bigger but still finite set, say Ω^+, which includes Ω_F. Then, by extending the notion of acceptability in an obvious way to Ω^+ he was able to prove that sequent systems for modal logic **K5** and **K5D** have Ω^+*-subformula property*. Decidability of these two logics is an immediate consequence of this result. It would be worthwhile and promising to pursue further considerations of "extended subformula property".

1.5 Cut Elimination

Along the same line, a semantical proof of cut elimination of the sequent system **GS4** is shown in this section. The idea is due to M. Takano [19]. Let **GS4**$^-$ be the system

[1] Quite recently, we proved in our joint work with T. Kowalski that subformula property implies analytic cut property in a certain general setting. Thus, Theorem 7 follows from Theorem 6.

GS4 without the cut rule. Let $\Gamma \Rightarrow \Delta$ be an arbitrary sequent which is not provable in **GS4**$^-$. Again, let Ω_F be the set $Sub(\Gamma \cup \Delta)$ of all subformulas of formulas in $\Gamma \cup \Delta$. Our goal is to show that $\Gamma \Rightarrow \Delta$ is false in a Kripke frame for **S4**.

Note first that since the system **GS4**$^-$ lacks the cut rule, the extension lemma for **GS4**$^-$ no longer holds. Hence, although Lindenbaum's lemma (restricted to Ω_F) still holds, the union of $\Sigma \cup \Theta$ is not always equal to Ω_F for a maximal **GS4**$^-$-consistent pair (Σ, Θ) in Ω_F. The existence of maximal **GS4**$^-$-consistent pairs are assured because Ω_F is finite.

Lemma 10 (Lindenbaum's lemma restricted to Ω_F) *For every pair (Σ, Θ) which is* **GS4**$^-$*-consistent in Ω_F, there exists a maximal* **GS4**$^-$*-consistent pair (Σ^*, Θ^*) in Ω_F such that $\Sigma \subseteq \Sigma^*$ and $\Theta \subseteq \Theta^*$.*

We define a structure $\langle W^c, R^c\rangle$ as follows.

- W^c is the set of all maximal **GS4**$^-$-consistent pairs (Σ, Θ) in Ω_F.
- For every $(\Sigma, \Theta), (\Lambda, \Pi) \in W^c$, the relation $(\Sigma, \Theta)R^c(\Lambda, \Pi)$ holds iff $\Sigma_\Box \subseteq \Lambda_\Box$.

The valuation V^c is defined by $V^c(p) = \{(\Sigma, \Theta) \in W^c; p \in \Sigma\}$, for every propositional variable $p \in \Omega_F$. We will show the following. (Here, $(\Sigma, \Theta) \models \delta$ is an abbreviation of $\mathscr{M}^c, (\Sigma, \Theta) \models \delta$, where the model $\mathscr{M}^c$ denotes the pair of $\langle W^c, R^c\rangle$ and V^c.)

Lemma 11 (partial truth lemma)
(1) The structure $\langle W^c, R^c\rangle$ is a finite Kripke frame for **S4**.
(2) For each formula $\alpha \in \Omega_F$ and each $(\Sigma, \Theta) \in W^c$,

- *if $\alpha \in \Sigma$ then $(\Sigma, \Theta) \models \alpha$,*
- *if $\alpha \in \Theta$ then $(\Sigma, \Theta) \not\models \alpha$.*

Proof Note that since the union of $\Sigma \cup \Theta$ is not always equal to Ω_F, the above (2) says that the truth lemma holds partially for **GS4**$^-$ (cf. Lemma 3 (2) for **S4**). Item (2) can be obtained by showing the following conditions I, II, and III for *downward saturation*, using induction. (For the simplicity's sake, $\vee$ is regarded here as a defined logical connective.)

I. The case where α (in Ω_F) is of the form $\beta \wedge \gamma$. It suffices to show that

(a) if $\beta \wedge \gamma \in \Sigma$ then both β and γ are in Σ,
(b) if $\beta \wedge \gamma \in \Theta$ then either β or γ are in Θ.

(a) It is easy to see that $(\{\beta, \gamma\} \cup \Sigma, \Theta)$ is **GS4**$^-$-consistent. Then, by the maximality of (Σ, Θ), both β and γ must belong to Σ.
(b) Suppose that $\beta \wedge \gamma \in \Theta$. If neither of $(\Sigma, \Theta \cup \{\beta\})$ and $(\Sigma, \Theta \cup \{\gamma\})$ is **GS4**$^-$-consistent, then both $\Sigma \Rightarrow \Theta, \beta$ and $\Sigma \Rightarrow \Theta, \gamma$ are **GS4**$^-$-provable. Thus, $\Sigma \Rightarrow \Theta, \beta \wedge \gamma$ is **GS4**$^-$-provable. But this leads to the conclusion that $\Sigma \Rightarrow \Theta$ is **GS4**$^-$-provable by using our assumption, which is contradictory. Thus, at least one of them must be **GS4**$^-$-consistent. By the maximality of (Σ, Θ), either β or γ belongs to Θ.

II. The case where α (in Ω_F) is of the form $\neg\beta$. It suffices to show that

(a) if $\neg\beta \in \Sigma$ then β is in Θ,
(b) if $\neg\beta \in \Theta$ then β is in Σ.

(a) Clearly, $(\Sigma, \Theta \cup \{\beta\})$ is **GS4**$^-$-consistent. Thus, β belongs to Θ.
(b) Similarly to (a).

III. The case where α (in Ω_F) is of the form $\Box\beta$. It suffices to show that

(a) if $\Box\beta \in \Sigma$ then $\beta \in \Lambda$ for each (Λ, Π) such that $(\Sigma, \Theta)R^c(\Lambda, \Pi)$,
(b) if $\Box\beta \in \Theta$ then $\beta \in \Pi$ for some (Λ, Π) such that $(\Sigma, \Theta)R^c(\Lambda, \Pi)$.

(a) Suppose that $(\Sigma, \Theta)R^c(\Lambda, \Pi)$, which means that $\Sigma_\Box \subseteq \Lambda_\Box$. Thus, $\Box\beta \in \Lambda$. Clearly, $(\{\beta\} \cup \Lambda, \Pi)$ is **GS4**$^-$-consistent (see the rule $(\Box \Rightarrow)$). Therefore, $\beta \in \Lambda$ by the maximality of (Λ, Π).
(b) Suppose that $\Box\beta \in \Theta$. Obviously $(\Sigma, \{\Box\beta\})$ is **GS4**$^-$-consistent, and hence so is $(\Box(\Sigma_\Box), \{\beta\})$ (see the rule $(\Rightarrow \Box 1)$). (Note that $\Box(\Sigma_\Box) \subseteq \Sigma$.) Thus, there exists a maximal **GS4**$^-$-consistent pair (Λ, Π) such that $\Box(\Sigma_\Box) \subseteq \Lambda$ and $\beta \in \Pi$. From the former, $\Sigma_\Box \subseteq \Lambda_\Box$ follows. Thus $\beta \in \Pi$ for $(\Sigma, \Theta)R^c(\Lambda, \Pi)$ follows.

Take any member (Σ, Θ) of W^c such that $\Gamma \subseteq \Sigma$ and $\Delta \subseteq \Theta$. Then by the above lemma, $(\Sigma, \Theta) \models \alpha$ holds for each formula $\alpha \in \Gamma$, and $(\Sigma, \Theta) \not\models \beta$ holds for each $\beta \in \Delta$. Therefore, $\Gamma \Rightarrow \Delta$ is false in this model. By taking the contraposition, we have the following.

Theorem 8 (cut elimination property) *If a sequent $\Gamma \Rightarrow \Delta$ is provable in* **GS4**, *there exists a proof of $\Gamma \Rightarrow \Delta$ in* **GS4** *without any application of cut rule.*

Similarly, cut elimination for **GK** and **GKT** can be shown. But, why does not the same method work well for **GS5**? To see this, let us consider (b) in the case III, but for **GS5**$^-$, i.e., **GS5** without cut rule. From the assumption that $\Box\beta \in \Theta$, we can infer also in this case that $(\Box(\Sigma_\Box), \Box(\Theta_\Box) \cup \{\beta\})$ is **GS5**$^-$-consistent in Ω_F (by the rule $(\Rightarrow \Box 2)$). So, there exists a maximal **GS5**$^-$-consistent pair (Λ, Π) in Ω_F such that $\Box(\Sigma_\Box) \subseteq \Lambda$ and $\Box(\Theta_\Box) \cup \{\beta\} \subseteq \Pi$. So far so good. But we cannot infer $\Sigma_\Box = \Lambda_\Box$ from this. This follows in fact whenever $\Sigma \cup \Theta = \Omega_F$ holds. Hence at this point, the argument for **GS5**$^-$ will break up.

As a matter of fact, the present proof of cut elimination is of its local form, since the notion of maximal consistency in Ω_F, instead of Ω, is used. In other words, what we have shown here is, precisely speaking, cut elimination property of the following stronger form.

Theorem 9 *If a sequent $\Gamma \Rightarrow \Delta$ is provable in* **GS4**, *there exists a proof of $\Gamma \Rightarrow \Delta$ in* **GS4** *with the* subformula property *which contains no applications of cut rule.*

Actually, the *global form* can be shown simply by replacing Ω_F by Ω in the above. In such a case, the existence of maximal **GS4**$^-$-consistent pairs mentioned in Lemma 10 can be ascertained by using a similar argument to the proof of Lemma 2 based on a given enumeration of all formulas. Of course, Θ may not always be

$\Omega \backslash \Sigma$ for a maximal **GS4**$^-$-consistent pair (Σ, Θ). It will be interesting also to compare our argument with discussions on *partial valuations* in their connection to cut elimination, in, e.g., Schütte [16] and Takeuti [21].

As we have seen, essential ingredients of Takano's method, are extension lemma and Lindenbaum's lemma, by which one can infer the required (partial) truth lemma. Though it looks different on the surface, the method has a close relation with Fitting's work based on *consistency property* in [8], as a consistency property is intended to describe conditions satisfied by the set of all maximal consistent pairs.

1.6 Quasi-embeddings and Downward Saturation

There has been a certain development of algebraic proofs of cut elimination in recent years, in particular for substructural logics (see, e.g., [2, 6, 9]). This algebraic method works well also for modal logics, and hence, for instance, the cut elimination for **GS4** can be derived algebraically (see [2]). In this section, we will present our attempt to clarify connections between the semantical proofs in the previous section and algebraic ones. Because of the lack of space, we cannot give the details of the proof. We assume a certain familiarity with terminologies and results in [2], in which an algebraic proof of cut elimination for some sequent systems for modal logics is outlined.

Suppose that **L** is a modal logic and $\mathcal{V}_{\mathbf{L}}$ is the corresponding variety, i.e., the variety of all **L**-modal algebras. Let **GL** be a given sequent system for **L**. Obviously, the cut elimination for **GL** is obtained if we can show that the sequent system **GL** *without the cut rule* is complete with respect to all algebras in $\mathcal{V}_{\mathbf{L}}$. An algebraic structure for **GL** *without the cut rule* is introduced and is called a *Gentzen structure* (or, a Gentzen matrix in [9]). Like standard proof of algebraic completeness for a given logic **L** using Lindenbaum algebras, we can show the following.

Lemma 12 *A sequent $\Sigma \Rightarrow \Theta$ is provable in* **GL** without the cut rule *iff it is valid in all Gentzen structures for* **GL** *without the cut rule.*

In fact, to show this lemma, the *absolutely free* Gentzen structure $\mathbf{B}_{\mathbf{GL}}$ for **GL** without the cut rule plays just the same role as Lindenbaum algebras. The underlying set of $\mathbf{B}_{\mathbf{GL}}$ is the set Ω of all formulas and its basic binary relation $\preceq$ on finite subsets of Ω is defined as follows.

- $\Sigma \preceq \Theta$ holds in $\mathbf{B}_{\mathbf{GL}}$ iff the sequent $\Sigma \Rightarrow \Theta$ is provable in **GL** *without the cut rule*, for all finite subsets Σ and Θ of Ω.

Now we will focus our attention only to the cut elimination for the sequent system **GS4**, as an example. As we mentioned above, to show the cut elimination it suffices to prove the completeness of **GS4**$^-$, i.e., **GS4** without the cut rule, with respect to **S4**-modal algebras. In fact, we can show the following basic theorem, although we omit the precise definition of *quasi-embeddings* (see [2] for further details).

Theorem 10 (quasi-embedding) *Every Gentzen structure* **B** *for* $\mathbf{GS4}^-$ *can be quasi-embedded into a complete modal algebra, called the* quasi-completion *of* **B**, *in* $\mathcal{V}_{\mathbf{S4}}$.

Since the nonvalidity of a given sequent is preserved under each quasi-embedding, the completeness of $\mathbf{GS4}^-$ with respect to **S4**-modal algebras follows from this theorem with Lemma 12. It should be noticed here that once we add the cut rule, each Gentzen structure for **GS4** (with the cut rule) will be an **S4**-modal algebra, and the quasi-embedding will be an embedding between modal algebras in the usual sense.

Before making a comparison of two approaches, we note that the algebraic proof outlined here is of the cut elimination in the global form while the proof in the previous section is of the local one (see Theorem 9). So, to make a precise comparison, it would be more suitable to take the algebraic proof of finite model property (see Sect. 7 of [2]), which is actually the local version of the algebraic proof of the cut elimination using a *finite* Gentzen structure. But because of the lack of space, we cannot discuss the problem here in details either.

Another point which we must keep in mind is that although *provability in* $\mathbf{GS4}^-$ (or, $\mathbf{GS4}^-$-consistency, in its negative form) is the basic notion in both approaches, the argument in the previous section is concerned mostly with *maximal* $\mathbf{GS4}^-$-*consistency*. We note here that for a given maximal $\mathbf{GS4}^-$-consistent pair (Σ, Θ) in Ω_F and for any formula α in Ω_F,

- $\alpha \notin \Sigma$ iff $\Sigma \cup \{\alpha\} \preceq \Theta$,
- $\alpha \notin \Theta$ iff $\Sigma \preceq \Theta \cup \{\alpha\}$.

Now we claim the following.

> *The existence of the quasi-embedding from the absolutely free Gentzen structure* $\mathbf{B}_{\mathbf{GS4}}$ *for* $\mathbf{GS4}^-$ *entails the downward saturation for maximal* $\mathbf{GS4}^-$*-consistent pairs.*

Here we give a brief explanation of this. For all finite subsets Φ, Π, Σ and Θ of Ω_F, $[\Sigma; \Theta]$ is the set of all pairs (Φ, Π) such that $\Sigma, \Phi \preceq \Pi, \Theta$ holds. We use ε for the empty set. For a formula $\alpha \in \Omega_F$, define a mapping k by $k(\alpha) = [\varepsilon; \{\alpha\}]$. If k is to be the quasi-embedding from $\mathbf{B}_{\mathbf{GS4}}$, it must satisfy the following condition for $\wedge$. (For other logical connectives, we omit conditions on k for the brevity's sake.)

- If $(\{\alpha\}, \varepsilon)$ is in $[\Sigma; \Theta]$ then $(\{\alpha \wedge \beta\}, \varepsilon)$ is in $[\Sigma; \Theta]$, and also if $(\{\beta\}, \varepsilon)$ is in $[\Sigma; \Theta]$ then $(\{\alpha \wedge \beta\}, \varepsilon)$ is in $[\Sigma; \Theta]$,
- $k(\alpha) \cap k(\beta) \subseteq k(\alpha \wedge \beta)$.

We show here that the condition I for $\wedge$ of downward saturation in the previous section follows from this, whenever (Σ, Θ) is maximal consistent. In fact, for (a) of I, if $\alpha \notin \Sigma$ then $\Sigma \cup \{\alpha\} \preceq \Theta$, i.e., $(\{\alpha\}, \varepsilon) \in [\Sigma; \Theta]$, and hence $(\{\alpha \wedge \beta\}, \varepsilon) \in [\Sigma; \Theta]$ by our assumption. This means $\Sigma \cup \{\alpha \wedge \beta\} \preceq \Theta$, and hence $\alpha \wedge \beta \notin \Sigma$. Similarly, we can show that $\beta \notin \Sigma$ implies $\alpha \wedge \beta \notin \Sigma$. For (b), if both $\alpha \notin \Theta$ and $\beta \notin \Theta$ then $(\Sigma, \Theta) \in k(\alpha) \cap k(\beta)$. Since $k(\alpha) \cap k(\beta) \subseteq k(\alpha \wedge \beta)$ by our assumption, $\Sigma \preceq \Theta \cup \{\alpha \wedge \beta\}$, and hence $\alpha \wedge \beta \notin \Theta$.

We have shown a certain relation between model-theoretic and algebraic ones, but not so satisfactorily yet. We think that this discrepancy will be partly of intrinsic character. Algebraic approaches developed so far are mostly for substructural logics that are not always distributive, while model-theoretic approaches to modal logics rely ultimately on the Jónsson-Tarski extension of Stone duality (cf. Theorem 5). For instance, in Sect. 1.3, we define essentially that a pair (Σ, Θ) of subsets Σ and Θ of Ω is **S4**-provable if for some $\alpha_1, \ldots, \alpha_m \in \Sigma$ and $\beta_1, \ldots, \beta_n \in \Theta$, the sequent $\alpha_1, \ldots, \alpha_m \Rightarrow \beta_1, \ldots, \beta_n$ is provable in **GS4**. But, we cannot adopt this kind of definition for logics lacking weakening rules. Thus, we cannot talk about maximal consistent pairs nor ultrafilters (in their algebraic form) for these logics. It might be better, then, to reconsider an algebraic framework which is more suitable for discussing cut elimination and subformula property in modal logics.

The basics to our approach consist of considering consistent pairs (or consistent sets) in a given logic (or a sequent system) and their maximal extensions (Lindenbaum's lemma), and showing that the Kripke model constructed by a set of maximal consistent pairs (or sets) satisfies the required logical property (truth lemma). As we have mentioned already, a related study has been done by M. Fitting in [8]. The notion *consistency property* of a collection of sets of formulas was introduced there, which describes conditions that the collection of all consistent sets in a given logic should have. Then, the *model existence theorem* was shown, which assures that if a given set of formulas is a member of the consistency property for a logic **L** it is satisfied in a model for **L**. From this model existence theorem, basic logical properties like Kripke completeness, cut elimination, Craig's interpolation theorem, etc. are derived. We take note that though *maximal consistent* pairs (or sets) are not discussed [8] in an explicit way, a consistency property is often assumed to be closed under chain unions in it. It means that the existence of maximal elements is in practice assumed by using Zorn's lemma. We note also that semantical proof of cut elimination by Fitting can be also applied to some intuitionistic modal logics (see [13]). Further discussions on connections of Fitting's approach with those in the present paper would be useful.

Many interesting problems on semantical approach to cut elimination and subformula property remain unsolved. They will be discussed in our future papers.

References

1. Amano, S.J.: The finite embeddability property for some modal algebras, Master thesis. Japan Advanced Institute of Science and Technology (2006)
2. Belardinelli, F., Jipsen, P., Ono, H.: Algebraic aspects of cut elimination. Stud. Log. **77**, 209–240 (2004)
3. Blackburn, P., de Rijke, M., Venema, Y.: Modal Logic. Cambridge Tracts in Theoretical Computer Science **53**, (2001)
4. Bull, R.: Some modal calculi based on IC. Formal systems and recursive functions. In: Crossley, J.N., Dummett, M.A.E., 3-7 (1965)
5. Chagrov, A., Zakharyaschev, M.: Modal Logic. Oxford Logic Guides, Clarendon Press, vol.35 (1997)

6. Ciabattoni, A., Galatos, N., Terui, K.: Algebraic proof theory for substructural logics: cut-elimination and completions. Ann. Pure Appl. Log. **163**, 266–290 (2012)
7. Curry, H.: The elimination theorem when modality is present. J. Symb. Log. **17**, 249–265 (1952)
8. Fitting, M.: Model existence theorems for modal and intuitionistic logics. J. Symb. Log. **38**, 613–627 (1973)
9. Galatos, N., Jipsen, P., Kowalski, T., Ono, H.: Residuated Lattices: an algebraic glimpse at substructural logics. Studies in Logic and the Foundations of Mathematics, Elsevier, vol. 151 (2007)
10. Gentzen, G.: Untersuchungen über das logische Schliessen I. II. Mathematische Zeitschrift **39**, (176-210, 405-431) (1934, 1935)
11. Ohnishi, M., Matsumoto, K.: Gentzen method in modal calculi, Osaka Math. J. **9**, 113-130 (1957) (Correction ibid. 10 (1958), p.147)
12. Okada, M., Terui, K.: The finite model property for various fragments of intuitionistic linear logic. J. Symb. Log. **64**, 790–802 (1999)
13. Ono, H.: On some intuitionistic modal logics, Publ. Res. Inst. Math. Sci. Kyoto University, **13**, 687–722 (1977)
14. Ono, H.: Proof-theoretic methods for nonclassical logic—an introduction. Theories of Types and Proofs (MSJ Memoirs 2). In: Takahashi, M., Okada, M., Dezani-Ciancaglini M. (eds.) Mathematical Society of Japan, 207-254 (1998)
15. Sato, M.: A study of Kripke-type models for some modal logic by Gentzen's sequential method. Publ. Res. Inst. Math. Sci. Kyoto University, **13**, 381-468 (1977)
16. Schütte, K.: Syntactical and semantical properties of simple type theory. J. Symb. Log. **25**, 305–326 (1960)
17. Schütte, K.: Vollständige Systeme modaler und intuitionistischer Logik. Ergebnisse der Mathematik und ihrer Grenzgebiete, Springer, vol. 42 (1968)
18. Takano, M.: Subformula property as a substitute for cut-elimination in modal propositional logics. Math. Jpn. **37**, 1145–1192 (1992)
19. Takano, M.: Semantical proofs of cut elimination and subformula property (in Japanese), abstract of talk at Japan Advanced Institute of Science and Technology (2000)
20. Takano, M.: A modified subformula property for the modal logics K5 and K5D. Bull. Sect. Log. **30**, 115–122 (2001)
21. Takeuti, G.: Proof Theory. Stud. Log. Found. Math. North-Holland, vol. 81 (1975)

Chapter 2
Ultraproducts of Admissible Models for Quantified Modal Logic

Robert Goldblatt

Abstract Admissible models for quantified modal logic have a restriction on which sets of worlds are admissible as propositions. They give an actualist interpretation of quantifiers that leads to very general completeness results: for any propositional modal logic S there is a quantificational proof system QS that is complete for validity in models whose algebra of admissible propositions validates S. In this paper, we construct ultraproducts of admissible models and use them to derive compactness theorems that combine with completeness to yield *strong* completeness: any QS-consistent set of formulas is satisfiable in a model whose admissible propositions validate S. The Barcan Formula is analysed separately and shown to axiomatise certain logics that are strongly complete over admissible models in which the quantifiers are given their standard Kripkean interpretation.

Keywords Admissible semantics · Quantified modal logic · Ultraproduct · Actualist quantification · Compactness · Strong completeness · Kripkean interpretation · Barcan formula

2.1 Introduction

A theory of *admissible semantics* for quantified modal logics was set out by the author in [5]. Its aim is to address the problem of *incompleteness* of some such logics under their Kripkean possible-worlds semantics. This includes cases where completeness for validity in Kripke frames holds at the propositional level but fails to lift to the quantificational setting.

An example of this failure concerns the Gödel-Löb logic GL, the normal propositional modal logic with the axiom $\Box(\Box A \to A) \to \Box A$. It axiomatises the interpretation of $\Box$ as "it is provable in Peano arithmetic that". GL is a decidable logic that is complete for validity in its Kripke frames. These Kripke frames validating GL have a natural mathematical description as the transitive inverse well-founded ones. But the set of formulas that are valid in the Kripkean *quantificational* models over

R. Goldblatt (✉)
Victoria University of Wellington, Wellington, New Zealand
e-mail: rob.goldblatt@msor.vuw.ac.nz

S.C.-M. Yang et al. (eds.), *Structural Analysis of Non-Classical Logics*,
Logic in Asia: Studia Logica Library, DOI 10.1007/978-3-662-48357-2_2

GL-frames is not recursively enumerable, so cannot be recursively axiomatised. What then are the prospects of developing a model theory that characterises logics defined proof-theoretically by adding standard axioms and inference rules for quantifiers to GL?

We answer this question by imposing a restriction on which sets of worlds count as propositions. Our models have a designated modal algebra *Prop* of sets of worlds, called the *admissible propositions*. Every formula is interpreted as an admissible proposition. For propositional modal languages such structures are called *general frames* and provide a complete semantics for any logic. In models for languages with quantification of individual variables, each world w of a general frame is assigned a subset Dw of some fixed universe U of possible individuals. Dw is the domain of individuals that exist, or are actual, in w.

In Kripkean models, a universal quantifier $\forall x$ is interpreted at w by taking the variable x to range over the domain Dw. This is the *actualist* interpretation of quantification, validating the Actual Instantiation scheme

AI: $\forall y(\forall x\varphi \to \varphi(y/x))$, where y is free for x in φ,

but not the Universal Instantiation scheme

UI: $\forall x\varphi \to \varphi(y/x)$, where y is free for x in φ

(because the value of y may not be actual in a particular world).

In an admissible model we take $\forall x\varphi$ to have the same meaning as the conjunction of the assertions "if a exists then $\varphi(a/x)$" for all $a \in U$. The conjunction operation is interpreted as the *meet*, or greatest lower bound, operation in the set $(Prop, \subseteq)$ of admissible propositions under the partial ordering $\subseteq$ of entailment (= set inclusion). In a Kripkean model, the meet of a set $\mathscr{X}$ of propositions is just its set-theoretic intersection $\bigcap \mathscr{X}$. But in an admissible model, the meet $\bigsqcap \mathscr{X}$ of $\mathscr{X}$ is the largest *admissible* subset of $\bigcap \mathscr{X}$. This can be understood as the weakest admissible proposition that entails every member of $\mathscr{X}$, and may have $\bigsqcap \mathscr{X} \subsetneq \bigcap \mathscr{X}$.

Using these ideas we have shown that for every propositional modal logic S there is a naturally axiomatised quantified logic QS (with axioms including AI and all instances of S-theorems), which is *complete* for validity in models whose underlying general frame of admissible propositions validates S. Completeness here means that every QS-consistent formula is satsifiable in a model of the kind just described. It is noteworthy that such models need not validate the commuting quantifiers axiom

$$\text{CQ}: \ \forall x\forall y\varphi \to \forall y\forall x\varphi,$$

which is valid in Kripkean models (see [6]).

In this paper, we take up the question of *strong* completeness, meaning that every consistent *set* of formulas is satisfiable in a model of the required kind. We introduce a definition of the ultraproduct $\mathscr{M}_\mu$ of a family $\{\mathscr{M}_i : i \in I\}$ of admissible models with respect to an ultrafilter μ on the index set I. We show that Łoś' Theorem, the so-called "fundamental theorem of ultraproducts", continues to hold for our admissible interpretation of the quantifier $\forall$. This theorem states that a formula is satisfiable in $\mathscr{M}_\mu$ iff it is satisfiable in "almost all" of the models $\mathscr{M}_i$. Armed with Łoś' Theorem it is then a matter of using standard arguments to derive a compactness theorem for admissible model theory and combine it with completeness to infer strong completeness for QS.

We then take up the question of the Barcan Formula BF: $\forall x\Box\varphi \to \Box\forall x\varphi$, and its converse CBF. In Kripkean models validity of BF is often identified with the condition of *contracting domains*: wRu implies $Dw \supseteq Du$. But admissible models can have contracting domains without validating BF. Perhaps surprisingly, every logic of the form QS is characterised by models with contracting domains. Imposition of this contracting domains condition on admissible models does not force the general validity of any non-theorems of QS. It is only in *Kripkean* models with contracting domains that validity of BF is guaranteed.

We apply our ultraproducts method to prove strong completeness of QS over contracting-domains models; of QS + CBF over models with *constant domains* (wRu implies $Dw = Du$); and of QS + CBF + CQ + BF over Kripkean constant-domain models. The proof for the last case works for arbitrarily large languages, overcoming a countability restriction on the original proof of completeness. The whole analysis reveals that the real role of BF in admissible model theory is to enable us to build models that give the quantifier $\forall$ its standard Kripkean interpretation.

Finally, we examine the universal instantiation axiom UI, which corresponds to the condition that a model has *one universal domain*: $D_w = U$ for all worlds w. The axioms CBF and CQ are derivable from UI. We show that QS + UI is strongly complete for validity in one-universal-domain admissible models whose underlying general frame validates S, and that QS + UI + BF is strongly complete over Kripkean models of this kind.

2.2 Admissible Models

Here, we set out the basic syntax of quantified modal logic, and its admissible semantics. Let $\{x_0, \dots, x_n, \dots\}$ be a fixed denumerable set of individual variables. The letters x, y will be used for arbitrary variables. Let $\mathscr{L}$ be a *signature*: a set of individual constants c, predicate symbols P, and function symbols F. An *$\mathscr{L}$-term* is any individual variable, any constant c, or inductively any expression $F\tau_1 \cdots \tau_n$ where F is an n-ary function symbol from $\mathscr{L}$, and $\tau_1, \dots, \tau_n$ are $\mathscr{L}$-terms.

An *atomic $\mathscr{L}$-formula* is any expression $P\tau_1 \cdots \tau_n$ where P is an n-ary predicate symbol from $\mathscr{L}$, and $\tau_1, \dots, \tau_n$ are $\mathscr{L}$-terms. The set of *$\mathscr{L}$-formulas* is generated from the atomic ones and a constant formula $\bot$ (Falsum) in the usual way, using the connectives $\wedge$ (conjunction), $\neg$ (negation), the modality $\Box$ and universal quantifiers $\forall x$ for each variable x.

A *model structure* is a system $\mathcal{S} = (W, R, Prop, U, D)$ such that:

- W is a non-empty set (of "worlds"), and R is a binary relation on W.
- *Prop* is a non-empty subset of the powerset $\wp W$ of W that is closed under binary intersections $X \cap Y$ and complements $-X$, hence under binary unions $X \cup Y$ and Boolean implications $X \Rightarrow Y = (-X) \cup Y$. Hence $\emptyset$, $W \in Prop$.
- *Prop* is closed under the operation $[R]$ defined by

$$[R]X = \{w \in W : \forall v \in W(wRv \text{ implies } v \in X)\}.$$

- U is a non-empty set, called the *universe* of $\mathcal{S}$; and
- D is a function assigning to each element w of W a subset Dw of U, called the *domain of* w.

A subset of W is *admissible* if it belongs to *Prop*. Members of *Prop* are also called the *admissible propositions* of $\mathcal{S}$.

The triple $(W, R, Prop)$ is sometimes called a *general frame*. Such a structure is used to provide semantics for propositional modal logic, in a manner that will be described in Sect. 2.6 below. When extracted from a model structure $\mathcal{S}$ as above it may be called the *underlying* general frame of $\mathcal{S}$.

An operation $\sqcap$ on collections of subsets of W is defined by putting, for each $\mathcal{X} \subseteq \wp W$,

$$\sqcap \mathcal{X} = \bigcup\{Y \in Prop : Y \subseteq \bigcap \mathcal{X}\}.$$

Thus $\sqcap \mathcal{X}$ is the union of all admissible subsets of $\bigcap \mathcal{X}$, hence $\sqcap \mathcal{X} \subseteq \bigcap \mathcal{X}$. It is not required that $\mathcal{X} \subseteq Prop$ in this definition: $\sqcap \mathcal{X}$ is defined for arbitrary $\mathcal{X} \subseteq \wp W$ and need not be admissible in general, even when $\mathcal{X} \subseteq Prop$. If we do have $\mathcal{X} \subseteq Prop$ and $\sqcap \mathcal{X} \in Prop$, then $\sqcap \mathcal{X}$ is the *greatest lower bound* of $\mathcal{X}$ in the partially ordered set $(Prop, \subseteq)$, i.e. the largest admissible set included in every member of $\mathcal{X}$. If $\bigcap \mathcal{X}$ is admissible, then $\sqcap \mathcal{X} = \bigcap \mathcal{X}$.

For each $a \in U$ we define $Ea = \{w \in W : a \in Dw\}$, representing the proposition "a exists". Sets of the form Ea may be referred to as "existence sets" or "existence propositions". They are not required to be admissible.

A *premodel* $\mathcal{M} = (\mathcal{S}, |-|^{\mathcal{M}})$ for signature $\mathcal{L}$, based on a model structure $\mathcal{S}$, is given by an interpretation function $|-|^{\mathcal{M}}$ on $\mathcal{L}$ that assigns:

- to each individual constant $\mathsf{c} \in \mathcal{L}$ an element $|\mathsf{c}|^{\mathcal{M}}$ of the universe U.
- to each n-ary function symbol $F \in \mathcal{L}$ an n-ary function $|F|^{\mathcal{M}}$ on the universe U, i.e. $|F|^{\mathcal{M}} : U^n \to U$.
- to each n-ary predicate symbol $P \in \mathcal{L}$ a function $|P|^{\mathcal{M}} : U^n \to \wp W$.

Intuitively, $|P|^{\mathcal{M}}(a_1, \ldots, a_n)$ represents the proposition that the predicate P holds of the n-tuple $(a_1, \ldots, a_n)$. A *variable-assignment* in a premodel is a function from the set ω of natural numbers into U. Thus, the set of variable-assignments is the set U^ω of all functions $f : \omega \to U$. The idea here is that f assigns the value fn to the variable x_n. Such an f then assigns to each $\mathcal{L}$-term τ a value $|\tau|^{\mathcal{M}} f \in U$, so overall $\mathcal{M}$ interprets τ as a function $|\tau|^{\mathcal{M}} : U^{\omega} \to U$. The inductive definition of $|\tau|^{\mathcal{M}} f$ is:

- $|x|^{\mathcal{M}} f = fn$, if x is the variable x_n.
- $|\mathsf{c}|^{\mathcal{M}} f = |\mathsf{c}|^{\mathcal{M}}$.
- $|F\tau_1 \cdots \tau_n|^{\mathcal{M}} f = |F|^{\mathcal{M}}(|\tau_1|^{\mathcal{M}} f, \ldots, |\tau_n|^{\mathcal{M}} f)$.

We write fx for fn when x is x_n, so we get $|x|^{\mathcal{M}} f = fx$. The notation $f[a/x]$ will be used for the function that *updates* f by assigning the value a to x and otherwise acting identically to f. Thus, $f[a/x]x = a$ and $f[a/x]y = fy$ if $y \neq x$.

A premodel gives an interpretation $|\varphi|^{\mathcal{M}} : U^{\omega} \to \wp W$ to each $\mathcal{L}$-formula. This interpretation is a *propositional function*, i.e. a function whose values are propositions (not necessarily admissible ones). For each assignment f, $|\varphi|^{\mathcal{M}} f$ is to be the *truth set* of all worlds at which φ is true under f. This is defined by induction on the formation of φ:

- $|P\tau_1 \cdots \tau_n|^{\mathcal{M}} f = |P|^{\mathcal{M}}(|\tau_1|^{\mathcal{M}} f, \ldots, |\tau_n|^{\mathcal{M}} f)$.
- $|\bot|^{\mathcal{M}} f = \emptyset$.
- $|\varphi \wedge \psi|^{\mathcal{M}} f = |\varphi|^{\mathcal{M}} f \cap |\psi|^{\mathcal{M}} f$.
- $|\neg\varphi|^{\mathcal{M}} f = W - |\varphi|^{\mathcal{M}} f$.
- $|\Box\varphi|^{\mathcal{M}} f = [R]|\varphi|^{\mathcal{M}} f$.
- $|\forall x\varphi|^{\mathcal{M}} f = \bigsqcap_{a\in U} \big(Ea \Rightarrow |\varphi|^{\mathcal{M}} f[a/x]\big)$.

Writing $\mathcal{M}, w, f \models \varphi$ to mean that $w \in |\varphi|^{\mathcal{M}} f$, we get the following standard clauses for this truth/satisfaction relation $\models$.

- $\mathcal{M}, w, f \models P\tau_1 \cdots \tau_n$ iff $w \in |P|^{\mathcal{M}}(|\tau_1|^{\mathcal{M}} f, \ldots, |\tau_n|^{\mathcal{M}} f)$.
- $\mathcal{M}, w, f \not\models \bot$.
- $\mathcal{M}, w, f \models \varphi \wedge \psi$ iff $\mathcal{M}, w, f \models \varphi$ and $\mathcal{M}, w, f \models \psi$.
- $\mathcal{M}, w, f \models \neg\varphi$ iff not $\mathcal{M}, w, f \models \varphi$.
- $\mathcal{M}, w, f \models \Box\varphi$ iff for all $v \in W(wRv$ implies $\mathcal{M}, v, f \models \varphi)$.

For the universal quantifier, the condition for $\mathcal{M}, w, f \models \forall x\varphi$ is that

$$\text{there is an } X \in Prop \text{ such that } w \in X \text{ and } X \subseteq \bigcap_{a\in U} \big(Ea \Rightarrow |\varphi|^{\mathcal{M}} f[a/x]\big). \quad (2.1)$$

Informally, this asserts that there is an admissible proposition X that is true at w and entails the assertions "if a exists then $\varphi(a/x)$" for all $a \in U$.

From (2.1) we see that

$$\mathcal{M}, w, f \models \forall x\varphi \ \textit{only if} \ \text{ for all } a \in Dw,\ \mathcal{M}, w, f[a/x] \models \varphi. \quad (2.2)$$

The converse need not hold [5, Example 1.6.6]. If it does hold, then $\mathcal{M}$ will be called *Kripkean*, because this means that $\forall$ gets the varying-domain semantics of [7]:

$$\mathcal{M}, w, f \models \forall x\varphi \ \text{ iff } \ \text{for all } a \in Dw,\ \mathcal{M}, w, f[a/x] \models \varphi.$$

Thus a Kripkean premodel is one that always has

$$|\forall x\varphi|^{\mathcal{M}} f = \bigcap_{a\in U} \big(Ea \Rightarrow |\varphi|^{\mathcal{M}} f[a/x]\big).$$

A formula φ is *valid in* premodel $\mathcal{M}$, written $\mathcal{M} \models \varphi$, if $|\varphi|^{\mathcal{M}} f = W$ for all f, i.e. if $\mathcal{M}, w, f \models \varphi$ for all $w \in W$ and $f \in U^{\omega}$.

An *admissible model*, or just *model*, for $\mathcal{L}$ is, by definition, a premodel in which every $\mathcal{L}$-formula φ is *admissible* in the sense that the function $|\varphi|^{\mathcal{M}}$ has the form $U^{\omega} \to \mathit{Prop}$, i.e. $|\varphi|^{\mathcal{M}} f \in \mathit{Prop}$ for all $f \in U^{\omega}$.

Informally, a model interprets a sentence $\forall x\varphi$ as the weakest admissible proposition that entails the assertions "if a exists then $\varphi(a/x)$" for all $a \in U$.

2.3 Ultraproducts of Premodels

Let μ be an ultrafilter on a set I. Recall that this means that μ is a collection of subsets of I such that $I \in \mu$; the complement $I - J$ of a subset $J \subseteq I$ belongs to μ iff $J \notin \mu$; and an intersection $J \cap K$ belongs to μ iff $J \in \mu$ and $K \in \mu$. Such a μ is closed under supersets: if $J \in \mu$ and $J \subseteq K$, then $K \in \mu$.

Each $J \in \mu$ is a "large" subset of I. We think of J as containing *almost all* members of I.

For any I-indexed collection $\{X_i : i \in I\}$ of sets, let $\prod_I X_i$ be the Cartesian product set whose points are the functions $f : I \to \bigcup_I X_i$ having $f(i) \in X_i$ for all $i \in I$. Define an equivalence relation $=_\mu$ on $\prod_I X_i$ by putting

$$f =_\mu g \quad \text{iff} \quad \{i \in I : f(i) = g(i)\} \in \mu.$$

The relation $f =_\mu g$ can be thought of as asserting that f and g agree at almost all $i \in I$. Let f_μ be the equivalence class $\{g \in \prod_I X_i : f =_\mu g\}$, and put $\prod_\mu X_i = \{f_\mu : f \in \prod_I X_i\}$. Then $\prod_\mu X_i$ is called the *ultraproduct* of the sets X_i with respect to μ. Many properties can be specified as holding of $\prod_\mu X_i$ iff they hold correspondingly of almost all of the X_i's.

Let $\{\mathcal{S}_i : i \in I\}$ be an I-indexed collection of model structures, with $\mathcal{S}_i = (W_i, R_i, \mathit{Prop}_i, U_i, D_i)$. We define the ultraproduct of the $\mathcal{S}_i$'s with respect to μ as a structure

$$\mathcal{S}_\mu = (W_\mu, R_\mu, \mathit{Prop}_\mu, U_\mu, D_\mu),$$

(which could also be denoted $\prod_\mu \mathcal{S}_i$). Here W_μ is the ultraproduct $\prod_\mu W_i$ of the W_i's and U_μ is the ultraproduct $\prod_\mu U_i$ of the U_i's. The binary relation R_μ on W_μ is well defined by putting, for all $f, g \in \prod_I W_i$,

$$f_\mu R_\mu g_\mu \quad \text{iff} \quad \{i \in I : f(i) R_i g(i)\} \in \mu.$$

The domain function $D_\mu : W_\mu \to \wp U_\mu$ is defined by putting

$$D_\mu f_\mu = \{g_\mu \in U_\mu : \{i : g(i) \in D_i f(i)\} \in \mu\}$$

for all $f \in \prod_I W_i$. This definition[1] can be seen as an example of the general procedure of lifting an operation to an ultraproduct by lifting it to the direct product and then transferring it to the $=_\mu$-equivalence classes. For this and other purposes it is convenient to lift the set membership relation to a relation $\in_\mu$ between any functions h, k with domain I by putting

$$h \in_\mu k \quad \text{iff} \quad \{i \in I : h(i) \in k(i)\} \in \mu.$$

Now the domain functions D_i induce the function $D_I : \prod_I W_i \to \prod_I \wp U_i$ where, for any $f \in \prod_I W_i$, the function $D_I f \in \prod_I \wp U_i$ is defined by putting, for each $i \in I$,

$$(D_I f)(i) = D_i f(i) \subseteq U_i.$$

Then the definition of D_μ becomes that

$$D_\mu f_\mu = \{g_\mu \in U_\mu : g \in_\mu D_I f\}.$$

We will write E_i for the existence operator in $\mathscr{S}_i$, so that $f(i) \in E_i g(i))$ iff $g(i) \in D_i g(i)$. The existence operator in $\mathscr{S}_\mu$ will be denoted E_μ, so that $f_\mu \in E_\mu g_\mu$ iff $g_\mu \in D_\mu f_\mu$. Thus

$$f_\mu \in E_\mu g_\mu \quad \text{iff} \quad \{i \in I : f(i) \in E_i g(i)\} \in \mu. \tag{2.3}$$

It remains to construct $Prop_\mu$ as a modal algebra of subsets of W_μ, closed under the Boolean set operations and the unary modal operator $[R_\mu]$ induced on $\wp W_\mu$ by the relation R_μ. This construction was carried out for general modal frames in [3], reproduced in [4]. It constructs $Prop_\mu$, not as the ultraproduct $\prod_\mu Prop_i$ of the modal algebras $Prop_i$, but as an algebra of subsets of W_μ that is isomorphic to $\prod_\mu Prop_i$. For each element σ of the Cartesian product $\prod_I Prop_i$, define a subset $S(\sigma)$ of W_μ by putting

$$S(\sigma) = \{f_\mu \in W_\mu : f \in_\mu \sigma\}.$$

Then, we put $Prop_\mu = \{S(\sigma) : \sigma \in \prod_I Prop_i\}$.

Now it can be shown that $S(\sigma)$ is well defined and that in general $\sigma =_\mu \sigma'$ iff $S(\sigma) = S(\sigma')$. Thus the map $\sigma_\mu \mapsto S(\sigma)$ is a bijection between $\prod_\mu Prop_i$ and $Prop_\mu$. Moreover, we have

---

[1] As with many operations on ultraproducts, it needs to be checked that D_μ is well defined, i.e. that $f_\mu = f'_\mu$ implies $D_\mu(f_\mu) = D_\mu(f'_\mu)$. Such checking is left to the reader in routine cases.

$$
\begin{aligned}
S(\sigma) \cap S(\sigma') &= S(\sigma \cap \sigma') \\
W_\mu - S(\sigma) &= S(-\sigma) \\
[R_\mu]S(\sigma) &= S([R_I]\sigma),
\end{aligned} \tag{2.4}
$$

where $\sigma \cap \sigma'$, $-\sigma$ and $[R_I]\sigma$ are the members of $\prod_I Prop_i$ defined pointwise by the corresponding operations on the algebras $Prop_i$, i.e. $(\sigma \cap \sigma')(i) = \sigma(i) \cap \sigma'(i)$, $(-\sigma)(i) = W_i - \sigma(i)$ and $([R_I]\sigma)(i) = [R_i]\sigma_i$.

It follows from (2.4) that $Prop_\mu$ is closed under $\cap$, $-$ and $[R_\mu]$. Full details of this construction can be found in [4, Sect. 1.7]. That completes the description of the ultraproduct of the the $\mathscr{S}_i$'s with respect to μ.

2.4 Łoś' Theorem

Given a collection $\{\mathscr{M}_i \in I\}$ of premodels for $\mathscr{L}$, with $\mathscr{M}_i = (\mathscr{S}_i, |-|^{\mathscr{M}_i})$, and an ultrafilter μ on I, we define a premodel $\mathscr{M}_\mu = (\mathscr{S}_\mu, |-|^{\mathscr{M}_\mu})$ on the ultraproduct $\mathscr{S}_\mu$ of the $\mathscr{S}_i$'s with respect to μ. We use *tuple* notation for functions here: a function f with domain I may be written as the tuple $\langle f(i) : i \in I\rangle$, and then $\langle f(i) : i \in I\rangle_\mu$ denotes f_μ.

The interpretation function $|-|^{\mathscr{M}_i}$ is defined as follows.

- For each individual constant $\mathsf{c} \in \mathscr{L}$, $|\mathsf{c}|^{\mathscr{M}_\mu} = \langle |\mathsf{c}|^{\mathscr{M}_i} : i \in I\rangle_\mu \in U_\mu$.
- For each n-ary function symbol $F \in \mathscr{L}$, the function $|F|^{\mathscr{M}_\mu} : {U_\mu}^n \to U_\mu$ is defined by putting, for all $f_1, \dots, f_n \in \prod_I U_i$,

$$
|F|^{\mathscr{M}_\mu}(f_{1\mu}, \dots, f_{n\mu}) = \langle |F|^{\mathscr{M}_i}(f_1(i), \dots, f_n(i)) : i \in I\rangle_\mu .
$$

- For each n-ary predicate symbol $P \in \mathscr{L}$, the function $|P|^{\mathscr{M}_\mu} : {U_\mu}^n \to \wp W_\mu$ is defined by putting, for all $f_1, \dots, f_n \in \prod_I U_i$ and all $g \in \prod_I W_i$,

$$
g_\mu \in |P|^{\mathscr{M}_\mu}(f_{1\mu}, \dots, f_{n\mu}) \ \text{ iff } \ \{i \in I : g(i) \in |P|^{\mathscr{M}_i}(f_1(i), \dots, f_n(i))\} \in \mu. \tag{2.5}
$$

The fundamental property of ultraproducts of models of (non-modal) first-order logic, due to Łoś, is in essence that a formula is satisfiable in an ultraproduct iff it is satisfiable in almost all of the component models. We now formulate the corresponding result for the admissible semantics of our premodels.

A sequence $f = \langle f_0, \dots, f_n, \dots\rangle \in (\prod_I U_i)^\omega$ of elements f_n of the Cartesian product $\prod_I U_i$ determines, for each $i \in I$, the sequence

$$
f \cdot i = \langle f_0(i), \dots, f_n(i), \dots\rangle \in U_i^\omega \tag{2.6}
$$

of elements of U_i. We write f_μ for the sequence $\langle f_{0\mu}, \ldots, f_{n\mu}, \ldots \rangle \in U_\mu{}^\omega$. Then it is straightforward to check that for any $g \in \prod_I U_i$ and any variable x we have

$$f_\mu[g_\mu/x] = f[g/x]_\mu, \tag{2.7}$$

and for each $i \in I$,

$$f[g/x] \cdot i = (f \cdot i)[g(i)/x]. \tag{2.8}$$

An induction on term formation shows that for any $\mathscr{L}$-term τ, the function $|\tau|^{\mathscr{M}_\mu} : U_\mu{}^\omega \to U_\mu$ has

$$|\tau|^{\mathscr{M}_\mu} f_\mu = \langle |\tau|^{\mathscr{M}_i} f \cdot i : i \in I \rangle_\mu. \tag{2.9}$$

The proof is essentially as in the classical case of ultraproducts of non-modal first-order logic [2, Sect. 4.1].

To formulate the fundamental theorem we define, for each formula φ and each $h \in \prod_I W_i$, the "truth set"

$$[\![h, \varphi, f]\!] = \{i \in I : h(i) \in |\varphi|^{\mathscr{M}_i} f \cdot i\} = \{i \in I : \mathscr{M}_i, h(i), f \cdot i \models \varphi\}.$$

Theorem 1 (Łoś' Theorem) *Let φ be any $\mathscr{L}$-formula. Then for all $f \in (\prod_I U_i)^\omega$ and $h \in \prod_I W_i$,*

$$h_\mu \in |\varphi|^{\mathscr{M}_\mu} f_\mu \quad \textit{iff} \quad [\![h, \varphi, f]\!] \in \mu.$$

In other words,

$$\mathscr{M}_\mu, h_\mu, f_\mu \models \varphi \quad \textit{iff} \quad \{i \in I : \mathscr{M}_i, h(i), f \cdot i \models \varphi\} \in \mu.$$

Proof This proceeds by induction on the formation of φ. For the case that φ is the atomic formula $P\tau_1 \cdots \tau_n$, the definition of $|P|^{\mathscr{M}_\mu}$ in (2.5) combines with (2.9) to show that $h_\mu \in |\varphi|^{\mathscr{M}_\mu} f_\mu$ iff the set

$$\{i \in I : h(i) \in |P|^{\mathscr{M}_i}(|\tau_1|^{\mathscr{M}_i} f \cdot i, \ldots, |\tau_n|^{\mathscr{M}_i} f \cdot i)\}$$

belongs to μ. But this set is $[\![h, P\tau_1 \cdots \tau_n, f]\!]$.

The case that φ is $\bot$ and the inductive steps for the connectives $\neg$, $\wedge$ and $\Box$ are as for propositional modal logic in [4, Sect. 1.7] (see also (2.13) below).

The really new case here is to show that the theorem holds for a formula $\forall x\varphi$ under the induction hypothesis that it holds for φ. Assume first that $[\![h, \forall x\varphi, f]\!] \in \mu$. To prove that $h_\mu \in |\forall x\varphi|^{\mathscr{M}_\mu} f_\mu$ we prove, in accordance with (2.1), that there exists some admissible set $S(\sigma) \in \mathit{Prop}_\mu$ such that $h_\mu \in S(\sigma)$ and

$$S(\sigma) \subseteq \bigcap_{g_\mu \in U_\mu} \big(E_\mu g_\mu \Rightarrow |\varphi|^{\mathscr{M}_\mu} f_\mu[g_\mu/x]\big). \tag{2.10}$$

Now if $i \in [\![h, \forall x\varphi, f]\!]$, then $h(i) \in |\forall x\varphi|^{\mathcal{M}_i} f \cdot i$, so applying (2.1) in $\mathcal{M}_i$, there is some admissible set $\sigma(i) \in Prop_i$ such that $h(i) \in \sigma(i)$ and

$$\sigma(i) \subseteq \bigcap_{d \in U_i} \big(E_i d \Rightarrow |\varphi|^{\mathcal{M}_i} f \cdot i[d/x]\big). \tag{2.11}$$

For $i \notin [\![h, \forall x\varphi, f]\!]$, put $\sigma(i) = \emptyset$. We have now defined a function $\sigma \in \prod_I Prop_i$ with $[\![h, \forall x\varphi, f]\!] \subseteq \{i : h(i) \in \sigma(i)\}$. Hence $\{i : h(i) \in \sigma(i)\} \in \mu$, so $h \in_\mu \sigma$ and therefore $h_\mu \in S(\sigma) \in Prop_\mu$. It remains to prove (2.10).

Take any $k_\mu \in S(\sigma)$, where $k \in \prod_I W_i$ and $k \in_\mu \sigma$. Let $g_\mu \in U_\mu$. If $k_\mu \in E_\mu g_\mu$, then the intersection

$$J = [\![h, \forall x\varphi, f]\!] \cap \{i : k(i) \in \sigma(i)\} \cap \{i : k(i) \in E_i g(i)\}$$

belongs to μ, since each of the three sets involved belongs to μ [cf. (2.3)]. But if $i \in J$, then (2.11) holds, and so as $k(i)$ belongs to $\sigma(i)$ and to $E_i g(i)$ we infer that it belongs to $|\varphi|^{\mathcal{M}_i} f \cdot i[g(i)/x]$, which is equal to $|\varphi|^{\mathcal{M}_i} f[g/x] \cdot i$ by (2.8). This shows that

$$J \subseteq \{i : k(i) \in |\varphi|^{\mathcal{M}_i} f[g/x] \cdot i\} = [\![k, \varphi, f[g/x]\,]\!].$$

Therefore, $[\![k, \varphi, f[g/x]\,]\!] \in \mu$, and so by the induction hypothesis on φ, k_μ belongs to $|\varphi|^{\mathcal{M}_\mu} f[g/x]_\mu$, which is equal to $|\varphi|^{\mathcal{M}_\mu} f_\mu[g_\mu/x]$ by (2.7). Altogether this proves that $k_\mu \in E_\mu g_\mu \Rightarrow |\varphi|^{\mathcal{M}_\mu} f_\mu[g_\mu/x]$, which completes the proof of (2.10), and hence the proof that $[\![h, \forall x\varphi, f]\!] \in \mu$ implies $h_\mu \in |\forall x\varphi|^{\mathcal{M}_\mu} f_\mu$.

For the converse, suppose that $[\![h, \forall x\varphi, f]\!] \notin \mu$. Then to show that $h_\mu \notin |\forall x\varphi|^{\mathcal{M}_\mu} f_\mu$ we take an arbitrary $S(\sigma) \in Prop_\mu$ such that $h_\mu \in S(\sigma)$, and will show that (2.10) fails. As μ is an ultrafilter we have $(I - [\![h, \forall x\varphi, f]\!]) \in \mu$, so as $h \in_\mu \sigma$ we get that the set

$$J' = (I - [\![h, \forall x\varphi, f]\!]) \cap \{i : h(i) \in \sigma(i)\}$$

belongs to μ. Now if $i \in J'$ we have $h(i) \notin |\forall x\varphi|^{\mathcal{M}_i} f \cdot i$ and $h(i) \in \sigma(i) \in Prop_i$, so (2.11) must fail. Hence there must be some $k(i) \in \sigma(i)$ and some $g(i) \in U_i$ with $k(i) \in E_i g(i) - |\varphi|^{\mathcal{M}_i} f \cdot i[g(i)/x]$. For $i \notin J'$ choose $k(i) \in W_i$ and $g(i) \in U_i$ arbitrarily. This defines $k \in \prod_I W_i$ and $g \in \prod_I U_i$.

Since $J' \subseteq \{i : k(i) \in E_i g(i)\}$ we get $k_\mu \in E_\mu g_\mu$. Whenever $i \in J'$ we have

$$k(i) \notin |\varphi|^{\mathcal{M}_i} f \cdot i[g(i)/x] = |\varphi|^{\mathcal{M}_i} f[g/x] \cdot i,$$

so $J' \cap [\![k, \varphi, f[g/x]\,]\!] = \emptyset$, and hence $[\![k, \varphi, f[g/x]\,]\!] \notin \mu$. The induction hypothesis on φ then gives

$$k_\mu \notin |\varphi|^{\mathcal{M}_\mu} f[g/x]_\mu = |\varphi|^{\mathcal{M}_\mu} f_\mu[g_\mu/x].$$

Altogether, this proves that $k_\mu \notin E_\mu g_\mu \Rightarrow |\varphi|^{\mathcal{M}_\mu} f_\mu[g_\mu/x]$, which shows that (2.10) fails, proving that $h_\mu \notin |\forall x\varphi|^{\mathcal{M}_\mu} f_\mu$, and hence completing the proof that the theorem holds for $\forall x\varphi$. □

Theorem 2 *$\mathcal{M}_\mu$ is a model if almost all of the $\mathcal{M}_i$'s are models.*

Proof Let $M = \{i \in I : \mathcal{M}_i$ is a model$\}$. Suppose that $M \in \mu$. To prove that $\mathcal{M}_\mu$ is a model we have to show that for any formula φ and any $f \in \prod_I U_i$, the set $|\varphi|^{\mathcal{M}_\mu} f_\mu$ is admissible in $\mathcal{M}_\mu$, i.e. belongs to $Prop_\mu$.

Define $\sigma \in \prod_I Prop_i$ by putting $\sigma(i) = |\varphi|^{\mathcal{M}_i} f \cdot i$ when $i \in M$, and $\sigma(i) = \emptyset$ otherwise. Note that when $i \in M$, φ is admissible in the model $\mathcal{M}_i$, so indeed $|\varphi|^{\mathcal{M}_i} f \cdot i \in Prop_i$. We will show that $|\varphi|^{\mathcal{M}_\mu} f_\mu = S(\sigma)$, giving the desired result that $|\varphi|^{\mathcal{M}_\mu} f_\mu \in Prop_\mu$.

Take any $h \in \prod_I W_i$. Then

$$[\![h, \varphi, f]\!] \cap M = \{i : h(i) \in \sigma(i)\} \cap M,$$

for if $i \in M$ then $|\varphi|^{\mathcal{M}_i} f \cdot i = \sigma(i)$, so $h(i) \in |\varphi|^{\mathcal{M}_i} f \cdot i$ iff $h(i) \in \sigma(i)$. Since $M \in \mu$ and μ is a filter, it follows that $[\![h, \varphi, f]\!] \in \mu$ iff $\{i : h(i) \in \sigma(i)\} \in \mu$. By Łoś' Theorem 1 and the definition of $S(\sigma)$, this says that $h_\mu \in |\varphi|^{\mathcal{M}_\mu} f_\mu$ iff $h_\mu \in S(\sigma)$, which gives the desired result. □

Theorem 3 *$\mathcal{M}_\mu$ is Kripkean if almost all of the $\mathcal{M}_i$'s are Kripkean.*

Proof Let $K = \{i \in I : \mathcal{M}_i$ is Kripkean$\}$. Suppose that $K \in \mu$. To prove that $\mathcal{M}_\mu$ is Kripkean we have to show that in general

$$|\forall x\varphi|^{\mathcal{M}_\mu} f_\mu = \bigcap_{g_\mu \in U_\mu} \left(E_\mu g_\mu \Rightarrow |\varphi|^{\mathcal{M}_\mu} f_\mu[g_\mu/x]\right). \tag{2.12}$$

Now from (2.2), which holds in any premodel, it follows that

$$|\forall x\varphi|^{\mathcal{M}_\mu} f_\mu \subseteq E_\mu g_\mu \Rightarrow |\varphi|^{\mathcal{M}_\mu} f_\mu[g_\mu/x]$$

for any $g_\mu \in U_\mu$. So the left to right inclusion of (2.12) holds. For the converse, suppose that $h_\mu \notin |\forall x\varphi|^{\mathcal{M}_\mu} f_\mu$. Then $[\![h, \forall x\varphi, f]\!] \notin \mu$ by Łoś' Theorem. Hence the set

$$J = (I - [\![h, \forall x\varphi, f]\!]) \cap K$$

belongs to μ. But if $i \in J$ then $h(i) \notin |\forall x\varphi|^{\mathcal{M}_i} f \cdot i$ and $\mathcal{M}_i$ is Kripkean, so there exists some element $g(i)$ of U_i with $h(i) \in E_i g(i) - |\varphi|^{\mathcal{M}_i} f \cdot i[g(i)/x]$. For $i \notin J$ choose $g(i) \in U_i$ arbitrarily. This defines $g \in \prod_I U_i$.

Since $J \subseteq \{i : h(i) \in E_i g(i)\}$ we get $h_\mu \in E_\mu g_\mu$. Whenever $i \in J$ we have

$$h(i) \notin |\varphi|^{\mathcal{M}_i} f \cdot i[g(i)/x] = |\varphi|^{\mathcal{M}_i} f[g/x] \cdot i,$$

so $J \cap [\![h, \varphi, f[g/x]]\!] = \emptyset$, and hence $[\![h, \varphi, f[g/x]]\!] \notin \mu$. Łoś' Theorem then gives

$$h_\mu \notin |\varphi|^{\mathcal{M}_\mu} f[g/x]_\mu = |\varphi|^{\mathcal{M}_\mu} f_\mu[g_\mu/x].$$

Altogether this proves that $h_\mu \notin E_\mu g_\mu \Rightarrow |\varphi|^{\mathcal{M}_\mu} f_\mu[g_\mu/x]$, showing that h_μ does not belong to the intersection on the right of (2.12), which completes the proof of (2.12). □

2.5 Compactness

We say that a formula φ is *satisfiable in* $\mathcal{M}$ if $|\varphi|^{\mathcal{M}} f \neq \emptyset$ for some $f \in U^\omega$, i.e. if $\mathcal{M}, w, f \models \varphi$ for some f and some $w \in W$. φ is *valid in* $\mathcal{M}$ if $\neg\varphi$ is not satisfiable in $\mathcal{M}$, which means that $\mathcal{M}, w, f \models \varphi$ for all $w \in W$ and $f \in U^\omega$.

If Γ is a set of formulas, we write $\mathcal{M}, w, f \models \Gamma$ to mean that for all $\varphi \in \Gamma$, $\mathcal{M}, w, f \models \varphi$. If this holds for some w and f then Γ is *satisfiable in* $\mathcal{M}$.

We say that a class $\mathbb{M}$ of premodels is *closed under ultraproducts* if, for all indexed subsets $\{\mathcal{M}_i : i \in I\}$ of $\mathbb{M}$ and all ultrafilters μ on I, the ultraproduct $\mathcal{M}_\mu$ belongs to $\mathbb{M}$.

Theorem 4 (Compactness) *Let $\mathcal{L}$ be any signature and $\mathbb{M}$ any class of premodels for $\mathcal{L}$ that is closed under utraproducts. For any set Γ of $\mathcal{L}$-formulas, if each finite subset of Γ is satisfiable in some member of $\mathbb{M}$, then Γ itself is satisfiable in some member of $\mathbb{M}$.*

Proof This follows the pattern of the standard ultraproducts proof of compactness for first-order logic.

Let $I = \{i \subseteq \Gamma : i \text{ is finite}\}$, and for each $i \in I$ put $J_i = \{i' \in I : i \subseteq i'\}$. Then the collection $\{J_i : i \in I\}$ has the finite intersection property, since for $i_1, \dots, i_n \in I$, the intersection $J_{i_1} \cap \dots \cap J_{i_n}$ contains $i_1 \cup \dots \cup i_n$. It follows that there is an ultrafilter μ on I such that $J_i \in \mu$ for all $i \in I$.

For each $i \in I$ there is by hypothesis a premodel $\mathcal{M}_i \in \mathbb{M}$ with set of worlds W_i and universe U_i such that $\mathcal{M}_i, w_i, f_i \models i$ for some $w_i \in W_i$ and some $f_i \in {U_i}^\omega$. Define a sequence $f = \langle f_0, \dots, f_n, \dots \rangle \in (\prod_I U_i)^\omega$ by putting $f_n(i) = f_i(n)$ for all $n < \omega$ and $i \in I$. Then for each $i \in I$, the sequence $f \cdot i \in U_i^\omega$ given by (2.6) is just f_i.

Now if $\varphi \in \Gamma$, consider $\{\varphi\} \in I$. For $i \in J_{\{\varphi\}}$, we have $\mathcal{M}_i, w_i, f_i \models \varphi$ as $\varphi \in i$. Hence

$$J_{\{\varphi\}} \subseteq \{i \in I : \mathcal{M}_i, w_i, f \cdot i \models \varphi\} = [\![h, \varphi, f]\!]$$

where $h(i) = w_i$ for all $i \in I$. Thus $[\![h, \varphi, f]\!] \in \mu$, and so $\mathcal{M}_\mu, h_\mu, f_\mu \models \varphi$ by Łoś' Theorem 1.

This shows that $\mathcal{M}_\mu, h_\mu, f_\mu \models \Gamma$, so Γ is satisfiable in the premodel $\mathcal{M}_\mu$, which belongs to the ultraproducts-closed class $\mathbb{M}$. □

Corollary 1 *For any set* Γ *of* $\mathscr{L}$*-formulas, if each finite subset of* Γ *is satisfiable in some (Kripkean)* $\mathscr{L}$*-model, then* Γ *is satisfiable in some (Kripkean)* $\mathscr{L}$*-model.*

Proof This follows from the theorem first by taking $\mathbb{M}$ to be the class of all $\mathscr{L}$-models, which is closed under ultraproducts by Theorem 2, and then by taking $\mathbb{M}$ to be the class of all Kripkean $\mathscr{L}$-models, which is closed under ultraproducts by Theorems 2 and 3. □

2.6 Propositional Logic

The formulas for propositional modal logic are generated from a denumerable list $\{p_n : n < \omega\}$ of propositional variables and the constant $\bot$ by using the connectives $\wedge$, $\neg$, and $\Box$. This language can be interpreted by models on a *general frame* $\mathscr{G} = (W, R, Prop)$, comprising a binary relation R on W and a set $Prop \subseteq \wp W$ closed under intersection $\cap$ complementation $-$ and the operation $[R]$, as in Sect. 2.2.

A model $\mathscr{M}$ on a general frame $\mathscr{G}$ is given by a variable assignment $|-|^{\mathscr{M}}$ such that $|p|^{\mathscr{M}} \in Prop$ for every propositional variable p. This assignment is then extended to define a truth set $|A|^{\mathscr{M}}$ for each propositional formula A, by induction on formula formation, as follows:

$$\begin{aligned} |\bot|^{\mathscr{M}} &= \emptyset. \\ |A \wedge B|^{\mathscr{M}} &= |A|^{\mathscr{M}} \cap |B|^{\mathscr{M}}. \\ |\neg A|^{\mathscr{M}} &= W - |A|^{\mathscr{M}}. \\ |\Box A|^{\mathscr{M}} &= [R]|A|^{\mathscr{M}}. \end{aligned}$$

The closure conditions on *Prop* then ensure that every formula is interpreted in $\mathscr{M}$ as an admissible proposition: $|A|^{\mathscr{M}} \in Prop$ for all propositional modal A. A formula A is *valid in the frame* $\mathscr{G}$, symbolised $\mathscr{G} \models A$, when $|A|^{\mathscr{M}} = W$ for all models $\mathscr{M}$ on $\mathscr{G}$. Thus $\mathscr{G} \models A$ when A is true at every point in every model on $\mathscr{G}$. A set S of propositional formulas is *valid in* $\mathscr{G}$, symbolised $\mathscr{G} \models$ S, when every member of S is valid in $\mathscr{G}$.

Let $\{\mathscr{G}_i : i \in I\}$ be a collection of general frames, with $\mathscr{G}_i = (W_i, R_i, Prop_i)$. If μ is an ultrafilter on the index set I, then we take the ultraproduct of the $\mathscr{G}_i$'s with respect to μ to be the structure

$$\mathscr{G}_\mu = (W_\mu, R_\mu, Prop_\mu)$$

who components were defined in Sect. 2.3. For any propositional modal formula A it can be shown [4, Corollary 1.7.13] that

$$\mathscr{G}_\mu \models A \quad \text{iff} \quad \{i \in I : \mathscr{G}_i \models A\} \in \mu. \tag{2.13}$$

In particular, if $\mathscr{G}_i \models A$ for all $i \in I$, then $\mathscr{G}_\mu \models A$. This implies

Theorem 5 *For any set* S *of propositional modal formulas, the class* $\{\mathscr{G} : \mathscr{G} \models \mathrm{S}\}$ *of all general frames validating* S *is closed under ultraproducts.* □

A *propositional modal logic* is a set S of propositional modal formulas that includes all such formulas that are Boolean tautologies or instances of the scheme

$$\mathrm{K}: \ \Box(A \to B) \to (\Box A \to \Box B),$$

and is closed under the rules of Modus Ponens (from A and $A \to B$ infer B) and Necessitation (from A infer $\Box A$).

For each general frame $\mathscr{G}$, the set $\mathrm{S}_{\mathscr{G}} = \{A : \mathscr{G} \models A\}$ of all propositional formulas valid in $\mathscr{G}$ is a propositional modal logic that is closed under the rule of uniform substitution for propositional variables. Conversely, there is a canonical frame construction showing that if a logic S is closed under uniform substitution, then it is equal to $S_{\mathscr{G}}$ for some general frame $\mathscr{G}$ (see [1, Sect. 5.5]).

2.7 Quantified Logics

For a given signature $\mathscr{L}$, a *quantified modal logic* is defined to be any set L of $\mathscr{L}$-formulas that includes all Boolean tautologies and instances of the axiom schemes listed in Fig. 2.1, and is closed under the inference rules of that Figure. A member φ of L is called an L-*theorem*, which we indicate by writing $\vdash_{\mathrm{L}} \varphi$.

Axiom schemes:

K: $\Box(\varphi \to \psi) \to (\Box\varphi \to \Box\psi)$
AI: $\forall y(\forall x\varphi \to \varphi(y/x))$, where y is free for x in φ.
UD: $\forall x(\varphi \to \psi) \to (\forall x\varphi \to \forall x\psi)$
VQ: $\varphi \to \forall x\varphi$, where x is not free in φ.

Inference rules:

MP: $\dfrac{\varphi, \ \varphi \to \psi}{\psi}$ *Modus Ponens*

N: $\dfrac{\varphi}{\Box\varphi}$ *Necessitation*

UG: $\dfrac{\varphi}{\forall x\varphi}$ *Universal Generalisation*

TI: $\dfrac{\varphi}{\varphi(\tau/x)}$, if τ is free for x in φ. *Term Instantiation*

GC: $\dfrac{\varphi(\mathsf{c}/x)}{\varphi}$, if c is not in φ. *Generalisation on Constants*

Fig. 2.1 Axioms and rules for quantified modal logics

A set Γ of $\mathscr{L}$-formulas is said to be L-consistent if there is no finite subset Γ_0 of Γ with $\vdash_L \neg \bigwedge \Gamma_0$, where $\bigwedge \Gamma_0$ is the conjunction of the members of Γ_0. In particular, a single formula φ is L-consistent iff $\{\varphi\}$ is L-consistent, which means that $\neg\varphi \notin$ L.

If S is any set of propositional modal formulas, we use the name QS for the *smallest* quantified modal logic that contains every $\mathscr{L}$-formula that is a substitution-instance of a member of S. In other words, QS is the intersection of all such quantified logics. If S is itself the smallest propositional modal logic that includes some set S_{ax} of propositional modal formulas, then QS $=$ QS_{ax} (see [5, Theorem 1.2.5], which also characterises QS-theorems in terms of derivability from substitution-instances of members of S by the axioms and rules of Fig. 2.1). Theorem 1.10.2 of [5] established the following characterisation:

> If S is any set of propositional modal formulas, then QS is characterised by validity in all models for $\mathscr{L}$ whose underlying general frame validates S.

The proof of this involves a canonical model construction that requires $\mathscr{L}$ to contain a denumerable infinity of individual constants. From now on we assume that all signatures have this property when required. It is a harmless assumption, since any logic can be conservatively extended by the addition of such constants.

The above characterisation of QS has two parts:

- *Soundness* If $\vdash_{QS} \varphi$, then φ is valid in all models for $\mathscr{L}$ whose underlying general frame validates the propositional logic S.
- *Completeness* If φ is valid in all models whose underlying general frame validates S, then $\vdash_{QS} \varphi$.

Since a formula φ is QS-consistent iff $\nvdash_{QS} \neg\varphi$, it is readily seen that completeness is equivalent to the statement

- Any QS-consistent formula φ is satisfiable in a model whose underlying general frame validates S.

Now a finite set of formulas is QS-consistent iff its conjunction is, and is satisfied at a point of a model iff its conjunction is. From this we see that completeness implies that

- Any *finite* QS-consistent set of formulas is satisfiable in a model whose underlying general frame validates S.

Strong completeness is the assertion that satisfiability holds for infinite consistent sets as well as finite ones. Here we can derive this stronger conclusion by combining completeness with an ultraproducts-based compactness argument.

Theorem 6 (Strong Completeness of QS) *If* S *is any set of propositional modal formulas, then for any signature* $\mathscr{L}$, *any* QS*-consistent set of* $\mathscr{L}$*-formulas is satisfiable in a model whose underlying general frame validates* S.

Proof Given S and $\mathscr{L}$, let $\mathbb{M}$ be the class of all models for $\mathscr{L}$ whose underlying general frame validates S. Now the property of being a model for $\mathscr{L}$ is preserved by ultraproducts (Theorem 2), as is the property of being a general frame that validates S (Theorem 5). Hence $\mathbb{M}$ is closed under ultraproducts.

Now if Γ is any QS-consistent set, then each finite subset of Γ is QS-consistent, and so is satisfiable in a member of $\mathbb{M}$ by the completeness of QS as stated above. Hence Theorem 4 implies that Γ is satisfiable in a member of $\mathbb{M}$, as required. $\square$

2.8 Strong Completeness with the Barcan Formulas

The Barcan Formula is the axiom scheme,

$$\text{BF:}\quad \forall x\Box\varphi \to \Box\forall x\varphi,$$

while the Converse Barcan Formula is

$$\text{CBF:}\quad \Box\forall x\varphi \to \forall x\Box\varphi.$$

We write L + BF and L + CBF for the least extensions of a quantified modal logic L that include BF and CBF, respectively.

Now validity of BF is often associated with the condition that a model structure has *contracting domains*: for all $w, u \in W$, wRu implies $Dw \supseteq Du$. Validity of CBF is often associated with *expanding domains*: wRu implies $Dw \subseteq Du$. However, these connections really only apply to models whose underlying frame is *full* in the sense that every set of worlds is admissible, i.e. $Prop = \wp W$. Full models are not adequate to characterise logics QS in general. *Admissible* models based on general frames are adequate, but in such models the relationship between contracting and expanding domains and the schemes BF and CBF is more complex. For instance, there exists admissible models that have contracting domains but do not validate BF. In fact there are such models that falsify BF and have *constant domains*: wRu implies $Dw = Du$. Admissible models rejecting BF even include ones with a single domain, having $D_w = U$ for all $w \in W$.

On the other hand, CBF is valid in all admissible models with expanding domains, and any logic of the form QS + CBF is characterised by models with expanding domains. But, perhaps surprisingly, these same logics are also characterised by models with *constant* domains. The class of expanding domain structures includes the constant domain ones, and these constant ones are sufficient to characterise QS + CBF, even when BF is not amongst its theorems.

What underlies these observations about QS + CBF is the perhaps more surprising fact that every logic of the form QS is characterised by models with *contracting* domains. In admissible models, imposition of this contracting domains condition does not force the general validity of any non-theorems of QS. Addition of the expanding domains condition to such models then compels the contracting domains to be constant. The work of Chap. 2 of [5] yields the following completeness results:

If S is any set of propositional modal formulas, then any finite QS-consistent set of formulas is satisfiable in a model whose underlying general frame validates S and has contracting domains.

Moreover, any finite QS + CBF-consistent set of formulas is satisfiable in a model whose underlying general frame validates S and has constant domains.

We now apply our ultraproduct construction to strengthen these facts to strong completeness results.

Theorem 7 *An ultraproduct* $\mathcal{M}_\mu = \prod_\mu \mathcal{M}_i$ *has contracting/expanding/constant domains if almost all of the* $\mathcal{M}_i$*'s have likewise.*

Proof Let $J = \{i \in I : \mathcal{M}_i$ has contracting domains$\}$ and suppose $J \in \mu$. We prove that $\mathcal{M}_\mu$ has contracting domains.

Let $f_\mu R_\mu g_\mu$. If $h_\mu \in D_\mu g_\mu$ then we have that the sets $\{i \in I : f(i) R_i g(i)\}$ and $\{i : h(i) \in D_i g(i)\}$ both belong to μ, and so the intersection

$$J \cap \{i \in I : f(i) R_i g(i)\} \cap \{i : h(i) \in D_i g(i)\}$$

belongs to μ. But the set $\{i : h(i) \in D_i f(i)\}$ includes this intersection, so it belongs to μ as well, showing that $h_\mu \in D_\mu f_\mu$. Hence $D_\mu f_\mu \supseteq D_\mu g_\mu$ as required.

The cases of expanding and constant domains, respectively, are similar. □

This theorem combines with the argument of Theorem 6, taking $\mathbb{M}$ to be the class of all contracting domains models whose underlying general frame validates S, and then restricting it those models with constant domains. In both cases, Theorem 7 implies that we get a class of models that is closed under ultraproducts. Given the above Completeness results we infer:

Theorem 8 (Contracting and Constant Domains Strong Completeness) *If* S *is any set of propositional modal formulas, then for any signature* $\mathcal{L}$*, any* QS*-consistent set of* $\mathcal{L}$*-formulas is satisfiable in a model whose underlying general frame validates* S *and has contracting domains. Moreover, any* QS + CBF*-consistent set of formulas is satisfiable in a model whose underlying general frame validates* S *and has constant domains.* □

Turning now to the Barcan Formula, we have already noted that it need not be valid in a contracting-domains model. In general it is only in *Kripkean* models with contracting domains that validity of BF is guaranteed. The real role of BF in admissible model theory is to enable us to build models that give the quantifier ∀ its standard Kripkean interpretation. In that context we also need to use the commuting quantifiers axiom

$$\text{CQ}: \quad \forall x \forall y \varphi \to \forall y \forall x \varphi$$

which is valid in Kripkean models, but not in general. In [5, Sect. 2.6] a canonical model construction was given that provides a completeness result for certain logics containing BF and which depends on the background signature being countable. The upshot is this:

If S is any set of propositional modal formulas, then for any countable signature $\mathscr{L}$, any finite QS + CBF + CQ + BF-consistent set of $\mathscr{L}$-formulas is satisfiable in a Kripkean constant-domains $\mathscr{L}$-model whose underlying general frame validates S.

We now lift this result to a strong completeness theorem, overcoming the countability restriction.

Theorem 9 (Strong Completeness for QS + CBF + CQ + BF) *If* S *is any set of propositional modal formulas, then for any signature* $\mathscr{L}$, *any* QS + CBF + CQ + BF*-consistent set of* $\mathscr{L}$*-formulas is satisfiable in a Kripkean constant-domains* $\mathscr{L}$*-model whose underlying general frame validates* S.

Proof Let $\mathbb{M}$ be the class of all Kripkean constant-domains $\mathscr{L}$-models whose underlying general frame validates S. Then $\mathbb{M}$ is closed under ultraproducts, by Theorems 2, 3, 5 and 7.

Let L be the logic QS + CBF + CQ + BF as a set of $\mathscr{L}$-formulas, and let Γ be an L-consistent set of $\mathscr{L}$-formulas. Put $I = \{i \subseteq \Gamma : i \text{ is finite}\}$. For each $i \in I$, let $\mathscr{L}_i$ be a countable subset of $\mathscr{L}$ that firstly includes all the (finitely many) members of $\mathscr{L}$ that occur in i; secondly has infinitely many constants, including some particular constant c_0; and thirdly for each positive integer n includes some particular n-ary function symbol F_n if $\mathscr{L}$ has n-ary function symbols. Then i is a set of $\mathscr{L}_i$-formulas.

Define L_i to be the logic QS + CBF + CQ + BF in the language $\mathscr{L}_i$. Then $\mathrm{L}_i \subseteq \mathrm{L}$, so if $\neg(\bigwedge i) \in \mathrm{L}_i$ we would have $\neg(\bigwedge i) \in \mathrm{L}$, contradicting the L-consistency of Γ. Hence $\neg(\bigwedge i) \notin \mathrm{L}_i$, showing i is L_i-consistent. Since the signature $\mathscr{L}_i$ is countable, the above completeness result for QS + CBF + CQ + BF implies that i is satisfiable in some Kripkean constant-domains $\mathscr{L}_i$-model $\mathcal{M}$ whose underlying general frame validates S.

Let $\mathscr{S} = (W, R, Prop, U, D)$ be the model structure of $\mathcal{M}$. We now expand $\mathcal{M}$ to an $\mathscr{L}$-premodel $\mathcal{M}'$ on $\mathscr{S}$ by declaring $\mathcal{M}'$ to be identical to $\mathcal{M}$ on $\mathscr{L}_i$, and for symbols ζ in $\mathscr{L} - \mathscr{L}_i$ putting $|\zeta|^{\mathcal{M}'} = |\mathsf{c}_0|^{\mathcal{M}}$ if ζ is a constant; $|\zeta|^{\mathcal{M}'} = |F_n|^{\mathcal{M}}$ if ζ is an n-ary function symbol; and if ζ is an n-ary predicate symbol, letting $|\zeta|^{\mathcal{M}'}$ be the n-ary function on U with constant value $\emptyset$.

For each $\mathscr{L}$-term τ, let τ' be the $\mathscr{L}_i$-term resulting from replacing any constant of τ not in $\mathscr{L}_i$ by c_0, and any n-ary function symbol of τ not in $\mathscr{L}_i$ by F_n. A routine induction on term-formation shows that in general $|\tau|^{\mathcal{M}'} = |\tau'|^{\mathcal{M}}$.

Then for each $\mathscr{L}$-formula φ, let φ' be the $\mathscr{L}_i$-formula resulting from replacing each atomic formula $P\tau_1 \cdots \tau_n$ within φ by $P\tau_1' \cdots \tau_n'$ if $P \in \mathscr{L}_i$, and by $\bot$ if $P \notin \mathscr{L}_i$. An induction on formula formation then shows that in general $|\varphi|^{\mathcal{M}'} = |\varphi'|^{\mathcal{M}}$.

It follows that $\mathcal{M}'$ is an $\mathscr{L}$-model: for any $\mathscr{L}$-formula φ and $f \in U^\omega$, $|\varphi|^{\mathcal{M}'} f = |\varphi'|^{\mathcal{M}} f \in Prop$ as $\mathcal{M}$ is an $\mathscr{L}_i$-model. So every $\mathscr{L}$-formula is admissible in $\mathcal{M}'$.

It also follows that $\mathcal{M}'$ is Kripkean. To see this, take any $\mathscr{L}$-formula φ, variable x, and $f \in U^\omega$, and let $\mathscr{X} = \{Ea \Rightarrow |\varphi|^{\mathcal{M}'} f[a/x] : a \in U\}$. So $|\forall x \varphi|^{\mathcal{M}'} f = \sqcap \mathscr{X}$. But

$$\bigcap \mathscr{X} = \bigcap_{a \in U} \left(Ea \Rightarrow |\varphi'|^{\mathcal{M}} f[a/x] : a \in U \right) = |\forall x(\varphi')|^{\mathcal{M}} f \in Prop,$$

because $\mathcal{M}$ is Kripkean and an $\mathcal{L}_i$-model. Thus $\bigcap \mathcal{X} \in Prop$, which implies that $\bigcap \mathcal{X} = \prod \mathcal{X}$. So $|\forall x\varphi|^{\mathcal{M}'} = \bigcap \mathcal{X}$, making $\mathcal{M}'$ a Kripkean model.

Now the underlying structure $\mathcal{S}$ of $\mathcal{M}'$ has constant domains and its general frame validates S. So $\mathcal{M}'$ belongs to the class $\mathbb{M}$ defined at the start of this proof. But each $\varphi \in i$ is an $\mathcal{L}_i$-formula so has $\varphi' = \varphi$, and hence $|\varphi|^{\mathcal{M}'} = |\varphi|^{\mathcal{M}}$. Since i is satisfiable in $\mathcal{M}$ it follows that it is satisfiable in $\mathcal{M}'$.

We have now established that any finite subset i of Γ is satisfiable in an $\mathcal{L}$-model belonging to the ultraproducts-closed class $\mathbb{M}$. Hence Theorem 4 implies that Γ is satisfiable in a member of $\mathbb{M}$, as required. □

2.9 One Universal Domain

A model structure $\mathcal{S}$ has *one universal domain* if $D_w = U$ for all w in $\mathcal{S}$. If this holds, then $Ea = W$ for all $a \in U$, and so $(Ea \Rightarrow X) = X$ in general. This implies that in any model $\mathcal{M}$ on $\mathcal{S}$ we have

$$|\forall x\varphi|^{\mathcal{M}} f = \prod_{a \in U} |\varphi|^{\mathcal{M}} f[a/x].$$

A model with one universal domain validates the universal instantiation axiom

$$\text{UI:}\ \ \forall x\varphi \to \varphi(\tau/x), \quad \text{where } \tau \text{ is free for } x \text{ in } \varphi.$$

It was shown in [5, Sect. 2.4] that any quantifed modal logic of the form QS + UI is complete for validity in one-universal-domain admissible models whose underlying general frame validates S.

Now it is readily seen that the property of having one universal domain is preserved by an ultraproduct $\mathcal{M}_\mu = \prod_\mu \mathcal{M}_i$. For if the set

$$J = \{i \in I : \mathcal{M}_i \text{ has one universal domain}\}$$

belongs to μ, then for any $f_u \in W_\mu$ and any $g_\mu \in U_\mu$, the set $\{i \in I : g(i) \in D_i(f(i))\}$ includes J and so belongs to μ. It follows that $g_\mu \in D_\mu f_\mu$. Hence $D_\mu f_\mu = U_\mu$, implying $\mathcal{M}_\mu$ has one universal domain.

Applying this observation to our earlier arguments, we can conclude that any logic of the form QS + UI is *strongly* complete for validity in one-universal-domain admissible models whose underlying general frame validates S.

A logic containing UI also contains the schemes CBF and CQ, but need not contain BF. For instance, BF is not derivable in QS4 + UI. Section 2.7 of [5] showed that any quantifed modal logic of the form QS + UI + BF in a countable signature is complete for validity in *Kripkean* one-universal-domain admissible models whose underlying general frame validates S. Here, our ultraproduct analysis allows us to strengthen this to conclude that, in arbitrary signatures, QS + UI + BF is *strongly* complete for validity in such models.

References

1. Blackburn, P., de Rijke, M., Venema, Y.: Modal Logic. Cambridge University Press, Cambridge (2001)
2. Chang, C.C., Keisler, H.J.: Model Theory. North-Holland, Amsterdam (1973)
3. Goldblatt, R.: Metamathematics of modal logic. Ph.D. thesis, Victoria University, Wellington (1974) (Included in [4])
4. Goldblatt, R.: Mathematics of Modality. CSLI Lecture Notes No. 43. CSLI Publications, Stanford University (1993)
5. Goldblatt, R.: Quantifiers, Propositions and Identity: Admissible Semantics for Quantified Modal and Substructural Logics. Number 38 in Lecture Notes in Logic. Cambridge University Press and the Association for Symbolic Logic (2011)
6. Goldblatt, R., Hodkinson, I.: Commutativity of quantifiers in varying-domain Kripke models. In: Makinson, D., Malinowski, J., Wansing, H. (eds.) Towards Mathematical Philosophy, vol. 28 of Trends in Logic, pp. 9–30. Springer, New York (2009)
7. Kripke, S.A.: Semantical considerations on modal logic. Acta Philosophica Fennica **16**, 83–94 (1963)

Chapter 3
Logic and/of Truthmaking

Jamin Asay

Abstract The purpose of this paper is to explore the question of how truthmaker theorists ought to think about their subject in relation to logic. Regarding logic *and* truthmaking, I defend the view that considerations drawn from advances in modal logic have little bearing on the legitimacy of truthmaker theory. To do so, I respond to objections Timothy Williamson has lodged against truthmaker theory. As for the logic *of* truthmaking, I show how the project of understanding the logical features of the truthmaking relation has led to an apparent impasse. I offer a new perspective on the logic of truthmaking that both explains the problem and offers a way out.

3.1 Introduction

What can logic teach us about truthmaking, and what can truthmaking teach us about logic? These are the questions I seek to address in this paper, which I intend to contribute to the more general ongoing discussion over the relationship between logic and metaphysics. I defend the view that while logic has no immediate implications for the theory of truthmaking (contrary to the view of several contemporary philosophers), addressing particular questions about the logic of truthmaking can help us better understand the metaphysical project that motivates and drives truthmaker theorists.

In the first main part of the paper, I explore some dimensions of the relationship between logic *and* truthmaking. Some have argued that key considerations drawn from logic all but refute the theory of truthmaking—such is the view defended by Williamson [28]. I defend truthmaker theory against such objections, and argue that, in principle, no such argument could be successfully developed. As a result, metaphysical inquiries such as truthmaker theory enjoy a limited kind of autonomy from logical investigation.

I then turn to the logic *of* truthmaking. If truthmaker theory can be rescued from the sorts of logical attacks I address in the first part of the paper, then the notion

J. Asay (✉)
The University of Hong Kong, Pokfulam, Hong Kong
e-mail: asay@hku.hk

S.C.-M. Yang et al. (eds.), *Structural Analysis of Non-Classical Logics*,
Logic in Asia: Studia Logica Library, DOI 10.1007/978-3-662-48357-2_3

of truthmaking is legitimate. Accordingly, we should seek to understand its logical features. The project of developing a theory of the logic of truthmaking has been underway for some time, but has led to a seemingly irresolvable deadlock. I offer a new perspective on the logic of truthmaking that both explains the impasse and offers a way out.

3.2 Truthmaker Theory

Before turning to the relationship between logic and truthmaker theory, it will be worthwhile to pause briefly on the nature of the latter. "Truthmaker theory" means a variety of things to a variety of people. As I shall understand it, truthmaker theory is a kind of metaphysical inquiry that subscribes to the belief that progress can be made in metaphysics by exploring what sorts of ontological posits are necessary in order to account for what is true. Truthmakers are the objects in reality in virtue of which truths are true. Because those truthmaking objects exist, the truths in question are true. So far, I shall suppose, all truthmaker theorists for the most part agree. Where they disagree is over which truths have truthmakers, what those truthmakers are, and how we are to account for the relation that obtains between a truth and its truthmakers. Sorting out those sorts of disputes is the bread and butter of those engaged in the truthmaking industry.

To take an example, consider that, necessarily, copper conducts electricity. Truthmaker theorists offer something in reality whose existence properly accounts for the truth in question. David Armstrong [2], for instance, argues that what makes it true that copper necessarily conducts electricity is (something along the lines of) a state of affairs composed of the universal *copper* and the universal *electrically conductive* standing in the second-order relational universal *necessitation.* There is a relation of necessitation, in other words, between the two properties of being made of copper and being electrically conductive. Because that state of affairs exists, anything composed of copper must also be electrically conductive, and hence it will be true that copper necessarily conducts electricity. Had *copper* failed to stand in the necessitation relation to *electrically conductive* (as it does to, say, *having atomic number 28*), then it would not be necessary that copper conducts electricity. Of course, many dispute Armstrong's particular metaphysical account of the truthmakers for laws of nature. But what everyone can acknowledge is that Armstrong, like all truthmaker theorists, is trying to come to terms with the proper ontological grounds that are necessary for understanding why certain claims are true. One can still sense the need for *something* to make true the laws of nature, even if one does not find Armstrong's own account compelling.

Truthmaker theorists, then, engage metaphysics by asking after what the truthmakers are for different truths. What makes counterfactuals true? Negative truths? Truths about possibility and necessity? Truthmaking questions can also extend into metaethics (what makes moral judgments true?), mathematics (what makes mathematical claims true?), and any other area of philosophy where metaphysical quandaries arise. (See, respectively, Asay [6] and Baron [7]).

3.3 Logic and Truthmaking

Not everyone is compelled by truthmaking as a metaphysical methodology. One particularly severe critic is Williamson [28], who has argued forcefully against the feasibility of truthmaker theory. In particular, he argues that certain compelling considerations drawn from modal logic demonstrate that truthmaker theory is incoherent. His objections are quite devastating if correct, and no one in the truthmaking literature has yet fully answered them or even really addressed them. In this section, I rebut Williamson's argument, and argue instead that no such argumentative strategy can succeed. Purely logical considerations cannot in and of themselves undermine metaphysical theories like truthmaking.

3.3.1 Williamson's Argument

Williamson's argument against truthmaker theory is simple and straightforward. He begins by articulating a thesis that he calls the "truthmaker principle," and then argues that it is inconsistent with the converse Barcan formula. But because the converse Barcan formula is true, the truthmaker principle (which is independently implausible anyway) must be false. Williamson's argumentative strategy is clear; he understands his argument as a contribution to "modal metaphysics disciplined by the rigour of modern logic" (1999: 253). In this particular conflict between a principle of logic and a principle of metaphysics, logic triumphs.

Let us examine Williamson's argument in more detail. First consider the principle he calls the "truthmaker principle." It is a form of *truthmaker maximalism*, the view that all truths have truthmakers. This thesis, while adopted by many truthmaker theorists, is not universally accepted in the truthmaking community. Some have argued, for instance, that negative existentials lack truthmakers (e.g., Bigelow [8] and Lewis [11]). Nonmaximalists could, in principle, accept Williamson's argument, as they agree that the maximalist truthmaker principle is false. But, as we shall see, Williamson's argument poses severe challenges to *any* truth with a truthmaker, regardless of whether or not all truths have truthmakers. Williamson presents the key principle under discussion as the view that, necessarily, if something is true, then there is something that exists whose existence, necessarily, guarantees that the truth in question is true. For example, since it is true that there are pandas, there must be an object that is such that, if it exists, it is true that there are pandas. Any particular panda lounging in the forests of Sichuan province would seem to provide the requisite credentials to be a truthmaker. Note that Williamson presents the view as placing only a necessary condition on truthmaking: if X is a truthmaker for Y, then

X's existence must necessitate the truth of Y. Whether or not it must do something else is of no concern to Williamson, since this minimal requirement is enough to fuel his argument.

The converse Barcan formula, meanwhile, asserts the following: if, necessarily, everything is F, then everything is necessarily F. As Williamson shows, it is a consequence of the converse Barcan formula that everything that exists exists necessarily. This result, combined with the truthmaker principle above, leads to what Williamson calls "modal collapse" (1999: 264). Suppose Penelope is one of those pandas in the forest. Since Penelope exists, she necessarily exists, by the converse Barcan formula. But the truthmaker principle holds that if Penelope makes it true that there are pandas, then in any possibility in which Penelope exists, it will be true that pandas exist. Penelope exists in every possibility, and so it turns out to be necessarily true that there are pandas. Moreover, if *every* truth has a necessitating truthmaker, and each of those truthmakers exist necessarily, then every truth is necessary: modal collapse. Any truth with a truthmaker turns out to be necessary; that is trouble enough for any truthmaker theorist, even one who rejects the view that all truths have truthmakers.

In response to the inconsistency, Williamson opts for the converse Barcan formula over the truthmaker principle. As for the former, he does not say much by way of positively defending it in the context of his anti-truthmaking argument; he relegates those arguments to elsewhere (e.g., Williamson [27]). Williamson does highlight some of the awkward consequences of denying it, and claims that supposed counterexamples to its necessary existence consequence (presumably, every object of ordinary experience, among others) can be resolved by attending to equivocation on the word "exist" (1999: 267). Furthermore, he points out that accepting the converse Barcan formula allows one to be more "bold" with one's quantified modal logic (1999: 264). Where Williamson devotes more time is in undermining the motivations for the truthmaking principle. If the choice is between an unmotivated principle of metaphysics and a highly plausible theorem of logic, then the superior alternative should be immediately obvious.

Williamson's anti-truthmaking strategy is to find an innocuous substitute principle that preserves the intent behind the truthmaker principle without succumbing to its problematic metaphysical consequences. The truthmaker principle he seeks to reject is formalized as follows:

$$\text{(TM)} \quad \Box(A \supset \exists x \Box(\exists y\, x = y \supset A))$$

Again, what (TM) says is that, necessarily, if some claim is true, then there is some object such that, necessarily, if that object exists, then the claim is true. This is one way of capturing the thought behind the words "if something is true, there must be *something* that makes it true," which Williamson accepts to be the platitudinous foundation of truthmaker theory. Williamson even allows that the platitude is true, at least on some reading. What Williamson makes a point of noticing is that the word "something" in the platitude is interpreted by truthmaker theorists as a kind of objectual quantification. Hence, (TM) requires that any time something is true, there must be some existing object whose existence guarantees the truth of the truth in question.

In response to this understanding of the idea behind truthmaking, Williamson poses a rhetorical question: "Why not treat the platitude as simply connecting the constant A in sentence position with a variable in sentence position?" (1999: 258). In other words, Williamson suggests precisifying the basic thought behind truthmaker theory without resort to objectual quantification, and offers instead the following:

$$(\text{TM}^*) \quad \Box(A \supset \exists p(p \,\&\, \Box(p \supset A)))$$

All (TM*) asserts is that, necessarily, if A is true, then there is "something" (in a nonobjectual sense) that is true and whose truth is sufficient, necessarily, for the truth of A. As Williamson points out, (TM*) is a logical truth, and does not carry the ontological implications of (TM). For example, suppose Penelope weighs 200 pounds. (TM) requires there to be some object whose existence necessitates the fact that Penelope weighs 200 pounds. Penelope herself is not such an object, since she might have existed and yet still have weighed somewhat more or less. (TM) requires a further object, such as a state of affairs or trope—objects that Williamson declares to be of "unobvious standing" (1999: 264)—that does guarantee that Penelope weighs 200 pounds. (See Armstrong [3] for a development of this style of argument.) By contrast, (TM*) requires no such ontological posit. Simply substitute "Penelope weighs two hundred pounds" for "p." After all, necessarily, if Penelope weighs 200 pounds, then Penelope weighs 200 pounds. According to Williamson, then, (TM*) captures the basic thought behind truthmaker theory without its ontological extravagances.

3.3.2 Objections to the Argument

We have now seen Williamson's anti-truthmaking argument in full. I offer two different sorts of rebuttal. First, I challenge a number of the premises of his argument. Second, I contest the overall rhetorical strategy of his argument, and its intention to discipline metaphysical inquiry by way of logical expertise.

Williamson's argument comes down to the inconsistency between (TM) and the converse Barcan formula, and the superiority of the latter. I shall focus my objections on the second pillar of the argument. As Williamson mostly relegates his support of the converse Barcan formula to elsewhere, so too will I mostly suppress my resistance to it. Any principle that entails that I am a necessary existent is extremely suspect, but I shall set aside that line of criticism for another time. I do note that there is no reason to believe that "boldness" in one's logic is more conducive to truth that being "bold" in one's metaphysical views. Williamson's preference for bold logic over bold metaphysics may well be indicative of his understanding of the relationship between logic and metaphysics, but it hardly counts as an independent argument in favor of one's logical system when it is under fire from competing views.

The bulk of my criticism is thus directed at Williamson's attack on (TM). Recall that he makes the familiar argumentative move of claiming that (TM) is an unwarranted attempt at capturing a simple platitude, given that it can be articulated by the more modest (TM*). However, there is no reason to think that (TM*) expresses anything like the basic idea driving truthmaker theory. Williamson does not offer any reason himself; he introduces (TM*) by way of the rhetorical question above, and proceeds as if the burden is on others to explain why (TM*) is insufficient as a truthmaker principle. Thankfully, that burden is rather easily met. Truthmaker theorists start from the idea that things get to be true by way of reality. Put another way, the truth-theoretic features of our world (i.e., which propositions, sentences, beliefs, or what have you are true or false) are dependent upon the nontruth-theoretic features of our world: what exists, and what properties those existing objects have. The truthmaking relation is then understood as one that obtains between a truth bearer and something from one's ontology. When truthmaker theorists ask after the truthmaker for the proposition that there are pandas, they are looking for an object—like Penelope—whose existence properly accounts for the truth of the proposition. (TM) captures this sentiment by requiring that when something is true, at the least there must be a sufficient ontological basis for it. Hence, truthmaker theorists adopt principles like (TM) and their use of objectual quantification. Truthmakers are the *objects* in reality that ground the truth values of truth bearers.

If truthmakers were not existing objects, but simply further truth-theoretic entities or facts, then the intended explanation of truth by way of ontology has not yet been given. (TM*), in stark contrast with (TM), claims that when something is true, there is something (read, again, nonobjectually) whose truth is sufficient for the initial truth. Truthmaker theorists agree, but maintain that this observation completely misses the point. One does not answer a truthmaking inquiry for a given truth by pointing to another (or the same, as (TM*) seems to allow) truth. Williamson has left completely unexplained how adopting (TM*) and ditching the appeal to objectual quantification can satisfy the idea that what is true depends upon what exists. He is correct to notice that truthmaker theorists use quantificational language in expressing the basic pull behind the idea of truthmaking; but it does not follow that any analysis of that quantification is sufficient for capturing the intended thought. (TM) satisfies the main goal of truthmaker theory by relating truths with objects in the world. By abandoning objectual quantification, (TM*) removes any possibility for doing the same.

The objectual quantification invoked by (TM) is, therefore, fundamental to the truthmaking enterprise, as it guarantees that truths are being accounted for by way of being. (TM) ensures ontological accountability. (TM*), by comparison, is ontologically silent. Consider again the fact that there are pandas. The advocate of (TM) notes that anyone with a clear ontological conscience who accepts this truth must also accept an ontology that properly grounds it, such as an ontology with pandas. (TM*) imposes no similar burden. Someone might agree that there are pandas, and cite other claims they agree with that entail that there are pandas (such as that there are pandas that live in Sichuan), in accordance with (TM*). But suppose this person has an ontological aversion to creatures like Penelope and her conspecifics. This person strikes all such things from his or her ontology. In fact, this person insists that *nothing* needs to exist in order for it to be true that there are pandas: one must just commit to some claim that entails that there are pandas. Truthmaker theorists see foul play here: one cannot accept that it is true that there are pandas and yet accept no panda into her ontology without succumbing to the worst sort of ontological bad faith. But such a person has fully respected (TM*), which, after all, says nothing about how truth is related to ontology. Should one insist that it is simply impossible or incoherent to accept the truth that there are pandas while rejecting pandas from one's ontology, this can only be because one is assuming that there are connections that must be drawn between truth and ontology, connections which (TM*) does not assert but which truthmaker theorists insist must be respected. (TM*) is an ontologically impotent principle. Williamson agrees, and finds this to be its key virtue. Yet (TM*), precisely because of its innocuousness, has no ability to account for the basic insight behind truthmaking. Perhaps Williamson feels no such pull; if so, he is not alone, as there is no shortage of critics of truthmaker theory. But to think that (TM*) speaks at all to the concerns of those who do feel truthmaking's appeal is simply indefensible.[1]

Williamson writes as if it is the *words* "Something makes a proposition true" that we know are true, though the thought expressed by the words is somehow ethereal and mysterious, such that it is spoils to the victor for whoever can defend the ontologically lightest version of what the sentence might plausibly express. But unless we have a fair grasp of what the words mean, there is nothing to find intuitive or compelling. A sentence can hardly be intuitive if we are quite unclear about what it expresses; at the least, finding an unclear sentence intuitive is worth very little weight in any rational inquiry. It is unfortunate that Williamson uncharitably reads his truthmaking opponents as being so unreflective regarding the basic concept motivating their project. Simply put, Williamson vastly underestimates truthmaker theorists' ability to articulate the basic idea that drives their metaphysical program. As a result, they are highly unlikely to take the bait Williamson offers with (TM*).

Hence, Williamson's claim that (TM*) offers a superior alternative to (TM) is baseless. If so, Williamson might still claim that (TM) is independently problematic, and so (TM*), while not offering a genuine replacement for (TM), is the best truthmaker theorists can have in a bad situation. Williamson's concern about (TM)—even setting aside its conflict with the converse Barcan formula—is that it leads truthmaker

[1] Williamson takes note of similar objections to the effect that (TM*) is not sufficiently ontologically weighty (1999: 262–264). His main response is to charge his critic with not allowing there to be a third, unexplained form of quantification that is neither objectual nor substitutional. The thrust of my comments is that it is quite obvious to all involved what sort of quantification is involved in truthmaking, and attempts to get truthmaking off the ground without it are doomed to fail. Williamson never even attempts to show how a nonontologically binding quantifier can provide the intended ontological import required by truthmaker theory. Later, Williamson will respond to this thought by charging truthmaker theorists with "ignorance or neglect of the possibilities for non-nominal quantification" (2013: 402). Williamson is unwilling to concede that truthmaker theorists have some insight into what the commitments of their guiding idea is, and that it is one that requires ontological implications. If Williamson thinks that nonobjectual quantification can save the day, he has not shown how, and so has not helped to dispel the ignorance he happily attributes to his colleagues.

theorists to the "postulation of such individuals of unobvious standing" (1999: 264). He has in mind here entities like states of affairs and tropes, the sorts of objects that truthmaker theorists posit in order to ground the truth of contingent predications, negative existentials, and others. Such entities are indeed controversial, and some have argued for more austere, nominalistically friendly accounts of truthmaking (e.g., Lewis [12] and Asay [5]). Furthermore, one might argue for nonmaximalist approaches to truthmaking that restrict the application of (TM), and similarly avoid postulating such entities (e.g., Lewis [11]).

In any event, truthmaker theorists fully admit that their posits are just that: ontological *posits* appealed to in order to fulfill a particular theoretical demand for which they have argued. So of course they are "unobvious;" that fact is not in dispute, and this does not come as news. But the reason why Williamson's charge falls particularly flat is that his ontological alternative is no less unobvious. According to Williamson, all beings—not just God, numbers, and propositions—are necessary beings. There are also some rather curious beings such as the thing that Wittgenstein could have but did not father (Williamson [27]: 258). Such a thing exists in the actual world, though not concretely, as it might have. Its existence is certainly no more obvious than the existence of the tropes that trope theorists say I'm looking at this very moment. Furthermore, in his attack on truthmaking, Williamson invokes "possible facts." As Williamson conceives them, possible facts are truthmakers for falsities. This is rather surprising, given that falsities do not have truthmakers; if they did, they would not be false. So falsities have no truthmakers, including entities called "possible facts." For Williamson, possible facts exist, and they stand in a truthmaking relationship with falsities, though not in such a way as to render those falsities true. I, by contrast, reject such objects as being theoretically unnecessary and ontologically suspect.

Williamson rejects my outright denial of possible facts because, he says, "We can sensibly ask 'How many possible truthmakers are there for [a given falsehood]?', in a sense in which the mere falsity of [that falsehood] does not answer our question" (1999: 268). In other words, Williamson here rejects the straightforward response that when something is false, nothing makes it true, and there literally is nothing that could have made it true. (If there were such a thing, it would have made the claim true, and so the falsity would not be false). On Williamson's alternative, there are things that could have made falsities true (raising the awkward question of why they do not), or there are things like mere possibilia, which in some sense exist and in some other sense do not. Williamson may well be happy to commit himself to a realm of entities that do not actually exist but still somehow manage to exist nonetheless. But there is absolutely no basis for the claim that these sorts of entities are obvious, when compared to truthmaker theorists' tropes and states of affairs. According to Williamson, his nonconcrete, nonspatiotemporal "possible facts" with their suppressed truthmaking powers are more ontologically obvious than, say, Armstrong's concrete, actual facts (which he calls "states of affairs"), which are located in space and time and constructed from the very materials given to us in empirical experience. Williamson's ontology may be correct, but he scores no points for obviousness.[2]

Hence, Williamson is in no position to claim that his preferred metaphysics is somehow less ontologically unobvious than the truthmaker theorists'. While this may be a rather small point, it does reveal a defect in Williamson's overall rhetorical strategy. Recall that he understands his argument to be an advance in metaphysics when shown the light by good attention to logic. But what closer inspection reveals is that his logic-first approach to metaphysics is already deeply metaphysically laden. This comes as no surprise to Williamson, of course, as he uses modal logic as a tool for developing and defending his preferred metaphysical views (e.g., Williamson [29] and [30]). Yet Williamson also believes himself to have shown that truthmaker theory is incoherent, because of the converse Barcan formula. In fact, however, the most he has shown is that anyone who accepts the converse Barcan formula must reject truthmaker theory as being incoherent. As a result, Williamson is guilty of dialectical overreach.[3] There are probably countless modal logics that are inconsistent with truthmaker theory (and other metaphysical theories). Truthmaker theorists should respond that such modal logics are not correct; they should say the same thing about the converse Barcan formula.

More generally, one's preferred modal logic is either neutral or committed with respect to the tenability of truthmaker theory. If the logic is neutral, then considerations drawn from it will have no bearing on the truth or falsity of truthmaker theory. If the logic is inconsistent with it, then the logic carries its own metaphysical baggage, and those metaphysical implications receive no special priority simply because they are associated with some particular logic. Anyone who wields a logic with the intent of attacking a metaphysical view is, to borrow Bradley's phrase, a "brother metaphysician." As a result, there seems to be no reason to think that logic has any special implications for metaphysical theories like truthmaker theory, or any other special status not belonging to any other realm of inquiry. Of course, if truthmaker theory contradicts some true theorem of logic, then truthmaker theory is false. But by the same token, if truthmaker theory contradicts some true claim of physics, then truthmaker theory is false. Logic has no privileged role to play in assessing truthmaker theory.[4]

---

[2] Without doubt, Williamson would take issue with my casual wielding of "exist," a word he oddly would prefer to be stricken from philosophy (1998: 259). That may be so, and attention to casual presuppositions concerning quantification in natural language is essential. But my purpose here is not to claim that the truthmaker theorist's view is true, or does not face problems of its own; it is simply to demonstrate that Williamson's implication that his requisite ontology is somehow more obvious is meritless.

[3] This is a charge he may well now accept. In his subsequent discussion of truthmaking (2013: 391–403), Williamson frames the discussion as why those who accept his metaphysical views must reject truthmakers, rather than as a direct assault on truthmaker theory itself. So perhaps he would now concede my objection. He does, in addition, repeat his arguments to the effect that truthmaker theory is unmotivated, though they suffer the same problems addressed above.

[4] Williamson, in later work [29], has developed a substantial metaphysical methodology that places enormous weight on considerations dealing with quantified modal logic, and it is not my intent here to claim to have undermined that much larger project. I certainly have offered no competing

3.4 The Logic of Truthmaker Theory

In the previous section, I argued that truthmaker theory's tenability is not immediately threatened by its inconsistency with particular logical views. Logic and metaphysics enjoy a kind of independence from one another: when conflicts arise, neither field enjoys a privileged position. Or, perhaps to put the point more accurately, logic and metaphysics are already intertwined with one another, and so neither emerges as an Archimedean point by which to judge the other. In this section, I turn to the logic *of* truthmaking. Given the viability of the notion of a truthmaker, we want to have an account of how best to reason with it. If an object is a truthmaker for some truth bearer, what sorts of further inferences may we draw? Research on this topic has been quite fruitful, but has led to a deadlock. My contention is that there is a deep lesson about the nature of truthmaking to be learned by attending to this seemingly irresolvable conflict about the correct logic of truthmaking. How one conceives of the logic of truthmaking is fundamentally connected to how one conceives of the very point and purpose of truthmaking.

3.4.1 Some (conflicting) Truthmaking Principles

One way to think about the logic of truthmaking is to consider some of the logical principles that help explain how the truthmaking relation works. Many of these have been articulated and defended in the literature. First consider this pair of disjunction principles:

(D_1) If T makes true <P ∨ Q>, then T makes true <P> or T makes true <Q>.[5]
(D_2) If T makes true <P> or T makes true <Q>, then T makes true <P ∨ Q>.

(D_2) has been with truthmaker theory from the beginning (see Russell [21]: 39). (D_1), as we shall see, is quite contentious. Consider also the similar pair of conjunction principles:

(C_1) If T makes true <P ∧ Q>, then T makes true <P> and T makes true <Q>.
(C_2) If T makes true <P> and T makes true <Q>, then T makes true <P ∧ Q>.

(Footnote 4 continued)
metaphysical methodology. My intent is merely to show why Williamson's purported refutation of truthmaker theory falls well short of the mark. Truthmaker theorists have no independent reason to accept the converse Barcan formula, and Williamson's challenges to the independent reasons to accept truthmaker theory are quite shallow. For direct criticism of Williamson's project, see Sullivan [25]. For an alternative view more sympathetic to truthmaking that also draws tight connections between logic and metaphysics, see Angere [1].

5 '<*p*>' is shorthand for 'the proposition that *p*'.

The second principle is again less controversial than the first. Notice that (C_1), like (D_2), follows from a more general principle, the entailment principle, which has also been much discussed:

(E) If T makes true <P> and <P> entails <Q>, then T makes true <Q>.

All of these principles have struck some in the truthmaking literature as fairly compelling. But it is well known that together they produce a devastating conclusion. (The argument is originally due to Restall [17].) According to standard models of entailment, every contingent truth entails every necessary truth, including the instances of the law of excluded middle. For example, <Pandas exist> entails <Gophers are amphibians or gophers are not amphibians> because it is impossible for the former to be true and the latter false (simply because it's impossible for the latter to be false). Suppose again that Penelope is a truthmaker for <Pandas exist>. By (E), she is also a truthmaker for <Gophers are amphibians or gophers are not amphibians>. By (D_1), we infer that Penelope is a truthmaker for either <Gophers are amphibians> or <Gophers are not amphibians>. We know that <Gophers are amphibians> is false, and has no truthmaker, so Penelope is a truthmaker for <Gophers are not amphibians>. Generalizing away, we see that every truthmaker is a truthmaker for every truth.

Responses to this argument run the gamut. One might reject (D_1): truthmakers for disjunctions are not necessarily truthmakers for the disjuncts (e.g., Read [16], López de Sa [13], and Tałasiewicz et al. [26]). One might accept (E), but only on a reading of entailment that denies that everything entails necessary truths (e.g., Restall [17] and Armstrong [3]). Gonzalo Rodriguez-Pereyra [19, 20] accepts (D_1) but rejects (E) outright, regardless of how entailment is understood (cf. O'Conaill and Tahko [15]). He has a number of reasons for doing so, most notably because (E) entails (C_1), which he thinks is false. (See Jago [9] for an argument that this combination of positions is unstable.) His view will provide the central focus of my discussion of how the logic of truthmaking can help us understand the nature of truthmaking.

Rodriguez-Pereyra's central contention is that (C_1) is open to counterexample. Take the conjunction <There are pandas and there are gophers>. Suppose Goober is a gopher. One plausible truthmaker for the conjunction is something along the lines of the mereological sum Penelope + Goober. However, Penelope + Goober is *not*, says Rodriguez-Pereyra, a truthmaker for either <There are pandas> or <There are gophers>, despite being a truthmaker for their conjunction. Neither proposition, he reasons, is true in virtue of that mereological sum. Indeed, they are true in virtue of parts of that sum, but not the complete sum. So the sum is not a truthmaker for the individual conjuncts. Hence, Rodriguez-Pereyra concludes that (C_1), and (E) along with it, are false.

A more common view of these kinds of cases is that while Penelope + Goober is not the *only*, or the most *minimal* truthmaker for the individual conjuncts, it is one of their truthmakers nevertheless.[6] After all, truths need not have just a single truthmaker, and the existence of the mereological sum metaphysically guarantees

[6] See O'Conaill and Tahko [15] for an account of minimal truthmakers.

the truth of both conjuncts. Against this reasoning, Rodriguez-Pereyra maintains that "a conjunctive fact is what a certain proposition is true in virtue of only if *all* the conjuncts contribute to the truth of the proposition. When some but not all the conjuncts of a conjunctive fact contribute to the truth of a certain proposition, the proposition is true in virtue of a part of the conjunctive fact, but not in virtue of the conjunctive fact itself" (2006: 972). The basic idea is that the mereological sum contains extraneous parts that are completely irrelevant to the truth of the proposition in question. Because truthmaking is a relation that accounts for what parts of reality genuinely make true a proposition, the inclusion of excess ontology disqualifies the entity from being a truthmaker. <There are pandas> is not true in virtue of Goober in any way at all, and so is not true in virtue of anything which includes Goober even as a part.

At this juncture, we may appear to be at an impasse, or simply a clash of intuitions. There are those who, like Armstrong and López de Sa, judge that Penelope + Goober is a truthmaker for <There are pandas>, and so see no problem with (C_1). And there is Rodriguez-Pereyra, who judges that it is not a truthmaker, and so rejects both (C_1) and (E). Both camps are aware of the extraneous parts belonging to Penelope + Goober. Where they disagree is whether or not that nullifies the truthmaking in question. It is unclear what further source of evidence one could consult to settle the matter, so it is tempting to conclude that there is nothing more to be said than that the two parties, equipped with irreconcilable judgments, must agree to disagree. I, however, find this response quite unsatisfying. In fact, I believe we can discern a fairly fundamental lesson for truthmaker theory here by analyzing the disagreement. The reason why the two camps diverge lies in what they conceive the main goals of truthmaker theory to be.

3.4.2 Two Approaches to Truthmaking

Rodriguez-Pereyra sees in the notion of truthmaking a special kind of matching. For any given truth, there are parts of reality that are relevant to its being true, and parts that are irrelevant. The goal of truthmaker theory, then, is to determine which truths match which parts of reality. Failing to discern the appropriate matching means that the truth in question is left unaccounted for. At the risk of deploying an overused and widely abused term, one way of describing Rodriguez-Pereyra's understanding of truthmaking is as of being a kind of *explanatory* project. Faced with some truth, that truth is to be explained by the parts of reality that are responsible for its truth. If a proffered truthmaker contains extraneous parts, we have given a bad explanation: the truth is not true in virtue of *that* slice of reality; it is some other portion that is responsible. So conceived, truthmaker theory seeks to give a special kind of ontological explanation to truths. The upshot is that truths and their truthmakers must fit together just right; there is little flexibility in the relationship between a truth and its ontological ground. The idea, it seems to me, is highly reminiscent of the traditional correspondence theory of truth, which also relied on a close kind of matching between truths and facts (or whatever the corresponding objects were supposed to be). Whether that matching was a kind of congruence between truth and object or some sort of correlation was up for debate. (See Kirkham [10]: 119–120.) The explanatory approach takes truthmaker theory's business to be offering a necessary kind of explanation of truths, much as the traditional correspondence theory did.[7]

Consider now a different entry into the idea of truthmaking. Armstrong reports that his initial attraction to the idea of a truthmaker came from his (and Charlie Martin's) assessment of the failings of metaphysical views like behaviorism and phenomenalism (2004: 1–3). These views happily committed to certain counterfactual truths like <If I were to go to the quad, I would have a sense impression of a tree> and <If I were asked the capital of Argentina, I would answer "Buenos Aires">; they might even "reduce" the existence of ontological posits like unperceived objects and mental states down to the truth of such counterfactuals. But to take such claims as true, but deny that there is any underlying reality that makes them true, is to treat the counterfactuals as brute truths—truths that "float free" of reality. The existence of such inexplicable truths is no improvement over the alternative of accepting the straightforward ontological commitments that accompany the counterfactuals. In the previous section, I highlighted the even less tenable view that accepts <There are pandas> as true while refusing to ontologically commit to any pandas. Truthmaker theorists find fault with anyone who is willing to commit to certain truths but unwilling to commit to a sufficient ontological basis for them. This way of thinking about truthmaking presents it as a kind of ontological *accounting*: the theories we accept as true impose crucial constraints on what sorts of ontologies we are entitled to accept. Truthmaking as accounting keeps us ontologically honest: we consider and commit to the right kind of ontology that can fund all the claims we take to be true. With the accounting idea in mind, it makes sense why adding extraneous parts to a truthmaker does not destroy its truthmaking capacities. If the truth of <There are pandas> is fully accounted for by Penelope, then it is fully accounted for by Penelope + Goober. Those who offer the mereological sum as a truthmaker for the conjunction have done their ontological due diligence; no one can accuse them of cheating on their ontological taxes, as it were.

My hypothesis for explaining the deadlock between theorists like Rodriguez-Pereyra and theorists like Armstrong and López de Sa is that because both conceptions of truthmaking are operant in the literature, and they have not been cleanly distinguished from each other, they inform our judgments about particular cases in multiple and sometimes conflicting ways. As a result, there is no universally agreed upon conception of why truthmaking is important, what its theoretical roles are, and how theories of truthmaking should be developed. To conclude my remarks,

---

[7] Which is not to say that all theories of truthmaking are attempts at theories of truth. On my view, explaining the nature of truth itself and the nature of truthmakers are independent philosophical projects, though they can come together (as they do in the traditional theories of truth). See Asay [4]: 125–127.

I would like to consider some of the issues raised by drawing this distinction between explanatory and accounting truthmaking, and how we might move forward from here.

3.4.3 Moving Forward

First, I would like to stress that my view is not simply that Rodriguez-Pereyra and Armstrong and the others are talking past one another. That they have different philosophical views about the nature of the truthmaking relation does not show that they're engaged merely in a verbal dispute. I am suggesting that the very clear disagreement they have—over the status of purported counterexamples to (C_1)—is best explained by presuppositions about the enterprise that have not been fully articulated. Now, the ideas behind both the explanatory and accounting notions of truthmaking are familiar and widespread; I am not suggesting that truthmaker theorists have failed to notice these underlying approaches. To the contrary, I believe that both ideas have made an impact on all truthmaker theorists. The discussion of truthmaking as being a kind of explanatory relation is quite robust. (See, e.g., Smith and Simon [24], Sanson and Caplan [22], and Schulte [23].) The notion of truthmaking as ontological accounting, on the other hand, fits well with the idea of truthmaking as a kind of "cheater catching" (as defended by Merricks [14]), though I do not care for the language of "cheating". What has not been noticed, I am suggesting, is that these two angles on truthmaker theory are potentially in conflict with one another, and thus there is an underlying tension in the truthmaking literature that needs to be addressed. The explanatory and accounting notions are both widely in play in contemporary truthmaker theory, and while for most intents and purposes they are complementary approaches, they do inevitably butt heads, as demonstrated by the argument over (C_1).

One question that inevitably arises from drawing the contrast is: supposing the two genuinely do conflict, which notion is the correct account of the truthmaking relation? In response, I am fairly wary of the idea that there is some privileged relation properly bearing the name "truthmaking," and that of our two candidates, at most one of them is deserving of it. I think that a better analysis of the situation is that there is one relation—call it 'T_E'—that Rodriguez-Pereyra detects between <There are pandas> and Penelope, but not between <There are pandas> and Penelope + Goober. And there is another relation—call it 'T_A'—that Armstrong and others find obtaining between <There are pandas> on the one hand, and both Penelope and Penelope + Goober on the other. For both relations, we can ask whether they are theoretically illuminating, whether they hold for all or only some truths, whether they can answer important explanatory questions, and whether they deserve philosophical investigation and analysis. We can ask, in other words, about which relation deserves our attention as theorists interested in the kinds of metaphysical questions that truthmaker theorists have been exploring. Rodriguez-Pereyra would answer that T_A is not a particularly interesting relation; it at least does not serve the purpose of explaining how truth bearers get to be true. Other theorists might respond that T_E

simply does not exist (there is not such a connection between truths and objects in the world), or that far fewer truths stand in it than theorists like Rodriguez-Pereyra suppose.

Though I cannot settle the matter here, I would like to voice a few considerations that suggest that truthmaker theory is better suited for embracing T_A as its core notion. First, taking T_E as the core truthmaking relation threatens to call into doubt some other paradigm instances of the truthmaking relation. For instance, Penelope is typically thought to stand in the truthmaking relation to <There are pandas>. What is unclear is how we can explain how Penelope stands in T_E to <There are pandas>. The proposition <There are pandas> does not appear to be true in virtue of *Penelope*. Certainly, Penelope's existence is not necessary for the truth of <There are pandas>. Similarly, it is odd to think that the truth of <There are pandas> depends upon the existence of Penelope. Penelope could never have existed, and yet that would have had no effect at all on the truth of <There are pandas>. That is some reason to think that there is no dependence at work here. Yet truthmaking, at least understood along the lines of T_E, is a kind of dependence: truths depend on their truthmakers for their truth. What the truth of <There are pandas> seems to depend on is there being some panda or other, not on Penelope or any other panda in particular. But "there being some panda or other" is not the name of an entity—not of any uncontentious entity, anyway—and so it is unclear why we should think that Penelope stands in T_E to <There are pandas>. By contrast, it is perfectly clear why Penelope stands in T_A to <There are pandas>. Her existence is metaphysically sufficient for the truth of the proposition. An ontological commitment to Penelope is more than enough to account for the truth of <There are pandas>. Theorists relying on T_A therefore have a much simpler time accounting for the judgment that Penelope is indeed a truthmaker for <There are pandas>. <There are pandas> might indeed stand in T_E to Penelope, but some work needs to be done to show why, and in a convincing and non-*ad hoc* way.

T_E theorists also face the challenge of articulating the kind of explanations that truthmakers are supposed to offer. Take, for instance, the fact that snow is white. Truthmaker theorists often make the claim that this fact (by which I mean "true truth bearer") has a truthmaker, and that this truthmaker explains the truth of the fact. But here is another explanation, quickly found on the Internet:

> Snow is a whole bunch of individual ice crystals arranged together. When a light photon enters a layer of snow, it goes through an ice crystal on the top, which changes its direction slightly and sends it on to a new ice crystal, which does the same thing. Basically, all the crystals bounce the light all around so that it comes right back out of the snow pile. It does the same thing to all the different light frequencies, so all colors of light are bounced back out. The "color" of all the frequencies in the visible spectrum combined in equal measure is white, so this is the color we see in snow, while it's not the color we see in the individual ice crystals that form snow.[8]

This explanation, of course, makes no reference to truthmakers. Those skeptical of truthmaker theory will wonder why such explanations are insufficient for explaining

[8]http://science.howstuffworks.com/nature/climate-weather/atmospheric/question524.htm (Accessed 28 Jan, 2015).

the truth of <Snow is white>. Truthmaker theorists might respond by insisting that there is a distinctive ontological kind of explanation that only truthmakers can speak to. In that case, we are owed an account of what this relation is, which must be something that goes above and beyond the T_A theorist's accounting demand. I do not intend to claim that no such account can be given (but see Tałasiewicz et al. [26]: 601–603), but rather that this is a substantial hurdle faced by the advocate of T_E and avoided by adopting T_A.

Another challenge for T_E is developing a sufficiently precise account of the "matching" that the relation supposes to hold between truths and their truthmakers. If adding Goober to Penelope is enough to nullify Penelope's being a truthmaker for <There are pandas>, the question arises as to how much one can add to or subtract from Penelope and still end up with a valid truthmaker. After all, one might consider Penelope herself to be a mereological sum, in which case we must ask whether she has any parts extraneous to the truth of <There are pandas>. Presumably, Penelope could shed all sorts of parts (some fur, a limb, the bamboo currently digesting in her stomach) without sacrificing the truth of <There are pandas>. But if so, then it seems that we should be tolerant of extraneous material belonging to Penelope. If Goober is indeed an extraneous addition gone too far, the T_E theorist owes us an explanation as to which parts, however negligible, disrupt or are required for the necessary matching to obtain. T_A theorists might face a similar question when it comes to accounting for an object's *minimal* truthmakers: how much of Penelope can one subtract while still having a truthmaker for <There are pandas>? But T_A theorists are not committed to the view that all truths have minimal truthmakers: some might not have them at all (see Armstrong [3]: 21–22). Nor is their central theoretical concern finding minimal truthmakers for every truth. Honest ontological accounting comes first; exploring further details is a worthwhile enterprise, but not a matter that puts pressure on understanding the core relation of the whole theory.

Finally, one theoretical disadvantage facing the T_E theorist is that it may be more difficult to defend a nonmaximalist truthmaker theory. Recall my suggestion that the tight connection that T_E assigns between truth and truthmaker is reminiscent of the traditional correspondence theory of truth. According to that theory, truths are explained by way of their standing in a particular relation of correspondence to parts of reality. The correspondence theory is a theory of *truth*; it takes the nature of truth to be something that requires a distinct kind of metaphysical explanation. That explanation is common for all truths: any and all truths are accounted for by way of their corresponding with reality. (The lack of a need for a common explanation of truths in this manner is the calling card of deflationary theories of truth.) There can be no "non-maximalist correspondence theory": if truth *is* correspondence with reality, then something cannot be true without corresponding with reality. I detect a similar thought behind Rodriguez-Pereyra's insistence that <There are pandas> needs *Penelope*, not Penelope + Goober, in order to be true. When truthmaking moves beyond simply keeping your ontological books up to date, it wanders into the territory of taking *truth* itself to be something in need of a unique kind of metaphysical explanation. If so, then taking some truths to lack truthmakers is at odds with the stronger truthmaking project represented by T_E. For such views, all truths need truthmakers because without them, the truth of truth bearers goes unexplained and unaccounted for.[9]

Maximalism is less necessary to truthmaking when equipped merely with T_A. If truthmaking is not out to explain the nature of truth itself, it is free to consider that when it comes to some truths, nothing ontologically is needed to properly ground them. The classic example is negative existential truths. It is true that there are no saber-toothed tigers left in 2015. As a negative existential, it makes a claim exclusively about what *does not* exist, and so it is at least nontrivial to claim that it needs something that *does* exist in order to be true. It is open, in principle, to the defender of T_A to think that some truths just do not need truthmakers. (Analytic truths are another potential example). Now, the way to think about negative existentials is a longstanding and much-disputed (if not the most disputed) topic in truthmaker theory. My claim is that T_A gives us more theoretical flexibility in our thinking about the ontological implications of negative truths, since it is not committed to maximalism from the outset, as T_E appears to be.

One final implication of taking T_A as central to truthmaker theory is that it may offer some resistance to the now seemingly universal adoption of the view that not all objects make true necessary truths. The Grand Canyon, so says common wisdom, necessitates the truth of <7+5=12>, but does not make it true. Most theorists accept this perspective on this and similar cases, and thus seek a hyperintensional account of the truthmaking relation. Even those who have developed the ontological accounting idea of truthmaking—notably Armstrong—feel the pull of the problem of necessary truths. But the problem is felt most keenly given T_E, as there's no apparent explanatory connection between America's most magnificent geological formation and Kant's favorite piece of arithmetic. If truthmaking is more about covering your ontological bases than it is about providing explanations of truth, then it becomes less obvious that necessary truths even need truthmakers. After all, many necessary truths appear not to depend on anything in order to be true—they would be true regardless of what does or does not exist.[10] In any event, the important observation is that even prominent voices in the truthmaking literature are pulled both by T_A and T_E. If my contention that we cannot have both is correct, then some of the developed consensus in the literature needs rethinking.

All in all, I am suggesting that developing truthmaker theory along the lines of T_A instead of T_E is theoretically advantageous, and may bypass some of the worries and objections that have been offered against various kinds of truthmaker theories over the years. Ultimately, my claim is that our thinking about truthmaking has been drawing

---

[9] As it turns out, Rodriguez-Pereyra at most commits himself to maximalism only with respect to some set of synthetic truths (2005: 18). I cannot say how he might respond to this line of reasoning that suggests an internal tension between his nonmaximalism and adoption of something like T_E, as he has not directly argued for his restriction of truthmaking to a certain class of synthetic truths.

[10] In my view, developed elsewhere, the distinction between analytic and synthetic truths is of greater relevance to the question of which truths have truthmakers than is the distinction between contingent and necessary truths. If there are synthetic necessary truths (e.g., <God exists>), then they would seem to depend upon the existence of certain (necessary) beings. But the same is not obviously true for analytically necessary truths.

on the notions behind both T_A and T_E, and that this mixed source of ideas explains a variety of judgments that are taken for granted in the truthmaking literature. Yet this diverse spring of inspiration leads to conflict, since it is not obvious how to reconcile the inconsistencies that dwell within it. Analogously, it seems that our moral thinking has both utilitarian and deontological dimensions to it; it is this mixed bag that leads to compelling counterexamples to *both* kinds of theories. For truthmaker theory to make progress, it must also recognize these conflicts; only by doing so can it start to develop a systematic metaphysical theory.

Acknowledgments Versions of this paper were presented at the Taiwan Philosophical Logic Colloquium at National Taiwan University in October 2014, and at the Korean Society for Analytic Philosophy and Pluralisms Global Research Network Workshop in Seoul in November 2014. My thanks go to the organizers and participants for their very constructive feedback, as well as to the referee for this volume, Maegan Fairchild, and Jack Yip for their helpful input and discussion of the material. The work described in this paper was substantially supported by a grant from the Research Grants Council of the Hong Kong Special Administrative Region, China (HKU 23400014).

References

1. Angere, S.: The logical structure of truthmaking. J. Philosl. Log. **44** (4), 351–374 (2015)
2. Armstrong, D.M.: What is a Law of Nature?. Cambridge University Press, Cambridge (1983)
3. Armstrong, D.M.: Truth and Truthmakers. Cambridge University Press, Cambridge (2004)
4. Asay, J.: The Primitivist Theory of Truth. Cambridge University Press, Cambridge (2013)
5. Asay, J.: Truthmaking for modal skeptics. Thought **2**, 303–312 (2013)
6. Asay, J.: Truthmaking, metaethics, and creeping minimalism. Philos. Stud. **163**, 213–232 (2013)
7. Baron, S.: A truthmaker indispensability argument. Synthese **190**, 2413–2427 (2013)
8. Bigelow, J.: The Reality of Numbers: A Physicalist's Philosophy of Mathematics. Clarendon Press, Oxford (1988)
9. Jago, M.: The conjunction and disjunction theses. Mind (New series) **118**, 411–415 (2009)
10. Kirkham, R.L.: Theories of Truth: A Critical Introduction. MIT Press, Cambridge (1992)
11. Lewis, D.: Truthmaking and difference-making. Noûs **35**, 602–615 (2001)
12. Lewis, D.: Things qua truthmakers. In: Real Metaphysics: Essays in Honour of D. H. Mellor (eds.) Hallvard Lillehammer and Gonzalo Rodriguez-Pereyra, London: Routledge 25–42 (2003)
13. López de Sa, D.: Disjunctions, conjunctions, and their truthmakers. Mind (New Series) **118**, 417–425 (2009)
14. Merricks, T.: Truth and Ontology. Clarendon Press, Oxford (2007)
15. O'Conaill D., Tahko, T.E.: Forthcoming. Minimal truthmakers. Pacific Philosophical Quarterly
16. Read, S.: Truthmakers and the disjunction thesis. Mind (New series) **109**, 67–80 (2000)
17. Restall, G.: Truthmakers, entailment and necessity. Australas. J. Philos. **74**, 331–340 (1996)
18. Rodriguez-Pereyra, G.: Why truthmakers. In: Beebee, H., Dodd, J. (eds.) Truthmakers: The Contemporary Debate, pp. 17–31. Clarendon Press, Oxford (2005)
19. Rodriguez-Pereyra, G.: Truthmaking, entailment, and the conjunction thesis. Mind (New series) **115**, 957–982 (2006)
20. Rodriguez-Pereyra, G.: The disjunction and conjunction theses. Mind (New series) **118**, 427–443 (2009)
21. Russell, B.: The philosophy of logical atomism (lectures 3–4). The Monist **29**, 32–63 (1919)
22. Sanson, D., Caplan, B.: The way things were. Philos. Phenomenolog. Res. **81**, 24–39 (2010)

23. Schulte, P.: Truthmakers: a tale of two explanatory projects. Synthese **181**, 413–431 (2011)
24. Smith, B., Simon, J.: Truthmaker explanations. In: Monnoyer, J.-M. (ed.) Metaphysics and Truthmakers, pp. 79–98. Ontos Verlag, Frankfurt (2007)
25. Sullivan, M.: Modal logic as methodology. Philos. Phenomenol. Res. **88**, 734–743 (2014)
26. Tałasiewicz, M., Odrowąż-Sypniewska, J., Wciórka, W., Wilkin, P.: Do we need a new theory of truthmaking? some comments on disjunction thesis, conjunction thesis, entailment principle and explanation. Philosophical Studies **165**, 591–604 (2013)
27. Williamson, T.: Bare possibilia. Erkenntnis **48**, 257–273 (1998)
28. Williamson, T.: Truthmakers and the converse barcan formula. Dialectica **53**, 253–270 (1999)
29. Williamson, T.: Modal Logic as Metaphysics. Oxford University Press, Oxford (2013)
30. Williamson, T.: Logic, metalogic and neutrality. Erkenntnis **79**, 211–231 (2014)

Chapter 4
Structural Models for Williamson's Modal Epistemology

Duen-Min Deng

Abstract In this paper, I examine Williamson's [15] counterfactual-based account of modal epistemology. I argue that such an account faces two serious problems—the cotenability problem and the gap problem. As I diagnose it, these problems somehow indicate that our standard way of understanding counterfactuals under the received possible-worlds semantics may have insufficient 'structures' to distinguish various different kinds of constraints on our counterfactual thinking. The remedy, I suggest, is to invoke the 'structural semantics' as developed by Pearl [10] and Halpern [4]. Based on this semantics, I offer some philosophical elucidation for various kinds of modality, and thereby provide a more satisfactory account of how our modal knowledge can be grounded in our knowledge of counterfactuals.

Keywords Structural models · Modal epistemology · Williamson · Causal necessity · Counterfactuals

4.1 Introduction

It seems undeniable that we have knowledge of many modal truths. We know, for instance, that the train could have travelled faster than it did, but it could not have travelled faster than light. We also know that water by nature *has to* be H_2O, and that gold by nature *has to* be the element with atomic number 79, etc. But how do we know these things? What could be the cognitive mechanism for such modal knowledge? To this question, Williamson [15] offers an ingenious answer by proposing a counterfactual-based account of modal epistemology. On this account, it is our cognitive capacity to handle counterfactual conditionals which provides what we need to handle modal claims. The idea, briefly, is that we can know something to be

I would like to thank the Ministry of Science and Technology of Taiwan (MOST) for the financial support (Project: 102-2410-H-002 -229 -MY2).

D.-M. Deng (✉)
National Taiwan University, Taipei, Taiwan
e-mail: dmdeng@ntu.edu.tw

S.C.-M. Yang et al. (eds.), *Structural Analysis of Non-Classical Logics*,
Logic in Asia: Studia Logica Library, DOI 10.1007/978-3-662-48357-2_4

impossible if our counterfactual development of its supposition yields a contradiction. As a result, the epistemology of metaphysical modality is just a special case of the epistemology of counterfactual conditionals.

However, it is not always clear how the account works when we consider some concrete cases. For example, consider

(G) Gold is the element with atomic number 79.

Many philosophers after Kripke regard (G) as a metaphysical necessity. On Williamson's account, this is to develop counterfactually the supposition that gold is *not* the element with atomic number 79, so as to see whether it yields a contradiction. But apparently, there is no contradiction thus engendered simply by this counterfactual development, and so the account needs to say something more. Being aware of the problem, Williamson suggests that it is part of our practice in evaluating a counterfactual conditional to *hold something fixed*, and so if we hold the right facts fixed (e.g. (G) itself), we can indeed get the required contradiction, and thus come to know the necessity of (G).

Whilst I am quite sympathetic to this general picture, I think Williamson's account fails to deal with cases like (G) by such a cotenability-based treatment. One of the main difficulties comes from the old problem of cotenability: it is not entirely clear *which* facts we should hold fixed and *when*. If we happen to hold (G) fixed as Williamson suggests, and thus come to know its necessity, this *modal* knowledge will then have no further ground beyond whatever is our reason for holding it fixed. This leads many commentators to regard Williamson's account as circular or unilluminating (see Sect. 4.2 below for discussions). But I think the problem is much deeper than this. Our reason for holding (G) fixed may be that (G) is what Williamson calls a *constitutive fact*, which represents a certain 'structure' of the world that should be kept invariant under various counterfactual thinking. But when we are to consider what would have happened *if gold were to have a different atomic number*, there seems to be no reason why we should continue to hold (G) fixed. We indeed hold (G) fixed in many counterfactual evaluations, but it also seems that we may allow (G) to break down in certain cases.

Take for another example the laws of nature and the corresponding *nomic* necessity. It is widely agreed that in evaluating ordinary *causal* counterfactuals we should hold the relevant laws fixed. But when we consider the laws themselves, inquiring *in what sense* these laws are said to be necessary, it would be quite implausible to say that a law is necessary simply because in envisaging its violation we are to hold *that very law* fixed. Knowledge of laws can indeed be a ground for knowledge of certain counterfactuals, and for knowledge of the corresponding causal possibilities; but knowledge of laws can hardly be a ground for our knowledge of their own necessity. Here again, a certain worry of circularity or self-groundedness seems to arise. But I think the problem goes much deeper. For there is indeed a sense in which laws are to be held fixed in evaluating counterfactuals, as they also represent a certain (causal) 'structure' of the world that should be kept invariant; but there is also a sense in which laws can be violated. This is why laws are sometimes felt to be 'necessary' and sometimes 'contingent'. The problem, then, is to provide a satisfactory account to accommodate both characters at once.

I think Williamson is right to ground our modal knowledge in our capacity to handle counterfactual conditionals. But the implicit problem is that our standard way of understanding these counterfactuals under the received Lewis–Stalnaker semantics presupposes a framework of possible worlds which is in itself quite neutral to what constraints are to be imposed on our counterfactual thinking. Such 'structural' facts like the essential constitution of things, the lawlike order of the world, and perhaps the relationship between determinates and determinables, etc., are not especially distinguished in this framework from other derived modal truths. It is therefore somewhat difficult to explicate the modal status of these very structural facts within the system, and thereby to make clear in what sense we are to hold them fixed and in what sense we may allow their violation. One possible way out, implicit in Lewis' own account and fully developed by Kment [6], is to impose the required constraints by a system of weighting for measuring the distances between worlds, such that structural facts get their special status by being incorporated into the weighting system. Whilst this may perhaps solve the problem here, I think a more promising approach is to give up the possible worlds semantics entirely, and to invoke an alternative framework where the 'structures' of various sorts are more appropriately represented.

At this point, I think it is quite helpful to consider the alternative semantics for counterfactuals developed by Pearl and Halpern [3, 4, 10]. For in this framework, the 'laws' are represented by the so-called 'structural equations', which get their special status by being constitutive of the frame for modelling causal counterfactuals. But at the same time, it is typical in causal modelling to allow such 'laws' to break down by surgically replacing the structural equations by some new ones directly assigning values. This makes structural semantics at least initially appealing, for we now have a richer resource to distinguish between various senses of 'holding fixed', and thus also to explicate the different modal status of the statements under consideration. The case for constitutive facts like (G) is slightly more complicated, for they are not directly represented by the structural equations. We need some way to encode information about a thing's essential nature, and to model the counterfactual supposition concerning the violation of the constitutive facts in question. In this paper, I shall provide such a treatment, which makes use of the structural models and the associated analysis of causal counterfactuals to interpret various sorts of modal claims, including those common examples of nomic, essentialist and metaphysical necessities. I think this can effectively supplement Williamson's account by retaining his basic intuition with a more appropriate semantic analysis to model how our capacity to handle counterfactuals may indeed ground our knowledge of various modal truths.

Here is the plan of the paper. In Sect. 4.2 I shall summarise Williamson's account and examine some of its problems. I shall argue that the main difficulty lies in its inability to answer the sceptical worry about metaphysical modality which it intends to answer. A solution to the worry will be suggested and outlined. In Sect. 4.3 I shall offer a formal characterisation of the structural semantics which I take to be more appropriate for dealing with the problem. The semantics is basically Halpern's [4].

But some crucial modifications will be made to accommodate cases involving *de re* modality. Section 4.4 will apply such a structural semantics to account for various sorts of modal claims. Based on this semantic analysis, I shall offer some further philosophical elucidation for the different kinds of necessity involved, explaining in what sense a law of nature is necessary, in what sense a thing has its constitutive nature necessarily, and in what sense a thing necessarily belongs to its category. Such elucidation will help to model how modal knowledge can be grounded in knowledge of counterfactuals.

4.2 Williamson's Account and Its Problems

As I said earlier, the central idea of Williamson's account is to take modal epistemology as a special case of the epistemology of counterfactual thinking. But why should we do so? One motivation is that this avoids invoking any mysterious faculty (e.g. intuition) for knowing such truths. For counterfactual reasoning, according to Williamson, is one of the basic cognitive capacities we frequently employ in our ordinary life and in science, which can be shown by its close connection with our *causal* thinking [15, p. 141]. As causal and counterfactual reasoning is so fundamental to our ordinary life, this gives us at least some *evolutionary* ground for modal knowledge. As he puts it, 'Humans evolved under no pressure to do philosophy.... Any cognitive capacity we have for philosophy is a more or less accidental byproduct of other developments' (p. 136). So if modal knowledge is in this way a by-product of counterfactual knowledge, which is evolutionarily basic, then it would be implausible to be sceptical of our capacity to handle it.

Now, to illustrate how we may acquire knowledge of counterfactuals, Williamson suggests a kind of 'simulation' account:

> We can still schematise a typical overall process of evaluating a counterfactual conditional thus: one supposes the antecedent and develops the supposition, adding further judgements within the supposition by reasoning, offline predictive mechanisms, and other offline judgements [15, pp. 152–3].

On this account, we evaluate the truth of a counterfactual conditional by counterfactually developing the supposition of its antecedent in mental simulation (ibid.). For example, suppose you see a rock sliding from the top of a mountain into a bush, and wonder where it would have ended if the bush had not been there. Williamson's suggestion is that you can know it by 'visualising the rock sliding without the bush there' (p. 142) and come to know the following truth:

(1) If the bush had not been there, the rock would have ended in the lake.

Although in this process we may appeal to our imaginative faculty (e.g. 'visualisation'), it is not essential. What is crucial, however, is our cognitive capacities to handle (separately) the *antecedent* and the *consequent*, for it is by some sort of 'offline' application of the same cognitive capacities that we may simulate and predict what would have happened next (pp. 147–150). In this rock-and-bush case (1), the offline evaluation of the antecedent (i.e. the bush's not being there) requires our imaginative faculty, but in other cases it may require some different cognitive capacities. The point is that on this account we *only* need whatever is required to evaluate sentences (i.e. the antecedent and the consequent) and then run it offline in our mental simulation; we do *not* need some special faculty of intuition to evaluate counterfactual conditionals.

This also gives us a hint about *modal* knowledge. For as Williamson observes, there is a close connection between statements of modality and counterfactual conditionals, which can be captured by the following formulas of equivalence (where '$\bot$' is the logical symbol for contradiction):

(2) $\Box A \equiv (\neg A \Box\!\!\rightarrow \bot)$
(3) $\Diamond A \equiv \neg(A \Box\!\!\rightarrow \bot)$

Now, if we combine these equivalences with the simulation account of counterfactual knowledge specified above, we will get an account of modal knowledge. More precisely, by (2) 'we assert $\Box A$ when our counterfactual development of the supposition $\neg A$ robustly yields a contradiction'; and by (3) 'we assert $\Diamond A$ when our counterfactual development of the supposition A does not robustly yield a contradiction' (p. 163). In this way, 'the capacity to handle metaphysical modality is an "accidental" byproduct of the cognitive mechanisms that provide our capacity to handle counterfactual conditionals' (p. 162).

As we have seen in Sect. 4.1, this account requires some complications when dealing with such cases as (G).

(G) Gold is the element with atomic number 79.

For in this case we need to 'hold something fixed' in our evaluation, for otherwise our counterfactual development of the negation of (G) will not yield any contradiction. Williamson therefore suggests that we hold the relevant *constitutive facts* fixed (e.g. the fact that gold is the element with atomic number 79, i.e. (G) itself), and thereby derive the required contradiction and assert the necessity of (G). As he puts it,

> If we know enough chemistry, our counterfactual development of the supposition that gold is [not] the element with atomic number 79 will generate a contradiction. The reason is not simply that we know that gold is the element with atomic number 79, for we can and must vary some items of our knowledge under counterfactual suppositions. Rather, part of the general way we develop counterfactual suppositions is to hold such constitutive facts fixed [15, p. 164].

However, such a suggestion was criticised by many commentators as circular or unilluminating. For it amounts to saying that we can know the necessity of (G) only if we hold (G) fixed in evaluating the corresponding counterfactual. But how do we know we should hold (G) fixed? The only reason seems to be that we hold (G) fixed because it is a *metaphysical necessity* ([13], p. 107; cf. [1], p. 490, fn.1). But that would be plainly circular: for in this way, in order to know the necessity of (G) we need to hold (G) fixed, but to hold (G) fixed we need to know (G) to be a metaphysical necessity. To avoid the circularity, we should not ground our holding (G) fixed in its modal status. But what else can be the ground? Williamson may be right in saying that we know we should hold (G) fixed if we know (G) to be a constitutive fact. But Williamson says quite little about how we can achieve such prior constitutive knowledge. It therefore appears that Williamson's account leaves a substantial part of our modal knowledge (i.e. the prior *constitutive* knowledge) unexplained, and is thus utterly unilluminating (cf. [11]).

Now, I do not think this criticism really touches the heart of the problem. For on the one hand, Williamson is quite clear to emphasise that, to evaluate the modal status of (G) by applying (2), what is required to know is *not* the *modal* truth that *(G) is metaphysically necessary*, but only a non-modal one which claims that (G) is a constitutive fact [16, p. 506]. This avoids the circularity. On the other hand, also implicitly in the passage quoted above, Williamson does offer a hint as to how we may achieve the required constitutive knowledge—i.e. by *knowing enough chemistry*. For it is by the relevant scientific theory that we may come to know the constitutive nature of gold. Constitutive facts (e.g. that water is H_2O, that gold is the element with atomic number 79, etc.) are known, not by some mysterious modal intuition, but by our usual inductive method of natural science. But once we acquire knowledge of such constitutive facts, there is no problem of holding them fixed in our evaluation of counterfactuals. For 'projecting constitutive matters such as atomic numbers into counterfactual supposition is part of our general way of assessing counterfactuals' [15, p. 170]. This is quite similar to the case about *laws of nature*. For laws are also known by inductive method of science but projectable into counterfactual supposition. Similarly, constitutive knowledge can be acquired by scientific method and projected into counterfactual supposition.

However, precisely at this point we may come to see more clearly what is the real problem for Williamson's account. For if constitutive knowledge is indeed acquired by inductive method just like knowledge of laws, then the counterfactuals they support can only be *causal* counterfactuals, and the necessity involved can only be a species of *causal* or *nomic* necessity.[1] That is to say, if it is indeed by 'knowing enough chemistry' that we come to know the constitutive nature of gold, we would no longer have the ground of holding-fixed when the counterfactual supposition we envisage is one where the relevant chemical theory fails to hold. As a result, the very necessity of (G) *that we know in this manner* is at best a kind of *nomic* necessity.

This presents a serious problem for Williamson. For Williamson intends his account to be able to answer the sceptical doubt concerning modal knowledge, and he tries to do this by taking modal knowledge as a special case of counterfactual knowledge. But there are different senses of counterfactual, just as there are different senses of modality—there is *causal* counterfactual concerning what could have been otherwise given our laws of nature; there is *metaphysical* counterfactual concerning what could have been otherwise metaphysically. Correspondingly, there are *causal* (or *nomic*) modality, *metaphysical* modality, etc. So even if Williamson is right to think that his account can defend modal knowledge by emphasising the evolutionary ground of counterfactual knowledge in our causal thinking, it does not

[1] In a recent paper E. J. Lowe raises a similar worry. See [9, pp. 932ff].

really answer the sceptical doubt concerning *metaphysical* modality. For one can be a sceptic only about metaphysical modality without being sceptical of causal modality. That is to say, one may grant that Williamson's account indeed shows that our capacity to handle (causal) counterfactuals does provide the required resource to handle *some* modal claims, but still denies that we can have any cognitive capacity to access a metaphysical reality that goes beyond empirical sciences. Williamson's account is unable to answer the sceptical doubt of that sort.

So how can we reply to the sceptical doubt in question, if Williamson's account does not really answer it? To this problem, I would suggest a *sceptical solution*: to grant with the sceptic that we indeed have no knowledge of metaphysical modality beyond what we can know from science, but to argue that such a sceptical conclusion is entirely harmless. That is to say, we may grant that we really have no cognitive access to a distinctively metaphysical reality, but this does not undermine our reasoning in science and in ordinary life. For what we need to be able to handle in science and ordinary life is but causal and nomic modalities, and almost all modal knowledge that we may acquire by scientific means is of this sort. This means that the solution I am offering here is in fact a 'regulative' solution, for it advices that, whenever we seem to have a case of knowing some metaphysically modal truth, we should try to find an explanation of it in *naturalistic* terms (e.g. as a species of causal modality). If this can be done, it will explain why the sceptical conclusion is harmless. For if all the modal truths we can clearly know can be accommodated in naturalistic terms, then the remaining cases of purely 'metaphysical' modality are really something beyond our cognitive access. We therefore have no difficulty in confessing that we have no knowledge of them.

Now, I think such a naturalising project should better be carried out with the structural models (as mentioned in Sect. 4.1). The reason is quite clear. For structural models are supposed to be more appropriate for representing *causal* counterfactuals, and in this sense they are quite suitable for expressing the requisite naturalistic explanation of modal knowledge. In the next section, I shall provide a formal characterisation of the structural models in question and the corresponding semantic analysis of counterfactuals. Based on such a semantic analysis, the naturalised account of modal knowledge will be offered in Sect. 4.4.

4.3 Structural Models

I now provide a formal characterisation of structural models. Following Halpern [4], I distinguish between a *signature* and the *models* over a given signature.[2] The distinction is crucial to my purpose, for, as we shall see below, variations in signatures and variations in models correspond to different kinds of modality. Roughly speaking, a signature represents a certain metaphysical framework within which the causal structure can be further characterised. But to make it even more perspicuous, I would

[2]My characterisation therefore differs from [3, 10, 17] or [18] in this respect.

add a further distinction between a *model* and the possible *states* assignable for a given model. So we have a three-level structure of ⟨signatures, models, states⟩, which will become very useful when we are to represent various species of modality.

Definition 1 (*Signature*) A *signature* is a quadruple

$$\mathcal{S} = \langle \mathcal{V}, \mathcal{R}, \mathcal{I}, \varrho \rangle,$$

where

(i) $\mathcal{V}$ is a set variables;
(ii) $\mathcal{R}$ is a function that assigns to each variable $X \in \mathcal{V}$ a non-empty set $\mathcal{R}(X)$ of possible values for X (i.e. the *range* of the values of X);
(iii) $\mathcal{I}$ is a set of *individuals*; and
(iv) $\varrho \subseteq \mathcal{I} \times \mathcal{V}$ is a relation between individuals and variables indicating their *relevancy*. (Intuitively, '$a \varrho X$', which abbreviates '$(a, X) \in \varrho$', means 'X is a variable relevant to the individual a'.)

In the causal modelling literature, usually the variables are divided into the *exogenous* and the *endogenous* ones, according to whether the variables in question are determined by factors outside or inside the model ([4], p. 318; [10], p. 203). Now, since I distinguish between a *signature* (which represents the shared metaphysical framework) and the *models* over the given signature (which represent the causal structure to be characterised and modelled within this framework), such a split of variables should therefore be relative to the *models*. For different models may take different variables as the target to be modelled by the associated structural equations (i.e. the endogenous ones), thus leaving different variables as the background factors determined outside (i.e. the exogenous ones). For this reason, the division should *not* be placed at the level of signature.[3] So here in my characterisation, we have only one set $\mathcal{V}$ in the signature as the set of all variables.

Another crucial point is that in the causal modelling literature, usually no special mention of *individuals* is needed. This is mainly because we can always use a single *variable* to represent what we intend to say about the individual. For example, to represent the temperature of the given gas, instead of saying that the temperature T of the gas g takes the value t, we can use a single variable T_g to represent *the temperature of the gas*. However, as my purpose here is to provide an account which can accommodate *essentialist* attributions such as (G), the separation of the set of individuals $\mathcal{I}$ within the framework is somehow mandatory for modelling *de re* modality, as we shall see in due course.

Since we have individuals in our framework, we can understand the variables as *properties* of individuals. More precisely, a variable is a *determinable trope* of its relevant individuals, and its values are the *determinate tropes* which fall under it.[4]

[3] In this sense, Halpern's characterisation of a signature as $\langle \mathcal{U}, \mathcal{V}, \mathcal{R} \rangle$ is somewhat misleading.

[4] The appeal to an ontology of tropes is convenient but not compulsory. If we want to avoid tropes, we may use some equivalent way to express the same idea, e.g. by taking a variable as the state of affair of the relevant individuals' instantiating some determinable universal.

For example, let T be the variable for the temperature of the given gas g, and t be one of its values, say, 50 °C. We may understand T as the determinable trope $\langle$*g's temperature*$\rangle$, and t as the determinate trope $\langle$*g's being at the temperature* $50\,^\circ C$ $\rangle$. Notice that each such trope may involve one or several individuals as its property-bearer(s), which are said to be the individuals *relevant to*, or *involved in*, the given trope. The relation ϱ is precisely postulated to capture such a relationship between them. In the example above, we say that the gas g is relevant to the variable T, or that g is involved in T, which we express in symbol as $g\varrho T$. But a variable may also involve more than one individual. For example, let X be the variable for the distance between two objects a and b, and x be one of its values, say, 20m. We may understand X as the two-place determinable trope $\langle$*the distance between a and b*$\rangle$, and x as the two-place determinate trope $\langle$*a andb 's being at the distance of* $20m$$\rangle$. In this case, we have $a\varrho X$ and $b\varrho X$, which says that the variable X involves the individuals a and b.

Now, the relation ϱ not only specifies the objectual contents of the *variables*, but also provides crucial information about the *individuals*. To capture this more clearly, it is helpful to make some definitions based on ϱ.

Definition 2 (*Degree, Category, andLogicalSpace*) Let $S = \langle \mathcal{V}, \mathcal{R}, \mathcal{I}, \varrho \rangle$ be a signature, we define three functions, δ, $\mathcal{C}$, and $\mathcal{D}$ as follows:

(i) $\delta : \mathcal{V} \to \mathbb{N} \cup \{0\}$ is the function that assigns to each variable the number of the individuals involved in it, called its *degree*; i.e. for each $X \in \mathcal{V}$, $\delta(X) =_{\text{def}} \|\{a \in \mathcal{I} \mid a\varrho X\}\|$ (where $\|A\|$ is the size of A).
(ii) $\mathcal{C} : \mathcal{I} \to \mathcal{P}(\mathcal{V})$ is the function that assigns to each individual the set of its relevant variables, called its *category*; i.e. for each $a \in \mathcal{I}$, $\mathcal{C}(a) =_{\text{def}} \{X \in \mathcal{V} \mid a\varrho X\}$.
(iii) $\mathcal{D}$ is the function that assigns to each individual the Cartesian product of the ranges of its relevant variables, called its *logical space*; i.e. for each $a \in \mathcal{I}$, $\mathcal{D}(a) =_{\text{def}} \prod_{X \in \mathcal{C}(a)} \mathcal{R}(X)$.

For each variable $X \in \mathcal{V}$, the degree of X is the number of individuals involved in it. This tells us what kind of variable X is. When $\delta(X) = 1$, the variable X is a monadic determinable trope (e.g. *temperature*,[5] *shape*, *colour*, etc.). When $\delta(X) = n \geq 2$, the variable X is an n-place relational determinable trope (e.g. *distance*, *mutual gravitational force*, etc.). A degenerate case is $\delta(X) = 0$. In this case, the variable X involves *no* individual at all, and thus it directly represents what it is intended to represent without being analysed into an object-property structure (e.g. the occurrence or non-occurrence of an event).[6]

[5]In fact it should be $\langle$*The temperature of a*$\rangle$ (for some individual a), as it is a trope rather than a universal. But for the sake of simplicity I shall just write *temperature* when $\langle$*The temperature of a*$\rangle$ can be clearly understood from the context. The same applies to other determinable tropes.

[6]It is not easy to find an example where the variable involves no individual whatsoever. But consider this. Let Y be the variable for whether the Big Bang has occurred, and suppose we do not want to take the Big Bang as an individual. Then in this case, it may be plausible to assume $\delta(Y) = 0$.

The category function $\mathcal{C}$ assigns to each individual the set of all determinable properties relevant to it. Now, fundamentally different kinds of things are associated with different sets of determinables. For example, for any material object m, $\mathcal{C}(m)$ should include *shape* and *colour* but not *intensity*[7]; for any field f, $\mathcal{C}(f)$ should include *intensity* but not *shape* or *colour*; for any wave w, $\mathcal{C}(w)$ should include *frequency* and *wavelength*, etc. Such a set of determinables delineates and defines the category of the given individual. For it generates the *logical space*[8] for the individual by taking the Cartesian product of the ranges of the associated variables. Given any individual $a \in \mathcal{I}$, since $\mathcal{C}(a)$ contains all the variables relevant to a (i.e. all the associable determinables of a), each possible way a might be can be represented by a unique point in its logical space $\mathcal{D}(a)$ according to the values assigned to the variables in $\mathcal{C}(a)$. As a result, $\mathcal{D}(a)$ delineates the possible ways a might be, and this provides substantial information about a's category.

As I said earlier, a signature represents a certain metaphysical framework. In this sense, its invariance under all structural models definable over it should be akin to a sort of metaphysical necessity. For example, 'For any $X \in \mathcal{V}$, the value of X can only be one amongst $\mathcal{R}(X)$' represents a certain structural truth which holds of necessity in a very strict sense. This will be explicated further in Sect. 4.4. But now, I shall provide a formal characterisation of the structural models and the possible states first.

Definition 3 (*Structural Models*) A *structural model* over a given signature $\mathcal{S} = \langle \mathcal{V}, \mathcal{R}, \mathcal{I}, \varrho \rangle$ is a triple

$$\mathcal{M} = \langle \mathcal{S}, \mathcal{V}_{\text{en}}, \mathcal{F} \rangle,$$

where

(i) $\mathcal{V}_{\text{en}}$ is a subset of $\mathcal{V}$, called the *endogenous variables*. We also define another subset, $\mathcal{V}_{\text{ex}} =_{\text{def}} \mathcal{V} \backslash \mathcal{V}_{\text{en}}$, called the *exogenous variables*; and

(ii) $\mathcal{F} = \{f_X \mid X \in \mathcal{V}_{\text{en}}\}$ is a set of functions, where each variable $X \in \mathcal{V}_{\text{en}}$ is associated with a unique function denoted by f_X whose arguments are $\mathcal{V} \backslash \{X\}$, such that $f_X : \prod_{Y \in \mathcal{V} \backslash \{X\}} \mathcal{R}(Y) \to \mathcal{R}(X)$ determines the value of X given the values of all other variables. We also define for each variable $X \in \mathcal{V}_{\text{en}}$ its *structural equation* as

$$X = f_X,$$

which takes $\mathcal{V} \backslash \{X\}$ as its independent variables and X as its dependent variable.

The *endogenous* variables $\mathcal{V}_{\text{en}}$ are the variables whose values are determined in the model $\mathcal{M}$ according to the associated structural equations. The *exogenous* variables $\mathcal{V}_{\text{ex}}$, by contrast, are the variables whose values are determined 'outside' the model [10, p. 203]. So there are no structural equations for exogenous variables, for nothing in the model can influence the values of the exogenous variables. Also for this reason,

---

[7]That is to say, there are such variables as ⟨*The shape of m*⟩ and ⟨ *The colour of m*⟩, but there is no such variable as ⟨ *The intensity of m*⟩. See footnote 5 above.

[8]The idea of *logical space* was proposed by van Fraassen [14] and developed by Stalnaker [12]; but my use of the notion differs quite substantially from theirs.

we should assume that the exogenous variables are all *independent* from each other. For if an exogenous variable *were* such that its value should depend upon some other variables, then we would have a structural equation specifying how its value is determined, and thus it should be an endogenous variable rather than an exogenous one.

Now, given a signature $\mathcal{S}$, intuitively each possible assignment of values to the variables in $\mathcal{V}$ represents a possible way the world might be. In fact, each such assignment also maps every individual $a \in \mathcal{I}$ to a unique point in its logical space $\mathcal{D}(a)$ (for it assigns values to all variables in $\mathcal{C}(a)$, thus locating a at some point in $\mathcal{D}(a)$). In this sense, such value-assignments for $\mathcal{V}$ are a kind of *location functions* that map the individuals into the logical space (cf. [14]), representing the various alternative ways the individuals might be. Their semantic role is therefore more or less akin to possible worlds [12, p. 348]. We may thus call each such value-assignment a *world-state* of the signature $\mathcal{S}$.

However, *not* every world-state is genuinely possible. For the values of the *endogenous* variables $\mathcal{V}_{\text{en}}$ should depend on some other variables according to the associated structural equations, and hence we cannot just arbitrarily assign values to them. Our value-assignment needs to satisfy the structural equations to be a genuinely *possible state* for the model $\mathcal{M}$. But the *exogenous* variables $\mathcal{V}_{\text{ex}}$, by contrast, have no such restriction. For the exogenous variables are all independent from each other, and hence we can always arbitrarily assign values to each of them without fear of conflict. Each such assignment, which we may call an *exogenous assignment*, represents a possible configuration of background factors for $\mathcal{M}$ against which the genuinely possible states of $\mathcal{M}$ are to be determined.

Definition 4 (*World-States and Exogenous Assignments*) Let $\mathcal{S} = \langle \mathcal{V}, \mathcal{R}, \mathcal{I}, \varrho \rangle$ be a signature, and $\mathcal{M} = \langle \mathcal{S}, \mathcal{V}_{\text{en}}, \mathcal{F} \rangle$ be a structural model over $\mathcal{S}$. A *world-state* of the *signature* $\mathcal{S}$ is a value-configuration of all the variables in $\mathcal{V}$. An *exogenous assignment* for the model $\mathcal{M}$ is a value-configuration of the exogenous variables. More precisely,

(i) A *world-state of* $\mathcal{S}$ is a function s which assigns to each variable $X \in \mathcal{V}$ some particular value $s(X) \in \mathcal{R}(X)$ as its assigned value.
(ii) An *exogenous assignment for* $\mathcal{M}$ is a function σ which assigns to each *exogenous* variable $X \in \mathcal{V}_{\text{ex}}$ some particular value $\sigma(X) \in \mathcal{R}(X)$ as its assigned value.

At this point, let me introduce some useful conventions and notations. Given a signature $\mathcal{S}$ and a model $\mathcal{M}$ over $\mathcal{S}$, we may assume that our variables $\mathcal{V}$ (and also $\mathcal{V}_{\text{ex}}$ and $\mathcal{V}_{\text{en}}$) are arranged in a certain order. So we may use a variable-vector $\boldsymbol{X}$ to denote these variables (in $\mathcal{V}$, $\mathcal{V}_{\text{en}}$, or $\mathcal{V}_{\text{ex}}$), and use a value-vector $\boldsymbol{x}$ to denote a corresponding value-configuration. In this way, each world-state s corresponds to a value-vector $\boldsymbol{x}$ such that $\boldsymbol{x} = s(\boldsymbol{X})$, and similarly for the exogenous assignments. When the set $\mathcal{V} = \{X_1, \ldots, X_n\}$ is finite, we may simply use an n-tuple $\boldsymbol{x} = \langle s(X_1), \ldots, s(X_n) \rangle$ to represent the world-state s in question. Similarly, when the set $\mathcal{V}_{\text{ex}} = \{X_1, \ldots, X_m\}$ is finite, we may use an m-tuple $\boldsymbol{u} = \langle \sigma(X_1), \ldots, \sigma(X_m) \rangle$ to represent the exogenous assignment σ in question.

Due to the associated structural equations, each model $\mathcal{M}$ under a particular exogenous assignment σ (written $\mathcal{M}(\sigma)$) imposes some constraints on what world-states are genuinely possible. If a world-state s satisfies the imposed constraints, we say that s is a *solution* to $\mathcal{M}(\sigma)$. Intuitively, each such solution represents a *possible state* for $\mathcal{M}$. This can be captured more precisely by the following definitions:

Definition 5 (*Solutions and Possible States*) Let $\mathcal{M} = \langle \mathcal{S}, \mathcal{V}_{\text{en}}, \mathcal{F} \rangle$ be a structural model over the signature $\mathcal{S} = \langle \mathcal{V}, \mathcal{R}, \mathcal{I}, \varrho \rangle$, and σ be an exogenous assignment for $\mathcal{M}$. Let $\boldsymbol{X}$ denote the variables in $\mathcal{V}$.

(i) Say that a world-state s (of the given signature $\mathcal{S}$) is a *solution* to $\mathcal{M}(\sigma)$, if and only (a) for each variable $U \in \mathcal{V}_{\text{ex}}$, $s(U) = \sigma(U)$, and (b) for each variable $Y \in \mathcal{V}_{\text{en}}$, s satisfies its structural equation, i.e. $s(Y) = f_Y(\boldsymbol{s_Y})$ (where $\boldsymbol{s_Y}$ is the vector resulting from removing the Y-component from the value-vector $\boldsymbol{x} = s(\boldsymbol{X})$).
(ii) Say that a world-state s is a *possible state for the model $\mathcal{M}$ under σ* if and only if s is a solution to $\mathcal{M}(\sigma)$.
(iii) Say that a world-state s is a *possible state for $\mathcal{M}$* if and only if there is an exogenous assignment τ such that s is a possible state for $\mathcal{M}(\tau)$.

Following [4], I allow that some structural models under some exogenous assignments may have more than one solution. In such cases, the background factors together with the constraints imposed by the causal relationships do not determine a unique state, but only a number of states which are equally possible. Philosophically, this captures the idea that our world may be causally underdetermined. But for those cases where a structural model may have *no solution* at all, it is more difficult to make good philosophical sense. So in this paper, I shall simply assume that all of our models under every exogenous assignment have at least one solution.[9]

Now, we may provide truth-conditions for some sentences. But to do this we need to specify our language first. Following [4], I also take as the *atomic formulas* of our formal language those sentences of the form $X = x$ (where X is a variable in $\mathcal{V}$ and x is a value in $\mathcal{R}(X)$, such that the sentence says the variable X has the value x).[10] By having individuals in our framework, this means that usually simple predications of individuals can be expressed by atomic formulas (e.g. 'a is red' can be expressed by the atomic formula which says the colour-variable for a has the red trope as its value).[11] The truth-conditions for these atomic formulas are quite straightforward:

[9] Such models are called '*solutionful*' in [17].

[10] Strictly speaking, such items as X and x should belong to the semantics. So it is slightly confusing to have them *also* in our language. But here I simply follow the established tradition by Pearl and Halpern in using such a language as to contain these items.

[11] Alternatively, we can take simple predications as atomic. This can be done by having names and predicates in our language instead of variables and values, such that each predicate is assigned a set of variables that all have the same value-range *plus* one of these values as its semantic value (e.g. 'x is red' is assigned ⟨the set of all colour-variables, red⟩ as its semantic value). Then we can stipulate the truth-conditions for these atomic sentences in terms of the assigned semantic values (e.g. given $\langle C, p \rangle$ and o as the semantic values of Px and a respectively, Pa is true in s iff for some unique $X \in C$, $o\varrho X$ and for this X, $s(X) = p$). This avoids the confusion mentioned in footnote 10.

$X = x$ is true in a world-state s if and only if $s(X) = x$. This generalises recursively to any Boolean combination of atomic formulas.

We now introduce '$\Diamond$' and '$\Box$' as two new operators into our language. Intuitively, when prefixed to a formula φ, '$\Diamond\varphi$' is intended to mean 'It is *naturally possible* that φ', and '$\Box\varphi$' to mean 'It is *naturally necessary* that φ'. To specify their truth-conditions, however, we need to notice that natural possibility (and necessity) should always be relative to the *models*. So, we use '$\mathcal{M}, s \models \Diamond\varphi$' (instead of '$s \models \Diamond\varphi$') to express the claim that $\Diamond\varphi$ is true in the world-state s relative to the model $\mathcal{M}$. Thus qualified, the truth-conditions for atomic formulas are as before, i.e. $\mathcal{M}, s \models X = x$ iff $s(X) = x$, and the truth-conditions for modal sentences can be given as follows.

Truth-Conditions 1 (Natural Modalities) *Let $\mathcal{M} = \langle \mathcal{S}, \mathcal{V}_{\text{en}}, \mathcal{F} \rangle$ be a structural model, s be a possible state for $\mathcal{M}$, and φ be a formula in our language. Let $s\restriction_{\mathcal{V}_{\text{ex}}}$ be the exogenous assignment resulting from restricting s on the exogenous variables (i.e. the exogenous assignment such that $s\restriction_{\mathcal{V}_{\text{ex}}}(X) = s(X)$ for all $X \in \mathcal{V}_{\text{ex}}$). Then we have the following truth-conditions:*

(i) $\mathcal{M}, s \models \Diamond\varphi$ *if and only if* $\mathcal{M}, t \models \varphi$ *for* some *possible state t of* $\mathcal{M}(s\restriction_{\mathcal{V}_{\text{ex}}})$.
(ii) $\mathcal{M}, s \models \Box\varphi$ *if and only if* $\mathcal{M}, t \models \varphi$ *for* all *possible states t of* $\mathcal{M}(s\restriction_{\mathcal{V}_{\text{ex}}})$.

Given that s is a possible state for $\mathcal{M}$, the truth-condition (i) says that $\Diamond\varphi$ is true (in s relative to $\mathcal{M}$) if and only if φ is true in *some* possible state of $\mathcal{M}$ under the exogenous assignment resulting from s, and (ii) says that $\Box\varphi$ is true if and only if φ is true in *all* such possible states. It is clear that these truth-conditions validate the equivalences $\Diamond\varphi \equiv \neg\Box\neg\varphi$ and $\Box\varphi \equiv \neg\Diamond\neg\varphi$. Moreover, if s is a possible state of $\mathcal{M}$, by definition s is already a possible state of $\mathcal{M}(s\restriction_{\mathcal{V}_{\text{ex}}})$, so the truth-conditions validate $\varphi \supset \Diamond\varphi$ (and hence also $\Box\varphi \supset \varphi$). Finally, it is easy to check that (a) if t is a possible state for $\mathcal{M}(s\restriction_{\mathcal{V}_{\text{ex}}})$ then s is a possible state for $\mathcal{M}(t\restriction_{\mathcal{V}_{\text{ex}}})$, and (b) if t is a possible state for $\mathcal{M}(s\restriction_{\mathcal{V}_{\text{ex}}})$, and r is a possible state for $\mathcal{M}(t\restriction_{\mathcal{V}_{\text{ex}}})$, then r is a possible state for $\mathcal{M}(s\restriction_{\mathcal{V}_{\text{ex}}})$. This means that our truth-conditions also validate $\varphi \supset \Box\Diamond\varphi$ and $\Box\varphi \supset \Box\Box\varphi$, and thus impose a modal system of an S5 structure.

Notice that usually we cannot determine whether a formula φ is true or false if we are given *only* an exogenous assignment σ for the model $\mathcal{M}$. The reason is that there can be more than one possible state for $\mathcal{M}(\sigma)$, such that φ may be true in one state and false in another. But given the S5 structure, even in cases where $\mathcal{M}(\sigma)$ has more than one possible state, a *modalised formula* (i.e. $\Diamond\varphi$ or $\Box\varphi$) should have the same truth-value in all these states, and so we can directly talk about *the* truth-values of such modalised formulas in $\mathcal{M}(\sigma)$ without any problem. This justifies our introducing the notation '$\mathcal{M}(\sigma) \models \Diamond\varphi$' (and '$\mathcal{M}(\sigma) \models \Box\varphi$') to mean '$\mathcal{M}, s \models \Diamond\varphi$' (and '$\mathcal{M}, s \models \Box\varphi$'), where s is any given possible state of $\mathcal{M}(\sigma)$'.[12] It then follows that $\mathcal{M}(\sigma) \models \Diamond\varphi$ iff $\mathcal{M}, s \models \varphi$ for some possible state s of $\mathcal{M}(\sigma)$, and $\mathcal{M}(\sigma) \models \Box\varphi$ iff $\mathcal{M}, s \models \varphi$ for all possible states s of $\mathcal{M}(\sigma)$.

---

[12]Here we assume that every model under every exogenous assignment has at least one solution.

We shall now offer truth-conditions for causal counterfactuals. To do this, we need to invoke the notions of submodels and extended/modified assignments to represent the counterfactual situations resulting from manipulatively setting certain values to some variables.

Definition 6 (*Submodels and Extended/Modified Assignments*) Let $\mathcal{M} = \langle \mathcal{S}, \mathcal{V}_{\text{en}}, \mathcal{F} \rangle$ be a structural model, and σ be an exogenous assignment for $\mathcal{M}$. Let $\boldsymbol{X} = \langle X_1, \ldots, X_n \rangle \in \mathcal{V}_{\text{en}}^n$ be an endogenous variable-vector, $\boldsymbol{x} = \langle x_1, \ldots, x_n \rangle$ be a value-vector for $\boldsymbol{X}$, $\boldsymbol{Y} = \langle Y_1, \ldots, Y_m \rangle \in \mathcal{V}_{\text{ex}}^m$ be an exogenous variable-vector, and $\boldsymbol{y} = \langle y_1, \ldots, y_m \rangle$ be a value-vector for $\boldsymbol{Y}$. We make the following definitions:

(i) A *submodel* of $\mathcal{M}$, denoted by $\mathcal{M}_{\boldsymbol{X}}$, is the structural model

$$\mathcal{M}_{\boldsymbol{X}} = \langle \mathcal{S}, \mathcal{V}_{\text{en}} \backslash \{X_i \mid 1 \leq i \leq n\}, \mathcal{F} \backslash \{f_{X_i} \mid 1 \leq i \leq n\} \rangle;$$

(ii) An *extended assignment* of σ, denoted by $\sigma_{\boldsymbol{X}=\boldsymbol{x}}$, is the exogenous assignment for the submodel $\mathcal{M}_{\boldsymbol{X}}$, such that $\sigma_{\boldsymbol{X}=\boldsymbol{x}} = \sigma \cup \{\langle X_i, x_i \rangle \mid 1 \leq i \leq n\}$; or more precisely,

$$\sigma_{\boldsymbol{X}=\boldsymbol{x}}(Z) = \begin{cases} \sigma(Z) & \text{if } Z \in \mathcal{V}_{\text{ex}}, \\ x_i & \text{if } Z = X_i \text{ for some } i. \end{cases}$$

(iii) A *modified assignment* of σ, denoted by $\sigma_{\boldsymbol{Y}/\boldsymbol{y}}$, is the exogenous assignment for the model $\mathcal{M}$, such that $\sigma_{\boldsymbol{Y}/\boldsymbol{y}}$ is exactly the same as σ except that for each i, $\sigma_{\boldsymbol{Y}/\boldsymbol{y}}(Y_i)$ is y_i rather than $\sigma(Y_i)$; or more precisely,

$$\sigma_{\boldsymbol{Y}/\boldsymbol{y}}(Z) = \begin{cases} \sigma(Z) & \text{if } Z \neq Y_i \text{ for any } i, \\ y_i & \text{if } Z = Y_i \text{ for some } i. \end{cases}$$

Intuitively, $\mathcal{M}_{\boldsymbol{X}}$ represents the causal structure which results from $\mathcal{M}$ by breaking any previously existing causal influence on each X_i (i.e. removing the structural function f_{X_i} from $\mathcal{F}$), so that each X_i becomes an independent variable to be relocated in $\mathcal{V}_{\text{ex}}$. Then we can arbitrarily assign values to X_i on top of σ without fear of conflict, and $\sigma_{\boldsymbol{X}=\boldsymbol{x}}$ is exactly such an assignment. Putting these together we get $\mathcal{M}_{\boldsymbol{X}}(\sigma_{\boldsymbol{X}=\boldsymbol{x}})$, whose solutions then represent those possible (counterfactual) situations where we 'surgically' set the value of each X_i to x_i.

On the other hand, since *exogenous* variables are already independent from each other, we may directly change their values without destroying any currently existing causal relationship, and $\sigma_{\boldsymbol{Y}/\boldsymbol{y}}$ is precisely postulated to serve this purpose. So, intuitively, the solutions to $\mathcal{M}(\sigma_{\boldsymbol{Y}/\boldsymbol{y}})$ represents those possible (counterfactual) situations where we directly set the value of each Y_i to y_i.

We now introduce '$\Diamond\!\!\rightarrow$' and '$\Box\!\!\rightarrow$' as two new sentence connectives into our language representing causal counterfactuals. However, we shall confine our language to contain *only* those counterfactuals whose antecedent is an *atomic for-*

mula or a conjunction of atomic formulas.[13] Thus qualified, a causal counterfactual of our language will be of the form $(X_1 = x_1 \wedge \cdots \wedge X_n = x_n) \diamond\!\!\rightarrow \psi$ or $(X_1 = x_1 \wedge \cdots \wedge X_n = x_n) \Box\!\!\rightarrow \psi$, where each X_i is a variable in $\mathcal{V}$, x_i a value in $\mathcal{R}(X_i)$, and ψ a formula of our language. Intuitively, '$\varphi \diamond\!\!\rightarrow \psi$' is intended to mean 'If we were to bring about that φ, then it might be the case that ψ', and '$\varphi \Box\!\!\rightarrow \psi$' to mean 'If we were to bring about that φ, then it would be the case that ψ'. The truth-conditions for causal counterfactuals can now be given.

Truth-Conditions 2 (Causal Counterfactuals) *Let $\mathcal{M} = \langle \mathcal{S}, \mathcal{V}_{\text{en}}, \mathcal{F} \rangle$ be a structural model, s be a possible state for $\mathcal{M}$, $\sigma = s \upharpoonright_{\mathcal{V}_{\text{ex}}}$ be the exogenous assignment resulting from s, and φ be a formula in our language. Let $\boldsymbol{X} = \langle X_1, \ldots, X_n \rangle \in \mathcal{V}_{\text{en}}^n$ be an endogenous variable-vector,[14] $\boldsymbol{x} = \langle x_1, \ldots, x_n \rangle$ be one of its value-vectors, $\boldsymbol{Y} = \langle Y_1, \ldots, Y_m \rangle \in \mathcal{V}_{\text{ex}}^m$ be an exogenous variable-vector, and $\boldsymbol{y} = \langle y_1, \ldots, y_m \rangle$ be one of its value-vectors. Then we have the following truth-conditions:*

(i) $\mathcal{M}, s \models (X_1 = x_1 \wedge \cdots \wedge X_n = x_n) \diamond\!\!\rightarrow \varphi$ *iff* $\mathcal{M}_{\boldsymbol{X}}(\sigma_{\boldsymbol{X}=\boldsymbol{x}}) \models \Diamond\varphi$, *and*
$\mathcal{M}, s \models (X_1 = x_1 \wedge \cdots \wedge X_n = x_n) \Box\!\!\rightarrow \varphi$ *iff* $\mathcal{M}_{\boldsymbol{X}}(\sigma_{\boldsymbol{X}=\boldsymbol{x}}) \models \Box\varphi$;
(ii) $\mathcal{M}, s \models (Y_1 = y_1 \wedge \cdots \wedge Y_m = y_m) \diamond\!\!\rightarrow \varphi$ *iff* $\mathcal{M}(\sigma_{\boldsymbol{Y}/\boldsymbol{y}}) \models \Diamond\varphi$, *and*
$\mathcal{M}, s \models (Y_1 = y_1 \wedge \cdots \wedge Y_m = y_m) \Box\!\!\rightarrow \varphi$ *iff* $\mathcal{M}(\sigma_{\boldsymbol{Y}/\boldsymbol{y}}) \models \Box\varphi$;
(iii) $\mathcal{M}, s \models (X_1 = x_1 \wedge \cdots \wedge X_n = x_n \wedge Y_1 = y_1 \wedge \cdots \wedge Y_m = y_m) \diamond\!\!\rightarrow \varphi$ *iff* $\mathcal{M}_{\boldsymbol{X}}(\sigma_{\boldsymbol{Y}/\boldsymbol{y};\boldsymbol{X}=\boldsymbol{x}}) \models \Diamond\varphi$, *and*
$\mathcal{M}, s \models (X_1 = x_1 \wedge \cdots \wedge X_n = x_n \wedge Y_1 = y_1 \wedge \cdots \wedge Y_m = y_m) \Box\!\!\rightarrow \varphi$ *iff* $\mathcal{M}_{\boldsymbol{X}}(\sigma_{\boldsymbol{Y}/\boldsymbol{y};\boldsymbol{X}=\boldsymbol{x}}) \models \Box\varphi$.

As explained earlier, intuitively $\mathcal{M}_{\boldsymbol{X}}(\sigma_{\boldsymbol{X}=\boldsymbol{x}})$ selects those possible situations where we surgically set the values of $\boldsymbol{X}$ to $\boldsymbol{x}$, whereas $\mathcal{M}(\sigma_{\boldsymbol{Y}/\boldsymbol{y}})$ selects those where we set $\boldsymbol{Y}$ to $\boldsymbol{y}$ and $\mathcal{M}_{\boldsymbol{X}}(\sigma_{\boldsymbol{Y}/\boldsymbol{y};\boldsymbol{X}=\boldsymbol{x}})$ selects those where we do both. These truth-conditions therefore capture the intuition that a might-counterfactual is true iff its consequent is true in at least one selected possible situation, whereas a would-counterfactual is true iff its consequent is true in all the selected situations.

Notice that for any counterfactual $(\varphi_1 \wedge \cdots \wedge \varphi_n) \diamond\!\!\rightarrow \psi$ or $(\varphi_1 \wedge \cdots \wedge \varphi_n) \Box\!\!\rightarrow \psi$ in our language (where each φ_i is an atomic formula), the order of φ_i in the antecedent has no effect on the truth-value of the counterfactual. So our (i)-(iii) indeed offers the truth-conditions for *all* counterfactuals in our language, as we can always rearrange the conjuncts in the antecedent according as the involved variables are endogenous or exogenous.

---

[13]Cf. Halpern [4]. But the language here is still richer than Halpern's, for I allow any formula to figure in the consequent of a causal counterfactual, whereas Halpern allows only a Boolean combination of atomics.

A formal characterisation of the language can now be given: (a) all sentences of the form $X = x$, called *atomic*, are wffs; (b) if φ and ψ are wffs, then so are $\neg\varphi$, $(\varphi \wedge \psi)$, $(\varphi \vee \psi)$, $(\varphi \supset \psi)$, $(\varphi \equiv \psi)$, $\Diamond\varphi$, and $\Box\varphi$; (c) if $\varphi_1, \ldots, \varphi_n$ are atomic formulas *containing no common variables* (footnote 14 explains the qualification), and ψ is a wff, then $(\varphi_1 \wedge \cdots \wedge \varphi_n) \diamond\!\!\rightarrow \psi$ and $(\varphi_1 \wedge \cdots \wedge \varphi_n) \Box\!\!\rightarrow \psi$ are wffs; and (d) no other expression is a wff.

[14]Here we must require that $X_i \neq X_j$ for any $i \neq j$. This is to avoid having such formulas as $(X_i = x_i \wedge X_i = x_i') \Box\!\!\rightarrow \psi$ (where $x_i \neq x_i'$) in our language, which do not make any sense as *causal* counterfactuals. (We cannot bring about both at the same time.)

This completes our formal characterisation of the structural semantics. We may now consider an example to illustrate how it works, before we apply it to our project of naturalising modal epistemology in terms of causal counterfactuals.

Example 1 (The Firing Squad[15]) Suppose our individuals include the court u, a captain c, two riflemen a and b, a prisoner d, and nothing else. Suppose we are considering the following cases, which are represented respectively as below:

whether the court u orders the execution ($U = 1$) or not ($U = 0$),
whether the captain c gives a signal ($C = 1$) or not ($C = 0$),
whether the rifleman a shoots ($A = 1$) or not ($A = 0$),
whether the rifleman b shoots ($B = 1$) or not ($B = 0$), and
whether the prisoner d dies ($D = 1$) or not ($D = 0$)

So we have the signature $\mathcal{S} = \langle \mathcal{V}, \mathcal{R}, \mathcal{I}, \varrho \rangle$, where $\mathcal{V} = \{U, C, A, B, D\}$, $\mathcal{R}(X) = \{0, 1\}$ for all $X \in \mathcal{V}$, $\mathcal{I} = \{u, c, a, b, d\}$, and $\varrho = \{\langle u, U \rangle, \langle c, C \rangle, \langle a, A \rangle, \langle b, B \rangle, \langle d, D \rangle\}$. Suppose our actual state s_1 is such that the court ordered the execution, the captain gave a signal, the two riflemen both shot, and the person died (i.e. $s_1 = \langle 1, 1, 1, 1, 1 \rangle$). Suppose the causal relationships between these variables are captured by the structural model $\mathcal{M} = \langle \mathcal{S}, \{C, A, B, D\}, \{f_C, f_A, f_B, f_D\} \rangle$, where

$$\begin{aligned} f_C &= U \\ f_A &= C \\ f_B &= C \\ f_D &= \max\{A, B\}. \end{aligned}$$

This can be represented by the following graph:

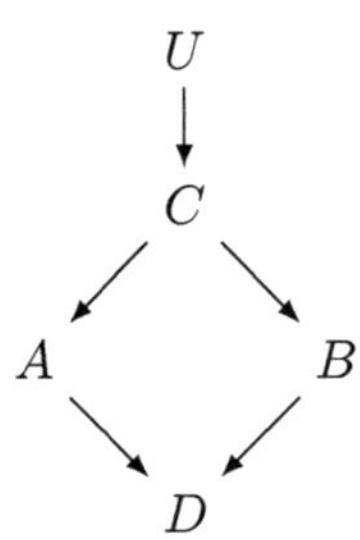

The model $\mathcal{M}$ has *two* possible states: $s_0 = \langle 0, 0, 0, 0, 0 \rangle$ and $s_1 = \langle 1, 1, 1, 1, 1 \rangle$. For there are two exogenous assignments (σ_0, which assigns 0 to the only exogenous variable U, and σ_1, which assigns 1 to U), and each of $\mathcal{M}(\sigma_0)$ and $\mathcal{M}(\sigma_1)$ has a unique solution. Given our actual state s_1, we may evaluate causal counterfactuals according to our truth-conditions. Consider the following statements:

(4) If we were to bring about that the rifleman a should not shoot, then the prisoner d would die.

[15]See [10, p. 207]. The case (6) below was provided by [2, p. 142].

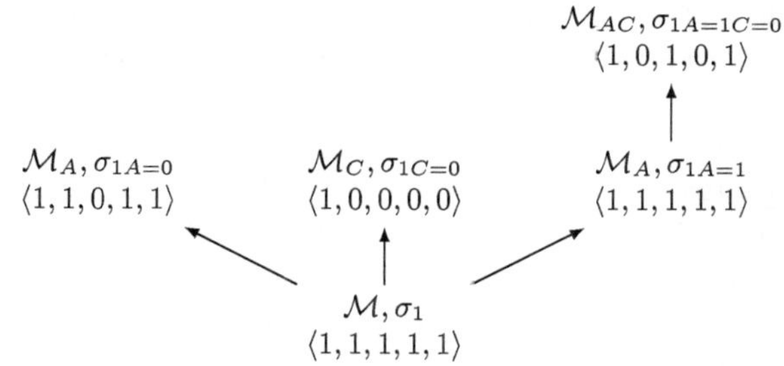

Fig. 4.1 Models and states for Example 1

(5) If we were to bring about that the captain c should give no signal, then the prisoner d would die.

(6) If we were to bring about that the rifleman a should shoot, then if we were to bring about that the captain c should give no signal, the prisoner d would die.

To evaluate (4), we need to consider the submodel $\mathcal{M}_A$ and the corresponding extended exogenous assignment $\sigma_{1A=0}$. Now $\mathcal{M}_A = \langle \mathcal{S}, \{C, B, D\}, \{f_C, f_B, f_D\} \rangle$, and $\sigma_{1A=0} = \{\langle U, 1 \rangle, \langle A, 0 \rangle\}$. It follows that $\mathcal{M}_A(\sigma_{1A=1})$ has $\langle 1, 1, 0, 1, 1 \rangle$ as its (unique) solution, in which $D = 1$ is true. As a result, $\mathcal{M}, s_1 \models A = 0 \,\Box\!\!\rightarrow D = 1$, and thus (4) is true.

The evaluation of (5) is similar. $\mathcal{M}_C = \langle \mathcal{S}, \{A, B, D\}, \{f_A, f_B, f_D\} \rangle$, and $\sigma_{1C=0} = \{\langle U, 1 \rangle, \langle C, 0 \rangle\}$. It follows that $\mathcal{M}_C(\sigma_{1C=0})$ has $\langle 1, 0, 0, 0, 0 \rangle$ as its (unique) solution, in which $D = 1$ is false. So $\mathcal{M}, s_1 \not\models C = 0 \,\Box\!\!\rightarrow D = 1$, and thus (5) is false.

Now, for the nested case (6), first we need to consider $\mathcal{M}_A$ and $\sigma_{1A=1}$ to see whether $C = 0 \,\Box\!\!\rightarrow D = 1$ holds in all possible states of $\mathcal{M}_A(\sigma_{1A=1})$. Now, since $\mathcal{M}_A = \langle \mathcal{S}, \{C, B, D\}, \{f_C, f_B, f_D\} \rangle$ and $\sigma_{1A=1} = \{\langle U, 1 \rangle, \langle A, 1 \rangle\}$, the (unique) solution to $\mathcal{M}_A(\sigma_{1A=1})$ is $\langle 1, 1, 1, 1, 1 \rangle$, which *coincides* with the actual state s_1. According to our truth-conditions, to determine whether (6) is true is to see whether $C = 0 \,\Box\!\!\rightarrow D = 1$ holds in s_1 in the model $\mathcal{M}_A$—i.e. the (only) possible state of $\mathcal{M}_A(\sigma_{1A=1})$. That is to say, we need to determine whether $\mathcal{M}_A, s_1 \models C = 0 \Box\!\!\rightarrow D = 1$ holds. To determine this, we need to consider $\mathcal{M}_A$'s submodel $\mathcal{M}_{AC} = \langle \mathcal{S}, \{B, D\}, \{f_B, f_D\} \rangle$, and the corresponding extended assignment $\sigma_{1A=1C=0}$. Since $\mathcal{M}_{AC}(\sigma_{1A=1C=0})$ has $\langle 1, 0, 1, 0, 1 \rangle$ as its (unique) solution, in which $D = 1$ is true, so $\mathcal{M}_A, s_1 \models C = 0 \,\Box\!\!\rightarrow D = 1$ holds. As a result, $\mathcal{M}, s_1 \models A = 1 \,\Box\!\!\rightarrow (C = 0 \,\Box\!\!\rightarrow D = 1)$ and thus (6) is true (Fig. 4.1).

(5) and (6) raise an interesting problem to the logic of causal counterfactuals. For in our model $\mathcal{M}$ and state s_1, both $A = 1 \Box\!\!\rightarrow (C = 0 \Box\!\!\rightarrow D = 1)$ and $A = 1$ are true but $C = 0 \Box\!\!\rightarrow D = 1$ is false. Briggs [2] takes this as showing that *modus ponens*,[16] and

[16]Some may find the example dubious on the basis that when φ *is* true usually we will not assert 'If φ then ψ' in the subjunctive as a *counterfactual* conditional. But notice that our '$\varphi \,\Box\!\!\rightarrow \psi$' is intended to mean not simply the subjunctive form of 'If φ then ψ', but 'If *we were to bring about* that φ, then it would be the case that ψ'. It is one thing to consider a situation where φ is true, but it is quite another to consider a situation where *we are surgically to bring about that* φ.

its underlying Lewisian assumption of *weak centering*,[17] can be violated if we enrich our language to include such nested counterfactuals as (6). This result has escaped the notice of the earlier advocates of structural semantics, who usually regard their logic to be approximately equivalent to Lewis's. For *weak centering* seems to be guaranteed by 'composition'—i.e. the fact that the actual state should be one of the possible states which result from surgically setting a variable to its *actual* value, or more succinctly, that setting a variable to its present value will not change anything about the present state. What has been ignored, however, is the fact that although setting a variable *surgically* to its present value will not change the present *state*, it can nevertheless change the *causal structure* of the model. By 'freezing' a variable at its present value, we will thereby block its prior causal influence, thus also break certain relations of counterfactual dependence (e.g. freezing A at 1 in our example breaks the dependency of D on C). This is how *weak centering* may fail for *causal* counterfactuals.[18]

We can see from the examples what extra resource the structural semantics may provide on top of the possible worlds semantics. First observe that what corresponds to a possible world in the structural semantics is *not* a world-state s nor a model $\mathcal{M}$, but a model–state pair $\langle \mathcal{M}, s \rangle$, as it is only with such a pair that we may evaluate the truth-value of a formula.[19] But such a model–state pair incorporates crucial information which is left out by its corresponding possible world: the *causal structure* represented by the structural equations.[20] Although the possible worlds semantics may still encode this information by adding a system assigning comparative similarities between worlds, the structural semantics rather takes it as *constitutive* of a (counterfactual) situation to include such information. This makes the structural semantics intuitively more appropriate for our project of naturalising modal epistemology in terms of causal counterfactuals, which I shall turn to in the next section.

4.4 Modal Epistemology Naturalised

As explained earlier, Williamson's modal epistemology has the advantage of being able to avoid invoking any mysterious faculty to explain how we acquire modal knowledge. The idea is to ground modal knowledge evolutionarily in our ordinary

[17]Weak centering is the assumption that for any φ and w, if φ is true at w then the selected worlds $f(\varphi, w)$ must include w. See [7] for more discussions.

[18]But notice that weak centering still holds for a special case: i.e. the case where the antecedent is an atomic formula concerning an *exogenous* variable. For in this case, our freezing the variable to its present value will change *neither* the state *nor* the causal structure of the model.

[19]Halpern and Pearl eventually make this clear in [5, p. 852]. Pearl calls such a pair a *causal world* [10, p. 207].

[20]This is why in our example we can have the same formula (i.e. $C = 0 \,\Box\!\!\rightarrow D = 1$) being false in the 'actual world' (i.e. $\langle \mathcal{M}, s_1 \rangle$) yet true in a world with the same state (i.e. $\langle \mathcal{M}_A, s_1 \rangle$). These two worlds are exactly alike in all the non-modal facts, but they still differ in causal structure. In other words, Humean supervenience fails.

capacity to handle (causal) counterfactuals, and then to explicate it as a capacity to perform some sort of 'mental simulation'.

On this account, when we are to evaluate a counterfactual conditional $A \Box\!\!\rightarrow B$, we may invoke all and only the cognitive resources we require for handling separately the antecedent A and the consequent B, and then apply them offline to simulate and predict what would have happened by counterfactually developing the supposition of the antecedent A, *with suitable background facts being held fixed.* Similarly, when we are to evaluate a modal claim $\Box A$, we evaluate the corresponding counterfactual conditional $\neg A \Box\!\!\rightarrow \bot$, and we do this by counterfactually developing the supposition $\neg A$, *with suitable background facts being held fixed*, to see whether it yields a contradiction. In this way, we may acquire modal knowledge without appealing to some mysterious faculty like intuition.

I think Williamson is on the right track in trying to reduce knowledge of modality to knowledge of counterfactuals. But his account, as I argued earlier, suffers from two problems: (1) the *cotenability* problem (i.e. that it is not always clear in our evaluation what background facts we should hold fixed, as it seems problematic just to hold A fixed in evaluating $\neg A \Box\!\!\rightarrow \bot$), and more seriously (2) the *gap* problem (i.e. that even granted the legitimacy of holding certain nomic and constitutive facts fixed which we learn from natural sciences, it still falls short of justifying knowledge of *metaphysical* modality). Earlier I suggested to solve the gap problem by restricting our modal knowledge to what can be explicated in naturalistic terms, such that we may quite harmlessly acknowledge our incapacity to know the alleged metaphysical modality that goes beyond our cognitive access. Now it is time to see how the structural semantics just characterised may help.

Let us consider the cotenability problem first. In a certain sense, we may also understand how the structural models work by a sort of 'simulation': to evaluate whether $\varphi \Box\!\!\rightarrow \psi$ is true, we simulate it by surgically setting some variables to certain values to bring about φ, *with suitable laws and facts being held fixed*, so as to see whether ψ would obtain. This is just like Williamson's simulation account, but here we have a more specific way to understand how we may 'counterfactually develop a supposition'—we simply set some variables to certain values, and then use the structural equations to calculate the possible values of our variables. But unlike Williamson's account, the structural semantics as I characterise it provides a handier way of expressing the distinction between what to vary and what to hold fixed. This can be considered in two categories: (i) the *laws of nature*, which are represented in our semantics by the *structural equations* of the model, and (ii) the *background facts*, which are represented in our semantics by the value-assignments to the *variables*. Now, to evaluate by such a 'simulation' whether a causal counterfactual $\varphi \Box\!\!\rightarrow \psi$ is true (where φ is a conjunction of the atomics $X_i = x_i$), we should hold some facts fixed and allow some others to vary, and also hold some facts fixed and allow some others to be violated. But here the distinction is readily made in the structural semantics. The variables X_i with their present values are precisely the facts we should vary, whereas all the *exogenous* variables (excluding X_i if any of them is exogenous) are what we should hold fixed. The remaining variables (i.e. those *endogenous* variables excluding X_i) we should neither vary nor hold fixed, but just leave them to be determined by this simulation. On the other hand, the structural equations $X_i = f_{X_i}$ (for $X_i \in \mathcal{V}_{\text{en}}$) are precisely the laws we should allow to break, whereas all the other structural equations of the model are the laws to be held fixed. So we have three different sets of variables, $\{X_i\}$, $\mathcal{V}_{\text{ex}} \backslash \{X_i\}$ and $\mathcal{V}_{\text{en}} \backslash \{X_i\}$, which correspond to a threefold division of all the facts into (a) what to be varied, (b) what to be held fixed and (c) what to be simulated. Similarly, we have two different sets of equations, $\{Y = f_Y \mid Y \in \mathcal{V}_{\text{en}} \cap \{X_i\}\}$ and $\{Y = f_Y \mid Y \in \mathcal{V}_{\text{en}} \backslash \{X_i\}\}$, which correspond to a division of all laws into what are to be violated and what are to be held fixed.

In a certain sense, our division of all variables into the exogenous and the endogenous ones is not entirely independent from our judgement about what to hold fixed. It is usually when we already have some intuitions about what we are to hold fixed as the background facts that we know more clearly how to make the exogenous–endogenous division. For example, in evaluating the counterfactual 'If I were to scratch the match, it would have lighted', we may want to take the aridity of the match as an exogenous variable because we have good reason to take it as a background factor to be held fixed.[21] But this is not an objection. For even in Lewis' possible worlds semantics, our reason for assigning a specific measure of comparative similarity rather than another may also appeal to certain pre-theoretical intuitions about what to hold fixed as the factual background. It merely indicates a very close conceptual connection between our evaluation of counterfactuals, our judgement about the factual background, our pre-theoretical understanding of the causal structure (including the exogenous–endogenous division) and our intuitions about comparative possibilities, such that it is almost impossible to have a theory for one without presupposing another. In this respect, the structural semantics is on a par with other semantics for counterfactuals.[22]

But there is still a difference. As we saw earlier, Williamson proposes that we evaluate a modal claim $\Box A$ through evaluating a corresponding counterfactual conditional $\neg A \Box\!\!\rightarrow \bot$. He then applies this to the case of constitutive facts such as (G), arguing that (G) is necessary because in holding (G) fixed the corresponding counterfactual conditional $\neg G \Box\!\!\rightarrow \bot$ should hold. Although this strikes us as counterintuitive, there is nevertheless nothing wrong with it in Lewis' semantics, provided that we have good reason for taking (G) as *cotenable*. For in Lewis' semantics, so long as (G) is necessary, it is indeed cotenable with any premise, its own negation included. This can be regarded as a degenerate case about cotenability. But in the structural semantics, it is *never* the case that a premise can be 'cotenable' with its own negation, whether it be necessary or not. We are never allowed to hold A fixed in evaluating what would happen had we brought about $\neg A$, simply because that would force us to take the same variables both as exogenous (so as to be held fixed) and as

---

[21] In that case, we cannot use the same structural model to evaluate the strengthened counterfactual 'If I were to soak the match in water and scratch it, it would have lighted', for here the aridity of the match is supposed to be something we are to simulate in the model, and thus should be taken as endogenous.

[22] That is, the 'ordering semantics' and the 'premise semantics' (see [8]).

endogenous (so as to be surgically brought about) at the same time. In this sense, the structural semantics helps to explain why Williamson's proposal is counter-intuitive.

But perhaps we may find some other facts to hold fixed? If so, there is still some hope that Williamson's proposal could work in the structural semantics. However, the problem is that Williamson's formula $\neg A \Box\!\!\rightarrow \bot$ is not even well-formed in our language. A smaller part of the problem is that in our language causal counterfactuals cannot have anything other than (conjunctions of) atomic formulas as antecedents. But this can be easily circumvented by using $A^* \Box\!\!\rightarrow \bot$ instead of $\neg A \Box\!\!\rightarrow \bot$, where A^* is a conjunction of atomics and is incompatible with A. So, instead of checking what would happen if gold *were not* the element with atomic number 79, we check what would happen if gold *were* the element with atomic number 78, etc. The more serious part of the problem concerns the precise meaning of having a contradiction $\bot$ in the consequent. If '$A \Box\!\!\rightarrow \bot$' means something like 'For some φ, $A \Box\!\!\rightarrow (\varphi \wedge \neg\varphi)$', then it is indeed well-formed in our language, but *trivially false* according to our semantics.[23] The reason is that our truth-conditions guarantee that $\varphi \wedge \neg\varphi$ should always be false in all possible states of any model, and thus $A \Box\!\!\rightarrow (\varphi \wedge \neg\varphi)$ has to be false. Another possible suggestion is to understand the symbol $\bot$ in the consequent as being used to represent the situation where our structural equations have no solution. On this interpretation, $(X_1 = x_1 \wedge \cdots \wedge X_n = x_n) \Box\!\!\rightarrow \bot$ holds in a possible state s of the model $\mathcal{M}$ when and only when $\mathcal{M}_{\mathbf{X}}(\sigma_{\mathbf{X}=\mathbf{x}})$ (where σ is the exogenous assignment $s \restriction_{\mathcal{V}_{\text{ex}}}$ derivable from s) has *no solution* at all. But as I remarked earlier, a structural model with no solution does not seem to make good philosophical sense. What is it supposed to mean when I surgically set X_i to x_i yet get no possible state at all because the structural equations have no solution? Or perhaps in that case I simply *couldn't* do such a setting? But why couldn't I do it? That seems to be in conflict with the basic idea of *causal* counterfactuals as interpreted in the structural semantics, where the antecedents are supposed to be something we can bring about by interventions. So this suggestion will not work either. As a result, we cannot invoke Williamson's equivalence $\Box A \equiv (\neg A \Box\!\!\rightarrow \bot)$ in the structural semantics to account for our modal knowledge.

What can we do then? I think even if we abandon Williamson's equivalence, we may still evaluate some modal claims about constitutive facts in terms of *causal* counterfactuals, provided that we have a good naturalistic way of understanding the modality involved. What does it mean by saying that a thing's constitutive nature (e.g. the atomic number of gold, the chemical structure of water, etc.) is *necessary* to it? My suggestion is that it means something like this: if we were surgically to change gold's atomic number, then it would no longer be gold. Notice that I am not saying '...then gold would not be gold' as if it would yield a contradiction (as is in Williamson's proposal). By contrast, my suggestion should be taken on a *de re* reading, saying about *the thing* which is actually gold that *it* would no longer be gold under the intervention in question. Another complication is that my suggestion

---

[23]This is a consequence of the structural semantics. It is also generally incorporated into the axiom systems (e.g. the 'existence' property in [10, p. 230]). Notice that [2] directly takes $\neg(A \Box\!\!\rightarrow \bot)$ as an axiom (p. 156).

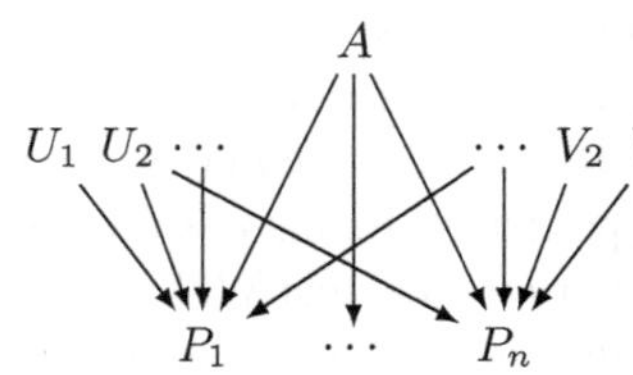

Fig. 4.2 Causal structure for gold

in fact requires us to have a criterion for determining whether something is or is not gold, for only so can we make our evaluation of the causal counterfactual in question. At this point, I would propose that we identify gold by a set of properties (e.g. being yellow, being malleable, having such and such a melting point, etc.), such that anything is gold if and only if it has most, or a weighted most, of these properties.[24] So suppose our identifying properties for gold are $p_1, \ldots, p_n$, then my suggestion is to understand the modal claim about gold's atomic number as this:

(7) For anything a which is actually gold, if we were to bring about that a has an atomic number 78 (or 80, etc.), then a would not have most (or a weighty most) of the properties $p_1, \ldots, p_n$.

Now, we can express (7) in the structural semantics. For instance, we may have a set of variables $\{A, P_1, \ldots, P_n\}$, representing a's atomic number and those determinable properties of a under which $p_1, \ldots, p_n$ falls, and we may have a model $\mathcal{M}$ with a causal structure like what is in Fig. 4.2. The model would be extremely complicated, and it should rightly be so. For to evaluate whether (7) is true, we need a lot of information about various causal relationships between P_i and various background factors, and we have to capture them in terms of the structural equations of the model. But this should be a virtue of my proposal rather than a vice, for it agrees with our intuition that knowing gold's atomic number *as constitutive* and *as necessary* should somehow involve a lot of scientific knowledge. It is not a result of some trivial conceptualisation. Such *modal* knowledge of constitutive facts is a highly complicated form of causal knowledge, as our models show. But it is still something we can accommodate in our structural semantics.

We may now come back to the gap problem and our naturalising project. In fact we have just provided a naturalised account for our modal knowledge of constitutive facts. We know that gold necessarily has the atomic number 79, because we know, with the help of certain scientific knowledge, that if we were to change gold's atomic number then it would no longer be gold. Similarly, we know that water necessarily has the chemical structure H_2O, because we know, with the help of certain chemical knowledge, that if we were to change water's chemical structure then it would no

[24]A possible objection might be that Kripke has already refuted such a cluster theory of names, on the basis that the theory could not allow the possibility that gold might lose all, or almost all, of the identifying properties in the set. My reply is that quite on the contrary, my proposal *allows* such a possibility, for my proposal does allow that something *which is actually gold* might have lost all of its identifying properties. But how about the possibility that something without any of these properties yet still be gold? I think for a case like that our intuition is very unclear.

longer be water. However, not all modal knowledge can be thus treated in terms of causal counterfactuals. Sometimes we may want to assert or evaluate the possibility or impossibility of something in a more straightforward sense, without considering what would happen if we were to change it this way or that way. For instance, we may want to assert that gold cannot possibly be both yellow and green, or to evaluate whether there is such a possibility that the law of gravitation might fail to hold. What can we do?

I think it is very helpful to distinguish between different species of modality in our semantics. We have already encountered one (i.e. natural modality), and we can now consider some more.

Truth-Conditions 3 (Natural and Metaphysical Modalities) *Let* $\mathcal{M} = \langle \mathcal{S}, \mathcal{V}_{\text{en}}, \mathcal{F} \rangle$ *be a structural model over the signature* $\mathcal{S}$*,* s *be a possible state of* $\mathcal{M}$*, and* φ *be a formula in our language. We define* $\Box\varphi$*,* $\boxdot\varphi$*, and* $\blacksquare\varphi$*, by the following truth-conditions.*

(i) $\mathcal{M}, s \models \Box\varphi$ *iff* $\mathcal{M}, t \models \varphi$ *for any possible state* t *of* $\mathcal{M}(s\restriction_{\mathcal{V}_{\text{ex}}})$.
(ii) $\mathcal{M}, s \models \boxdot\varphi$ *iff* $\mathcal{M}, t \models \varphi$ *for any possible state* t *of the model* $\mathcal{M}$.
(iii) $\mathcal{M}, s \models \blacksquare\varphi$ *iff* $\mathcal{N}, t \models \varphi$ *for any possible state* t *of any model* $\mathcal{N}$ *over* $\mathcal{S}$.

As explained earlier, $\Box\varphi$ in our semantics represents *natural* necessity. According to our truth-condition (i), something is naturally necessary if and only if it is true in all possible states of the model *with all the actual laws and background factors being held fixed*. This is the sense in which we may say it is naturally impossible that one can get from London to Cambridge in less than five minutes. However, there is still a sense in which this is 'naturally' possible—i.e. that it does not violate the actual laws of nature. For a convenient terminology, I call this sense of modality *nomic* (denoted by '$\boxdot$'), and use the truth-condition (ii) to capture it. Accordingly, something is *nomically* necessary if and only if it is true in all possible states of the model *with all the actual laws being held fixed* (but with background factors being allowed to vary). So, in this sense, getting from London to Cambridge in less than five minutes is *nomically* possible (relative to a setting of background factors to include the availability of some extremely high-speed aircraft), but travelling faster than light is *nomically* impossible.

But how about *metaphysical* modality? Here I define in our structural semantics a modal operator '$\blacksquare$' to capture some of the our uses of metaphysical necessity. Accordingly, something is metaphysically necessary if and only if it is true in all possible states of any model whatsoever over the given signature. So, in this sense, we may say that gold's being both yellow and green is metaphysically impossible, for in our semantics no variable (gold's colour included) can take two different values at once (i.e. in the same possible state). Also in this sense, we may regard the relation between a thing and its category (i.e. $a \in \mathcal{C}(a)$), or any other truth about the basic setting in the *signature*, as a matter of metaphysical necessity. On these definitions, the *laws of nature* will be nomically necessary but metaphysically contingent, for given any model $\mathcal{M}$, its structural equations should be satisfied in all its possible states, but we can always find such a model $\mathcal{N}$ where they fail to hold (e.g. a model with no endogenous variables, such that we may assign arbitrary values to all the variables).

Now, if metaphysical modality is thus understood, how does it fit the naturalising picture I propose? In a certain sense, knowledge of metaphysical modality is indeed quite different from knowledge of causal counterfactuals. That is part of the reason why earlier I cast some doubt on the idea of reducing the former to the latter, and present it as a gap problem for Williamson's account. But this is not to deny that we may still ground modal knowledge in our capacity to know causal counterfactuals. For our evaluation of causal counterfactuals has to presuppose some 'framework' like the *signature* of our semantics, and so if we have the capacity to handle causal counterfactuals, we should thereby have the capacity to handle truths about the presupposed framework. It is in this sense that knowledge of metaphysical modality is grounded in knowledge of counterfactuals. But the reason is not that we can reduce the former to the latter by some equivalence as Williamson proposes. The reason is rather that our capacity to handle the latter provides all we need to handle the former.

But notice that this will cover *only* a very restricted range of the so-called metaphysical necessities. For only those 'structural' truths about the signature (e.g. about the basic settings of variables and their value-ranges, and of individuals and their categories, etc.) can be accommodated in this way as part of our presupposition for counterfactual knowledge. Other modal claims in the metaphysics literature, which are alleged to involve '*metaphysical*' modality, may still be ungrounded. To these modal claims, I remain sceptical. We still have no such 'knowledge' concerning, say, whether zombies are metaphysically possible, or whether atomless gunk is metaphysically possible. We may have good philosophical arguments for or against such possibilities, but that does not seem to be settled as 'knowledge'. The cases for the constitutive truths and for the 'structural' truths just considered are quite different. For, if my argument in this paper is correct, these are the truths for which our modal knowledge can be grounded in our capacities to handle causal counterfactuals.

References

1. Boghossian, P.: Williamson on the a priori and the analytic. Philos. Phenomenol. Res. **82**(2), 488–497 (2011)
2. Briggs, R.: Interventionist counterfactuals. Philos. Stud. **160**(1), 139–166 (2012)
3. Galles, D., Pearl, J.: An axiomatic characterization of causal counterfactuals. Found. Sci. **3**(1), 151–182 (1998)
4. Halpern, J.: Axiomatizing causal reasoning. J. Artif. Intell. Res. **12**, 317–337 (2000)
5. Halpern, J., Pearl, J.: Causes and explanations: a structural-model approach. Part I: causes. Br. J. Philos. Sci. **56**(4), 843–887 (2005)
6. Kment, B.: Counterfactuals and the analysis of necessity. Philos. Perspect. **20**(1), 237–302 (2006)
7. Lewis, D.: Counterfactuals. Blackwell, Oxford (1973)
8. Lewis, D.: Ordering semantics and premise semantics for counterfactuals. J. Philos. Log. **10**(2), 217–234 (1981)

9. Lowe, E.J.: What is the source of our knowledge of modal truths? Mind **121**(484), 919–950 (2012)
10. Pearl, J.: Causality: Models, Reasoning and Inference, 2nd edn. Cambridge University Press, Cambridge (2009)
11. Roca-Royes, S.: Modal knowledge and counterfactual knowledge. Log. Anal. **54**(216), 537–552 (2011)
12. Stalnaker, R.: Anti-essentialism. Midwest studies. Philosophy **4**(1), 343–355 (1979)
13. Tahko, T.E.: Counterfactuals and modal epistemology. Grazer Philos. Stud. **86**(1), 93–115 (2012)
14. van Fraassen, B.C.: Meaning relations among predicates. Noûs **1**(2), 161–179 (1967)
15. Williamson, T.: The Philosophy of Philosophy. Blackwell, Oxford (2007)
16. Williamson, T.: Reply to Boghossian. Philos. Phenomenol. Res. **82**(2), 498–506 (2011)
17. Zhang, J.: A Lewisian logic of causal counterfactuals. Minds Mach. **23**(1), 77–93 (2013)
18. Zhang, J., Lam, W.-Y., De Clercq, R.: A peculiarity in Pearl's logic of interventionist counterfactuals. J. Philos. Log. **42**(5), 783–794 (2013)

Chapter 5
Motivating the Causal Modeling Semantics of Counterfactuals, or, Why We Should Favor the Causal Modeling Semantics over the Possible-Worlds Semantics

Kok Yong Lee

Abstract Philosophers have long analyzed the truth-condition of counterfactual conditionals in terms of the possible-worlds semantics advanced by Lewis [13] and Stalnaker [23]. In this paper, I argue that, from the perspective of philosophical semantics, the causal modeling semantics proposed by Pearl [17] and others (e.g., Briggs [3]) is more plausible than the Lewis-Stalnaker possible-worlds semantics. I offer two reasons. First, the possible-worlds semantics has suffered from a specific type of counterexamples. While the causal modeling semantics can handle such examples with ease, the only way for the possible-worlds semantics to do so seems to cost it its distinctive status as a philosophical semantics. Second, the causal modeling semantics, but not the possible-worlds semantics, has the resources enough for accounting for both forward-tracking and backtracking counterfactual conditionals.

Keywords Causal model · Causal modeling semantics · Counterfactual conditional · Possible-worlds semantics · Backtracking · Intervention

5.1 Introduction

Traditionally, philosophers have analyzed the truth-condition of counterfactual conditionals (hereafter "counterfactuals") in terms of the *possible-worlds semantics* advanced by David Lewis [13] and Robert Stalnaker [23]. In this paper, I argue that, from the perspective of philosophical semantics, it is better to give up the possible-worlds semantics and opt for the *causal modeling semantics* proposed by Judea Pearl [17] and others (cf., e.g., Briggs [3]). I will make an important modification to the orthodox causal modeling semantics though.

K.Y. Lee (✉)
Department of Philosophy, National Chung Cheng University,
Chia-yi County 621, Min-hsiung, Taiwan
e-mail: kokyonglee.mu@gmail.com

S.C.-M. Yang et al. (eds.), *Structural Analysis of Non-Classical Logics*,
Logic in Asia: Studia Logica Library, DOI 10.1007/978-3-662-48357-2_5

I offer two reasons for favoring the causal modeling semantics over the possible-worlds semantics. First, the possible-worlds semantics has suffered from a specific type of counterexamples. While the causal modeling semantics can handle such examples with ease, the only way for the possible-worlds semantics to do so seems to cost it its distinctive status as a philosophical semantics. Second, the possible-worlds semantics is incomplete at best, since it fails to take backtracking counterfactuals into account. The causal modeling semantics, by contrast, has the resources enough for accounting for both forward-tracking and backtracking counterfactuals.

The following consists of seven sections. In Sect. 5.2, I will review the possible-worlds semantics of counterfactuals, in particular, the notion of comparative similarity among worlds. In Sect. 5.3, I will discuss two counterexamples to the possible-worlds semantics, which indicate that the similarity of worlds needs to be characterized in terms of causal dependence. In Sect. 5.4, I will point out that the possible-worlds semantics fails to take backtracking counterfactuals into account. I will discuss and reject Lewis' reasons for dismissing backtracking counterfactuals. In Sect. 5.5, I will introduce a new causal modeling semantics. In Sect. 5.6, I will demonstrate that the distinction between forward-tracking and backtracking counterfactuals can be explained naturally by the new causal modeling semantics. In Sect. 5.7, I will show that the new causal modeling semantics is immune to the counterexamples mentioned in Sect. 5.3. In Sect. 5.8, I will summarize the main findings.

5.2 The Possible-Worlds Semantics

Let ">" stand for the counterfactual conditional connective, "A > C" for the counterfactual conditional *If A had obtained, then C would have obtained.*[1] Intuitively, when determining whether "A > C" is true, we first envisage a (counterfactual) scenario s′ such that (i) "A" is true in s′, and that (ii) s′ is as similar to the (actual) scenario s as "A" being truth permits it to (cf. Lewis [13], 1). We then determine whether "C" is true in s′. "A > C" is true in s if, and only if, "C" is true in s′. We may define a *selection function f* as a function that selects a set of situations s′ based on A and s.[2] The *i*ntuitive *p*icture of the truth-condition of counterfactuals is as follows:

(IP) "A > C" is true in s if and only if "C" is true in each s′ ∈ f(A, s). (Cf. Briggs [3], 140–1)

IP is just a framework. To further develop it, some substantial contents must be given to the selection function.

Let "A-world" stand for the world in which "A" is true. The possible-worlds semantics interprets the selection function as a function of the comparative *s*imilarity among possible *w*orlds:

[1]Throughout this paper, propositions (or events) are denoted by italics sentences.

[2]The selection function was first introduced by Stalnaker [23]. I am using the notion in a broader sense.

(SW) "A > C" is (non-vacuously) true in w_i if and only if some A-worlds w_j in which "C" is true are more similar to w_i than any A-world w_l in which "C" is false is.

The similarity talk is somehow intuitive as ordinary people employ something similar when determining the truth-values of counterfactuals. Still, SW is just a first step; a lot more needs to be said in order for it to be instructive.

How should we interpret the notion of similarity among worlds? Arguably, the similarity in play cannot be *overall similarity* [6]. In his [14], Lewis proposes a complex system of weights of similarity among worlds. On this system, when evaluating the degree of similarity among worlds:

(L1) It is of the first importance to avoid big miracles or big quasi-miracles.
(L2) It is of the second importance to maximize the region of perfect match.
(L3) It is of the third importance to avoid small miracles or small quasi-miracles.
(L4) It is of the fourth importance to maximize the region of imperfect match. (For the sake of discussion, I adopt Schaffer's formulation (cf. Schaffer [20]))

Call L1-L4 "System L" and the possible-worlds semantics equipped with System L "the L-possible-worlds semantics." Some clarifications are called for. *Miracles* here mean violations of physical laws. Taking violations of laws as events, we may talk about the "size" of miracles based on the number of violations involved. Suppose that physical laws are indeterministic. An indeterministic event is counted as a *quasi-miracle* if it seems to "conspire to produce a pattern" (Lewis [15], 60). A quasi-miracle is an event "which is both low in probability and which has a pattern which is, by our lights, *remarkable*" (Hawthorne [9], 398, original italics). *Perfect match* indicates molecule-to-molecule identity, while *imperfect match*, overall similarity.

5.3 Troubling Cases and Refinements

The L-possible-worlds semantics suffers from a specific type of counterexamples, which reveals one of its deepest problems. That is, it fails to take into account the notion of causal dependence, which plays a crucial role in ordinary people's determination of the truth-values of counterfactuals.[3]

Consider Ryan Wasserman's example:

> *Bomb.* Imagine a deterministic world … that is much like our own in its distribution of objects and qualities, but which contains a black box in the middle of the Milky Way. In the black box there is a beetle and a button. If the button is pushed, a signal will run along a wire and out of the box. Beyond the wire, there are no causal avenues running out of the black box—whatever happens in the box stays in the box. The wire is connected to a "mega-bomb" which is lightening fast and deadly powerful—if the mega-bomb explodes, everything in the future light cone of the bomb will be destroyed. But the universe is spared. The beetle does not strike, the bomb does not destroy. Let us suppose, finally, that the black box and all of its contents is [sic] destroyed in a lawful manner shortly after *t*. (Wasserman [25], 59)

[3]There are other criticisms (cf., e.g., Pruss [2]). For simplicity's sake, I will leave them aside.

Let "Push > Destroy" stand for *If the beetle had pushed the button, the universe would have been destroyed*. Intuitively, "Push > Destroy" is true in *Bomb*, but the L-possible-worlds semantics yields the wrong result.

Let w_1 be the world of *Bomb* in which the beetle does not push the button, and the universe is not destroyed, w_2 be the counterfactual world in which the beetle, due to a small miracle, pushes the button, and the universe is destroyed, and w_3 be the counterfactual world in which the beetle, due to a small miracle, pushes the button, but, due to yet another small miracle, the signal does not transmit to the mega-bomb—hence the universe is not destroyed.

According to the L-possible-worlds semantics, w_3 is more similar to w_1 than w_2 is, since (i) while w_3 contains more small miracles than w_2 does, it also has a larger region of perfect match than w_2 does, and (ii) it is more important to maximize the region of perfect match than to avoid small miracles when determining the degrees of similarity among worlds. It follows that "Push > Destroy" is false in w_1. Counterintuitive.

Bomb happens in a deterministic world. Yet parallel counterexamples can be constructed out of an indeterministic setting. Michael Slote once reported Sidney Morgenbesser's example:

> *Bet*. Imagine a completely underdetermined random coin. Your friend offers you good odds that it will not come up heads; you decline to bet, he flips, and the coin comes out heads. He then says: "you see; if you had bet (heads), you would have won." (Slote [22], 27, Footnote 33)

Let "Bet > Win" stand for *If the hearer had bet (heads), she would have won*. Intuitively, "Bet > Win" is true in *Bet*, but the L-possible-worlds semantics yields the wrong result again.

Let w_4 be the world of *Bet* in which the hearer does not bet (heads), the coin lands heads, and thus the hearer does not win, w_5 be the counterfactual world in which the hearer, due to a small miracle, bets (heads), the coin lands heads, and thus the hearer wins, and w_6 be the counterfactual world in which the hearer, due to a small miracle, bets (heads), the coin lands tails, and thus the hearer does not win.

According to the L-possible-worlds semantics, w_5 is no more similar to w_6 than w_8 is, since (i) both w_5 and w_6 contain the same small miracle, and (ii) w_5 contains the imperfect match that the coin lands heads, while w_6 contains the imperfect match that the hearer does not win the bet—hence, w_5 and w_6 are seemingly equally similar to w_4. It follows that "Bet > Win" is not true in *Bet*. Counterintuitive.

What is wrong with the L-possible-worlds semantics? The problem, as many have pointed out (cf. Schaffer [20]; Edgington ([5], 20)), is this: when determining the truth-values of counterfactuals, System L fails to take into account the different ways a possible world may obtain the region of (im)perfect match. For instance, in *Bomb*, the region of perfect match between w_1 and w_3—that the universe is not destroyed—is *causally dependent* on Push, the antecedent of the counterfactual in question. Intuitively, when determining the truth-values of counterfactuals, maximizing the region of perfect match of this sort should be weighed less (if at all) than avoiding small miracles. Similarly, in *Bet*, the region of *mismatch* between w_4 and w_5—that the hearer wins in w_5 but not in w_4—is *causally dependent* on whether or not Bet, the antecedent of the counterfactual in question, obtains, while the *mismatch* between w_4 and w_6—the coin lands heads in w_4 but lands tails in w_6—is *causally independent* on whether or not Bet obtains. Intuitively, when determining the truth-values of counterfactuals, minimizing mismatch of the former sort should be weighed less (if at all) than minimizing mismatch of the latter sort.

Jonathan Schaffer thinks that the L-possible-worlds semantics is remediable. He proposes to refine System L as follows. When evaluating the degrees of similarity among worlds:

(S1) It is of the first importance to avoid big miracles or big quasi-miracles.
(S2) It is of the second importance to maximize the region of perfect match, *from those regions causally independent of whether or not the antecedent obtains.*
(S3) It is of the third importance to avoid small miracles or small quasi-miracles.
(S4) It is of the fourth importance to maximize the region of imperfect match, *from those regions causally independent of whether or not the antecedent obtains.* (Schaffer [20], 305, original italics)

Call S1-S4 "System S," and the possible-worlds semantics equipped with System S "the S-possible-worlds semantics." System S takes into account the different ways a possible worlds may obtain the region of (im)perfect match, which play a crucial role in determining the truth-values of counterfactuals. That is, when determining the degree of similarity among worlds, only the region of (im)perfect match *causally independent* of whether or not the antecedent of the counterfactual in question obtains should be regarded as important.

The S-possible-worlds semantics is able to handle cases like *Bomb* and *Bet*. Consider *Bomb*. According to the S-possible-worlds semantics, w_2 is more similar to w_1 than w_3 is, since (i) w_2 contains fewer small miracles than w_3 does (w_3's region of perfect match counts for nothing now, since it is causally dependent on Push), and (ii) it is important to avoid small miracles when determining the similarity among worlds. It follows that "Push > Destroy" is true in *Bomb*, as desired. Likewise, in *Bet*, w_5 is more similar to w_4 than w_6 is, since (i) w_5 contains a larger region of imperfect match than w_6 does (w_6's region of imperfect match counts for nothing now, since it is causally dependent on Bet), and (ii) it is important to maximize the region of imperfect match when determining the similarity among worlds. It follows that "Bet > Win" is true in *Bet*, as desired.

However, there is still a flaw in Schaffer's refinement. Like System L, when determining the truth-values of counterfactuals, System S regards the different ways of avoiding miracles as equally important. This is mistaken. Consider:

Power. John and Linda are drinking wine in John's apartment. They finish the last bottle and long for some more. John looks at the glass of water in front of them, and says to Linda, "If I had the power of Jesus, I would have served you more wine."

Let "Power > Wine" stand for *If John had the power of Jesus, he would have served Linda more wine*. Intuitively, "Power > Wine" is true in *Power*. The S-possible-worlds semantics, however, fails to give the correct verdict.

Let w_7 be the world of *Power* in which John does not have the power of Jesus and does not serve Linda more wine, w_8 be the counterfactual world in which John has the power of Jesus, John executes his power to transform the glass of water before him into a glass of wine (which is a big miracle), and he then serves it to Linda, and w_9 be the counterfactual world in which John has the power of Jesus, but, due to a small miracle in his brain, he changes his mind and does not execute his power. Thus Linda does not get more wine.

According to System S (and System L, too), w_9 is more similar to w_7 than w_8 is, since (i) while w_9 contains a small miracle in John's brain, w_8 contains a big miracle of turning water into wine, and (ii) it is more important to avoid big miracles than to avoid small miracles when determining the similarity among worlds. It follows that "Power > Wine" is false in *Power*. Counterintuitive.

Power poses a problem to the S-possible-worlds semantics in as much the same way as *Bomb* and *Bet* do to the L-possible-worlds semantics. System L regards the different ways of obtaining the region of (im)perfect match as equally important, which is problematic since the region of (im)perfect match causally dependent on the antecedent of the counterfactual in question should play no significant role in determining the truth-values of counterfactuals. Similarly, System S regards the different ways of avoiding miracles as equally important, which is problematic since miracles causally dependent on the antecedent of the counterfactual in question should play no significant role in determining the truth-values of counterfactuals.

Still, System S is remediable. Following the spirit of Schaffer's refinement of System L, we may replace S1 and S3 with the following respectively:

(S1′) It is of the first importance to avoid big miracle or big quasi-miracles, *for miracles causally independent of whether or not the antecedent obtains.*

(S3′) It is of the third importance to avoid small miracles or small quasi-miracles, *for miracles causally independent of whether or not the antecedent obtains.*

Call the resulting account "System S′," and the possible-worlds semantics equipped with System S′ "the S′-possible-worlds semantics."

The S′-possible-worlds semantics handles *Power* nicely, for now w_8 is regarded as more similar to w_7 than w_9 is, since (i) w_8 contains fewer small miracles than w_9 does (w_8's big miracle counts for nothing now, since it is causally dependent on Power), and (ii) it is important to avoid small miracles when determining the degrees of similarity among worlds. It follows that "Power > Wine" is true in *Power*, as desired.[4]

[4]Some might complain that cases like *Power* were illegitimate for involving supernatural power, or that counterfactuals with a physically impossible antecedent such as "Power > Wine" should receive a different semantic treatment. However, I see no inherent problem for counterfactuals involving supernatural power. Nor do I think that the difference between "Power > Wine" and, say, "Bet > Win" warrants different semantic treatments.

Perhaps even System S′ is not immune to criticisms.[5] But let us not pursue the issue further. For the present purposes, it is important to highlight the general direction for which System S and System S′ are heading. As noted, the possible-worlds semantics proposes a similarity-of-worlds interpretation of the selection function. In order for the possible-worlds semantics to gain its distinctive status as a philosophical semantics, it would be better if the notion of similarity is not reducible to some other notions, such as the notion of causal dependence, which is central to the causal-modeling-semantics interpretation of the selection function (see Sect. 5.5). Otherwise, the status of the possible-worlds semantics as a genuine alternative to the causal modeling semantics would become doubtful.

System L is doing just fine. The similarity of worlds, according to System L, is determined by the conditions of avoiding miracles and maximizing (im)perfect match, which are defined independently of the notion of causal dependence. System L, however, has suffered from a series of counterexamples. To refine, System S and System S′ suggest that the two conditions of avoiding miracles and maximizing (im)perfect match should be further confined by certain causal constraints. The general idea, as specified by S2 and S4, is to define the similarity of worlds in such a way that events causally independent of the antecedent are preserved as much as possible. The same goes for events causally determined by the antecedent, as specified by S1′ and S3′. Defined in this way, the term "similarity" loses any of its intuitive meaning and may better be understood as a placeholder for something essentially causal. The problem is that such a similarity interpretation of the selection function appears alarmingly like a version of the causal interpretation offered by the causal modeling semantics (see Sect. 5.5 for more on the latter).

In other words, System S and System S′'s interpretation of the selection function seems to be a causal interpretation in disguise. If so, the possible-worlds semantics is deprived of its status as a distinctive philosophical semantics. For as long as the notion of the similarity of worlds relies heavily on the notion of causal dependence,

---

[5]James Woodward has offered a counterexample to Lewis' idea that avoiding big miracles is always more important than avoiding small miracles:

> Consider a simple example ... C is a deterministic direct (type) cause of E but also deterministically causes E indirectly by means of n causal routes that go through $C_1, \ldots, C_n$. Consider the counterfactual (1) "If $C_1, \ldots, C_n$ had not occurred, E would not have occurred." (Woodward 2013, Endnote 4)

Intuitively, (1) seems false, but the System S′ fails to give the correct verdict. Let w_{10} be the world in which C, $C_1, \ldots, C_n$, and E hold, w_{11} be the world in which, due to a small miracle, C does not hold, and $C_1, \ldots, C_n$, and E do not hold, and w_{12} be the world in which C holds, but due to a big miracle $C_1, \ldots, C_n$ do not hold, but E still holds.

Suppose that C is within the *immediate past* of $C_1, \ldots, C_n$. That C is within the immediate past of C_i means that C had to have obtained if C_i were to obtain (as we will see in Sect. 5.4, Lewis allows backtracking counterfactualization in immediate past). It follows that, according to the S′-possible-worlds semantics, w_{11} is more similar to w_{10} than w_{12} is, since w_{12} contains a big miracle while w_{11} does not. Hence, (1) turns out to be true. Counterintuitive.

Thanks for an anonymous reviewer for correcting a serious mistake in the original draft.

the possible-worlds semantics seems to devolve into a cumbersome causal modeling semantics.

Of course, the possible-worlds semantics and the causal modeling semantics are still different in other aspects. For instance, the possible-worlds semantics takes propositions to be true in a possible world, which is a global scenario including infinitely many events, while the causal modeling semantics opts for causal models, which, as we will see, are local scenarios consisting of a finite number of events. But the difference does not show that the two are distinctively different, since the framework of the possible-worlds semantics is consistent with the idea of propositions being true in local scenarios (or something less globally encompassing than possible worlds). And the causal modeling semantics, in principle, can work with possible worlds as well.

There is still room for discussion. Perhaps, it could be shown that the similarity interpretation of the selection function is not just a causal interpretation in disguise. Perhaps, there could be something interesting in the notion of similarity of worlds, which is not exhausted by causal dependence. But the burden of proof is now on the proponents of the possible-worlds semantics.

5.4 Backtracking Counterfactuals

The possible-worlds semantics also faces the general problem of not being able to account for backtracking counterfactualization (i.e., to counterfactualize back in time, and then forward again (cf. Bennett [2], 208)). To be fair, the problem of backtracking counterfactuals is not specific to the possible-worlds semantics, as many accounts of the causal modeling semantics are vulnerable to the same problem. Still, the problem indicates that the possible-worlds semantics is at best incomplete.

The following is a famous example that illustrates the distinction between forward-tracking and backtracking counterfactuals:

> *Ask*. Jack had a quarrel with Jim yesterday, and Jack is still mad at Jim. When Jack is not mad, he is a generous person. He will help his friend if asked for a favor. Jim, on the other hand, is a prideful person, who will not ask someone for help after having a quarrel with this person. As a result, Jim does not ask Jack for help. (cf. Lewis [14], 456; also see Downing [4])

Let “Ask > Help” stand for *If Jim had asked Jack for help, Jack would have helped him*. “Ask > Help” seems false in *Ask*, but only under what we may call *forward-tracking* counterfactualization: if Jim were to ask Jack for help, he would have been rejected since Jack is mad at him, and Jack is not generous when he is mad. Under what we may call *backtracking* counterfactualization, however, “Ask > Help” seems true in *Ask*: Jim is a prideful person; he would not have asked Jim for help after having such a quarrel with him yesterday. Hence, if Jim were to ask Jack for help, it must be that they did not quarrel yesterday. If so, Jack would not be mad at Jim and would have helped him.

The possible-worlds semantics, at least in its orthodox form, is insensitive to the distinction between forward-tracking and backtracking counterfactuals. More precisely, the possible-worlds semantics has no resources for handling backtracking counterfactuals. The possible-worlds semantics always gives a definite verdict on the truth-values of counterfactuals like "Ask > Help," usually the one in accord with forward-tracking counterfactuals.

Consider *Ask*. Let w_{10} be the world of *Ask* in which Jim did not ask Jack for help and was not rejected, w_{11} be the counterfactual world in which, due to a small miracle, Jim and Jack did not quarrel yesterday, and Jim asked Jack for help and was not rejected, and w_{12} be the counterfactual world in which, due to a small miracle, Jim asked Jack for help and was rejected.

According to both System L and System S (and System S' for that matter), w_{12} is more similar to w_{10} than w_{11} is, since (i) while both w_{11} and w_{12} contain a small miracle, w_{12} contains a larger region of perfect match (causally independent of Ask), and (ii) it is important to maximize the region of perfect match, other things being equal. It follows that, according to the orthodox possible-worlds semantics, "Ask > Help" is false in w_{10}. This verdict is not so much wrong as it is incomplete, since it is in complete disregard of backtracking reading of "Ask > Help."

That the possible-worlds semantics does not square well with backtracking counterfactuals is nothing new.[6] Lewis is aware of the problem, but he quickly spares the semantics the difficulty by dismissing backtracking counterfactuals as illegitimate. Since Lewis' view is by no means uncommon it is worth examining Lewis' reasons closely.

First, Lewis argues that backtracking counterfactuals are *nonstandard* since ordinary counterfactuals are not backtracking in character:

> We *ordinarily* resolve the vagueness of counterfactuals in such a way that counterfactual dependence is asymmetric (except perhaps in cases of time travel or the like). Under this *standard* resolution, backtracking arguments are mistaken: if the present were different the past would be the same, but the same past causes would fail somehow to cause the same present effects. If Jim asked Jack for help today, somehow Jim would have overcome his pride and asked despite yesterday's quarrel. (Lewis [14], 458, my italics)

This quotation seems to suggest that backtracking counterfactuals are nonstandard because *ordinary* counterfactuals are non-backtracking in character. But what does "ordinary" mean here? Presumably, it does not mean that forward-tracking interpretation of counterfactuals are used more frequently than backtracking ones, since frequency is a contingent matter—there could well be a society in which backtracking counterfactuals are used more often instead.

Lewis also notes that backtracking counterfactuals "will not be clearly true or clearly false," if taken "out of context" (Lewis [14], 485). But it cannot be the case that ordinary counterfactuals are not backtracking in character because the truth-values

[6] That the possible-worlds semantics fails to account for backtracking counterfactuals is the reason why the semantics also has difficulties dealing with backward counterfactuals (counterfactuals whose antecedent happens after its consequent) (cf. Northcott [16]) and backward causation (cf. Tooley 2002).

of backtracking counterfactuals are context-dependent, for clearly the truth-values of forward-tracking counterfactuals are no less context-dependent.

Lewis also points out that backtracking counterfactuals are marked by a syntactic peculiarity. For instance, it would be more natural to say, in *Ask*, "If Jim asked Jack for help today, there would have to have been no quarrel yesterday" (Lewis [14], 458). However, such a syntactic peculiarity should have nothing to do with counting backtracking counterfactuals as non-ordinary either, since not all languages have different syntactic structures for backtracking and forward-tracking counterfactuals. Mandarin, for one, uses the same syntactic structure for backtracking and forward-tracking counterfactuals.[7] Yet, as far as I can tell, Mandarin speakers' understanding of counterfactuals does not differ significantly from English speakers'.

At any rate, I think it is incorrect to take backtracking counterfactuals as non-ordinary. But even if backtracking counterfactual were non-ordinary, it still did not follow that they are illegitimate or mistaken. Lewis' quotation above clearly conflates the distinction between ordinariness and correctness. Just because backtracking counterfactualization is a non-ordinary interpretation of counterfactuals it does not follow that it is mistaken. Given the fact that we are not very good at making probabilistically correct judgments (cf. Kahneman [12]), it is safe to say that ordinary probabilistic judgments are not based on probability theory. But this does not show that probabilistic judgments based on probability theory are mistaken.

Lewis' second, and perhaps more powerful, reason against backtracking counterfactuals is his view on counterfactual dependence:

> The way the future depends counterfactually on the way the present is. ... Likewise the present depends counterfactually on the past, and in general the way things are later depends on the way things were earlier.
> Not so in reverse. Seldom, if ever, can we find a clearly true counterfactual about how the past would be different if the present were somehow different. (Lewis [14], 455)

Counterfactual dependence, in Lewis' opinion, is temporally asymmetric: temporally later events counterfactually depend on temporally earlier events but not the other way around. Obviously, if counterfactual dependence is temporally asymmetric in this way, backtracking counterfactuals, according to which an earlier event counterfactually depends on a later event, are illegitimate.

There is, however, a serious flaw in Lewis' contention of the temporal asymmetry of counterfactual dependence. That is, the contention is not even tenable in Lewis' own account of forward-tracking counterfactuals [1]. Suppose that A obtains at t_1 and C obtains at t_2 (t_1 is before t_2). According to the standard view, which Lewis also endorses, in evaluating whether or not "A > C" is true in w, we first imagine a world w′ that are exactly identical to w until t_0 (t_0 is before t_1 and is supposed to be as close to t_1 as possible). At t_0 a miracle happens in w′ that causes A to obtain at t_1 (call this event D). We then determine whether C would have obtained at t_2 in w′. This story, quite natural on its own, does not satisfy the temporal asymmetry of counterfactual dependence: whether or not D obtains at t_0 depends on whether or not

[7]In fact, Mandarin does not even syntactically distinguish counterfactual conditionals from indicative conditionals.

A obtains in t_1—in Lewis' terms, the antecedent causally determines what would have happened in its "immediate past." However, if counterfactual dependence were temporally asymmetric, it is very puzzling how A could have causally determined its immediate past.

Worse, what counts as "immediate" past in Lewis' account may not be temporally close to the time at which the antecedent obtains. In other words, backtracking counterfactualizing to the "immediate past" could be virtually indistinguishable from backtracking counterfactualizing to the "non-immediate past." For instance, "If in 1933 there had been twice as many Jews in Germany as there actually were, there would have been an even larger holocaust" seems true (cf. Bennett [1], 79). On Lewis' account, it seems natural that the miraculous event that causes the number of Jews in Germany in 1933 to be twice as many as there actually were must happen for quite a long period of time before 1933. For instance, over a long period of time before 1933, many Jewish parents in Germany would have to have more children than they actually had. If the range of immediate past can extend to years, the term "immediate past" loses any of its intuitive meaning. It seems that what counts as immediate past is simply the one that causes the antecedent of a forward-tracking counterfactual to obtain. If so, it is *ad hoc* to allow backtracking counterfactualization only to the immediate past, but not beyond.

To sum up, it seems that there is no convincing reason for the dismissal of backtracking counterfactuals. A complete semantics of counterfactuals should account for both forward-tracking and backtracking counterfactuals. The possible-worlds semantics, at least in its orthodox form, is not in a position to offer such a complete semantics. I think the causal modeling semantics can do better. While the prominent causal modeling semantics still falls short of being a complete semantics, the notion of causal models gives us what we need in order to construct a complete semantics of counterfactuals, or so I will argue.

5.5 The Causal Modeling Semantics

Let us first introduce the notion of causal models. A causal model is a mathematical object that represents (or is supposed to represent) the causal relations of the events in a scenario. To elaborate, it is useful to begin with an example. Let us then construct a causal model K for *Ask*.

A causal model M is a triple $<V, S, A>$.[8] V is a finite set of variables $\{V_1, V_2, \ldots, V_n\}$. These are *variables for events* in the scenario that M is supposed to represent. K's V naturally contains the following variables:

[8]The causal modeling semantics has been developed by Jude Pearl and many others (cf. Pearl [17]; also see Galles and Pearl [7]). The following formulation has been influenced by Briggs [3]. Hiddleston [10] has constructed a different type of causal modeling semantics. For more on Hiddleston's account, see Footnote 23.

QUARREL represents whether or not Jim and Jack quarreled yesterday.
MAD represents whether or not Jack is mad at Jim.
PRIDE represents whether or not Jim is a prideful person.
ASK represents whether or not Jim asks Jack for help.
HELP represents whether or not Jack helps Jim.

In general, each variable $V_i \in V$ admits a range of values, but, for simplicity's sake, we will only deal with *binary variables*. That is, all V_i's discussed below only take on two possible values, i.e., "Yes" or "No".

It is customary to use "$V_i = v_i$" to stand for *The variable* V_i *takes on the value* v_i. For binary variables such as ASK and MAD, we may use "1" and "0" to stand for "Yes" and "No" respectively. For instance, "ASK = 1" means that Jim asks Jack for help, while "MAD = 0" means that Jack is not mad at Jim.

The second element of a causal model, *S*, is a set of structural equations, which specifies the relationships of causal dependence among variables. The causal dependence in play may be deterministic and indeterministic, although I will focus on deterministic causal relations for the time being. For each $V_i \in V$, *S* contains at most one structural equation of the following form:

$$V_i \Leftarrow f_i(PA_i).$$

The meaning of the symbol "$\Leftarrow$" is twofold. On the one hand, "$X \Leftarrow Y$" means that X is causally dependent on Y, i.e., whether X obtains or not is causally dependent on whether Y obtains or not. On the other hand, "$X \Leftarrow Y$" indicates that X will take on the value of Y. PA_i, which is a subset of *V* is the set of V_i's *parents* (V_i is called PA_i's *child)*. Parenthood is essentially a causal relation: the parents of an event are its causes, and the children of an event are its effects. F_i is a function that maps PA_i to {0, 1}, for we only deal with binary variables here. We may further regard f_i as truth-functional with truth and falsity being represented by 1 and 0 respectively. We will also treat variables on the right-hand side of the equation as propositions such that "Y" means Y = 1, and "$\sim$Y" means Y = 0.

Naturally, K's *S* contains the following structural equations:

MAD $\Leftarrow$ QUARREL
ASK $\Leftarrow$ ($\sim$PRIDE $\vee$ $\sim$QUARREL)
HELP $\Leftarrow$ (ASK $\wedge$ $\sim$MAD)

In words, "MAD $\Leftarrow$ QUARREL" means that whether or not Jack is mad at Jim depends causally on whether or not they had a quarrel yesterday. Jim will be mad at Jim *if and only if* they had a quarrel yesterday.[9] "ASK $\Leftarrow$ ($\sim$PRIDE $\vee$ $\sim$QUARREL)" means that whether or not Jim will ask Jack for help depends causally on whether

[9] According to *Ask*, Jim will be mad at Jim *if and only if* they had a quarrel yesterday. We assume that none of the conditions sabotaging the if direction of the biconditional (such as Jack has suffered from amnesia) holds. Nor does any of the conditions sabotaging the only-if direction (such as Jack has a burst of anger) hold. The same goes for other structural equations. In Galles and Pearl's [7] term, these conditions are called "inhibiting" and "triggering abnormalities" respectively. Implicit in each structural equation is the assumption that such abnormalities do not hold.

or not Jim is a prideful person and on whether or not they had a quarrel yesterday. Jim will ask Jack for help if and only if either Jim is not a prideful person or they did not have a quarrel yesterday. "HELP ⇐ (ASK ∧ ∼MAD)" means that whether or not Jack will help Jim depends causally on whether or not Jim asks Jack for help and on whether or not Jack is mad at Jim. Jack will help Jim if and only if Jim asks Jack for help and Jack is not mad at Jim.

There is no structural equation for QUARREL and PRIDE; their parents are not specified by K. We thus distinguish two types of variables: *exogenous* variables, whose parents are not specified by the causal model, and *endogenous* variables, whose parents are so specified. In K, QUARREL and PRIDE are exogenous, while the rest are endogenous. The values of exogenous variables are given to a causal model; they are presupposed, so to speak.

The third element of a causal model, *A*, is a function that assigns values to all variables in the model.[10] For each exogenous variable $V_i \in V$, *A* assigns the value v_i to V_i. For each endogenous variable $V_i \in V$, *A* assigns the value v_i to V_i based on the values of exogenous variables and the set of structural equations *S*. For instance, K's *A* is as follows:

A(ASK) = *A*(HELP) = 0, and
A(QUARREL) =*A*(PRIDE) = *A*(MAD) = 1.[11]

In words, in *Ask*, Jim and Jack had a quarrel yesterday, Jack is mad at Jim, Jim is a prideful person, Jim does not ask Jack for help, and Jack does not help Jim.

It is useful to illustrate a causal model in terms of a *directed acyclic graph* (*DAG*). A DAG consists of a set of nodes, which stand for the variables in *V*, and a set of directed acyclic arrows, which captures the *parental* relationships among variables. Specifically, if V_i is a parent of V_j (or, equivalently, V_j is a child of V_i), then there is an arrow pointing from the former to the latter. For binary variables, we use shaded nodes to indicate that the corresponding variables have the value of "1"; otherwise, the value of "0". Figure 5.1 is the DAG of K.

With the notion of causal model at hand, we are in a position to construct the causal modeling semantics, which is also based on IP:

(IP) "A > C" is true in a scenario s if and only if "C" is true in all s′ ∈ f(A, s).

Specifically, scenarios are interpreted as causal models and the selection function as a function that maps the antecedent A and a causal model M to certain *submodels M′*. Informally, a submodel M′ is a causal model generated by causally manipulating M in a certain way. The truth-condition of counterfactuals is specified as follows:

(CM) "A > C" is true in a causal model M if and only if "C" is true in some submodels M′.

[10]For the assignment function, cf. Hiddleston [10] and Briggs [3].

[11]Calculation: QUARREL = 1 and PRIDE = 1 (by assumption). If QUARREL = 1, then MAD = 1 (by MAD ⇐ QUARREL). If QUARREL = 1 and PRIDE = 1, then ASK = 0 (by ASK ⇐ (∼PRIDE ∨ ∼QUARREL)). If MAD = 1, then HELP = 0 (by HELP ⇐ (ASK ∧ ∼MAD)).

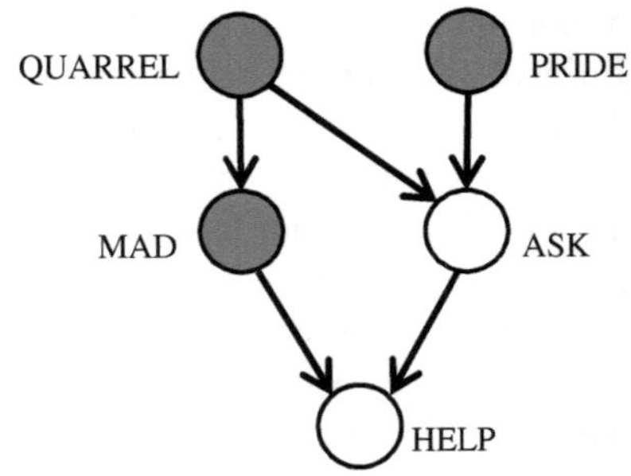

Fig. 5.1 DAG of K

The general idea behind CM is quite intuitive. Given that a causal model M represents a scenario s, a submodel M′ thus represents a "counterfactual" scenario s′ with respect to s, generated by causally manipulating s in a certain way. The task now is to specify the notion of submodel.

My claim is that there are essentially two kinds of submodels, since there are two distinct ways to manipulate a causal model. That is, one may manipulate M by changing either the set of structural equations *S* or the value assignment *A*. I call them "intervention" and "extrapolation" respectively.

Let us start with intervention, which has been featured in the prominent accounts of the causal modeling semantics (cf., e.g., Galles and Pearl [7]; Pearl [17]; Briggs [3]). Let M (=<*V*, *S*, *A*>) be a causal model, B be a sentence of the form "$C_1 = c_1 \wedge \ldots \wedge C_m = c_m$",[12] *VB* be the set of variables that are in B. An *intervention in M with respect to B* generates a *submodel* M_B (=<V_B, S_B, A_B>) of M such that:

(i) $V_B = V$.
(ii) $S_B = S$ except that for each $C_i \in VB$, S_B replaces the structural equation $C_i = f_i(PA_i)$ of *S* with the structural equation $C_i \Leftarrow c_i$, if C_i is endogenous.
(iii) $A_B = A$ except that (a) for each $C_i \in VB$, A_B sets the value of C_i to c_i if C_i is exogenous, and that (b) for each $V_i \in (V_B \backslash VB)$, A_B assigns the value v_i to V_i based on the value of C_i and S_B.[13]

In words, to intervene in a causal model M with respect to B (i.e., $C_1 = c_1 \wedge \ldots \vee C_m = c_m$) is to replace the original structural equation of $C_i \in VB$ with the new structural equation $C_i \Leftarrow c_i$. If C_i is exogenous, intervention simply sets the value

[12]Galles and Pearl's [7] original semantics has limited expressive power. In particular, they consider only counterfactuals of the form "$(A_1 \wedge \ldots \wedge A_n) > (C_1 \wedge \ldots \wedge C_m)$" while A_i and C_j have the form "$A_i = a_i$' and '$C_j = c_j$" respectively. Halpern [8] has developed a semantics for "A > C" with A taking the form "$A_1 \wedge \ldots \wedge A_n$" (like Pearl's), while C being any Boolean combination of sentences of the form "$C_i = c_i$." Briggs [3] further extends the semantics to deal with "A > C" with A to be any Boolean combination of sentences of the form "$A_i = a_i$." For simplicity's sake, I will here focus on a language with less expressive power. That is, I will follow Pearl in assuming that the sentences involved in intervention (and extrapolation) consist only of conjunctions.

[13]Thanks for an anonymous reviewer for pointing out some problems in the original formulation. Also see the definition of extrapolation below.

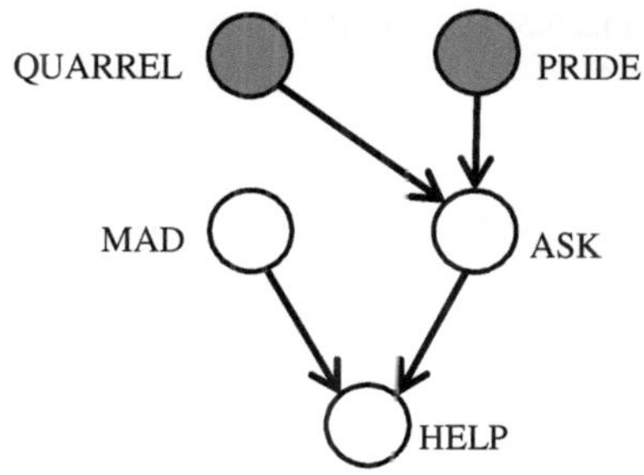

Fig. 5.2 DAG of $K_{(MAD=0)}$

of C_i to be c_i. The value assignment A_B assigns the value c_i to C_i. The values of the rest of the variables are calculated based on the value of C_i and S_B.

Suppose that we intervene in K with respect to (MAD = 0). The intervention generates the submodel $K_{(MAD=0)}$, whose set of variables is identical to K's. $K_{(MAD=0)}$'s $S_{(MAD=0)}$, by contrast, consists of the following:

MAD ⇐ 0
ASK ⇐ (~PRIDE ∨ ~QUARREL)
HELP ⇐ (ASK ∧ ~MAD)

The meaning of "MAD ⇐ 0" is twofold. On the one hand, it means that MAD is no longer causally dependent on other variables in the model. That is, whether Jack is mad at Jim no longer depends on whether or not they had a quarrel yesterday. On the other hand, it means that MAD is to take on the value of "0", i.e., Jack is not mad at Jim.

Accordingly, $A_{(MAD=0)}$ is as follows:

$A_{(MAD=0)}$(MAD) = $A_{(MAD=0)}$(ASK) = $A_{(MAD=0)}$(HELP) = 0, and
$A_{(MAD=0)}$(QUARREL) = $A_{(MAD=0)}$(PRIDE) = 1.[14]

Figure 5.2 is the DAG of $K_{(MAD=0)}$. Comparing Fig. 5.1 with Fig. 5.2, we can see that intervention "mutilates" (cf. Pearl [18]) the arrows in the original DAG, thereby canceling the parental relationships of some variables. Intervention allows, but does not imply different value assignments.

Let us move on to extrapolation, which, by contrast, has generally been assigned a marginal role (if at all). Let M (=<V, S, A>) be a causal model, B be a sentence of the form "$C_1 = c_1 \wedge \ldots \wedge C_m = c_m$," and *VB* be the set of variables that are in B. An *extrapolation on M with respect to B* generates a *submodel* M^B (=<V^B, S^B, A^B>) of M such that:

(i) $V^B = V$.
(ii) $S^B = S$.

[14]Calculation: QUARREL = 1 and PRIDE = 1 (by assumption). MAD = 0 (by Intervention). If QUARREL = 1 and PRIDE = 1, then ASK = 0 (by ASK ⇐ (~PRIDE ∨ ~QUARREL)). If ASK = 0, then HELP = 0 (by HELP ⇐ (ASK ∧ ~MAD)).

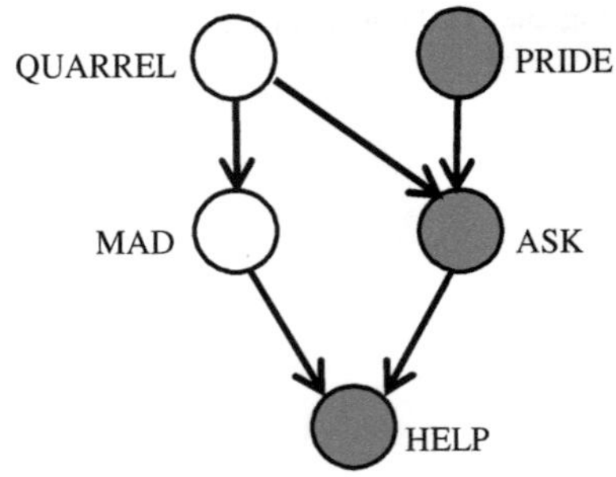

Fig. 5.3 DAG of $K^{(MAD=0)}$

(iii) $A^B = A$ except that (a) for each $C_i \in VB$, A sets the value of C_i to c_i, and that (b) for each $V_i \in (V^B \backslash VB)$, A^B assigns the value v_i to V_i based on the value of C_i and S^B.

In words, to extrapolate a causal model M with respect to B (i.e., $C_1 = c_1 \wedge ... \vee C_m = c_m$) is to set the value of $C_i \in VB$ to be c_i, and calculate the values of the variables causally related (directly or indirectly) to C_i based on the value of C_i and SB.

Suppose that we extrapolate K with respect to (MAD = 0). The extrapolation gives rise to the submodel K(MAD = 0). K and $K^{(MAD=0)}$ have the same sets of variables and structural equations. $A^{(MAD=0)}$ is as follows:

$A^{(MAD=0)}$(QUARREL) = $A^{(MAD=0)}$(MAD) = 0, and

$A^{(MAD=0)}$(PRIDE) = $A^{(MAD=0)}$(ASK) = $A^{(MAD=0)}$(HELP) = 1.[15]

Figure 5.3 is the DAG of $K^{(MAD=0)}$.

Sometimes, an extrapolation may fail to determine a unique submodel.[16] To elaborate, suppose that a causal model M consists of four variables X_1, X_2, X_3 and X_4. The structural equations are:

$$X_3 \Leftarrow\sim X_1 \vee X_2$$

$$X_4 \Leftarrow\sim X_2 \wedge X_3$$

Suppose that A is as follows:

$$A(X_2) = A(X_3) = A(X_4) = 0, \text{ and}$$
$$A(X_1) = 1.$$

---

[15]Calculation: MAD = 0 (by extrapolation). PRIDE = 1 (by assumption). If MAD = 0, then QUARREL = 0 (by MAD ⇐ QUARREL). If QUARREL = 0, then ASK = 1 (by ASK ⇐ (~PRIDE ∨ ~QUARREL)). If MAD = 0 and ASK = 1, then HELP = 1 (by HELP ⇐ (ASK ∧ ~MAD)).

[16]This point was originally addressed in a footnote. Thanks for an anonymous reviewer for urging me to address it in the main text.

Let us extrapolate M with respect to ($X_3 = 1$). It seems that this extrapolation gives rise to two equally good submodels $M^{(X3=1)(a)}$ and $M^{(X3=1)(b)}$, whose value assignments are as follows:

$A^{(X3=1)(a)}(X_4) = 0$, and
$A^{(X3=1)(a)}(X_1) = A^{(X3=1)(a)}(X_2) = A^{(X3=1)(a)}(X_3) = 1$;[17]
$A^{(X3=1)(b)}(X_1) = A^{(X3=1)(b)}(X_2) = 0$, and
$A^{(X3=1)(b)}(X_3) = A^{(X3=1)(b)}(X_4) = 1$.[18]

In particular, "$X_4 = 1$" is true in $M^{(X3=1)(b)}$ but false in $M^{(X3=1)(a)}$. The difference between these two submodels consists in the values of the variables we hold fixed when extrapolating M with respect to ($X_3 = 1$). If we hold the value of X_1 fixed (i.e., $X_1 = 1$), then we get $M^{(X3=1)(a)}$. $M^{(X3=1)(b)}$, by contrast, is the result of holding fixed the value of X_2 (i.e., $X_2 = 0$).

What this shows is that extrapolation is *context-sensitive*. To extrapolate a causal model with respect to ($C_i = c_i$) presupposes holding something fixed, and what should be held fixed is always a matter determined by the context. We may call the submodels M′ determined by the context *the relevant submodels*.[19] To use the previous example, if $M^{(X3=1)(b)}$ is the relevant submodel, then "$X_3 = 1 > X_4 = 1$" is true in M, while the same counterfactual is false in M if $M^{(X3=1)(a)}$ is relevant.

I propose that intervention and extrapolation give rise to different kinds of relevant submodel(s).[20] Hence, CM should be *disambiguated* into:

(CM_{IN}) "A > C" is $true_{IN}$ in a causal model M if and only if "C" is true in the relevant submodels M_A.[21]
(CM_{EX}) "A > C" is $true_{EX}$ in a causal model M if and only if "C" is true in the relevant submodels M^A.[22]

---

[17]Calculation: $X_3 = 1$ (by extrapolation). $X_1 = 1$ (by assumption). If $X_3 = 1$ and $X_1 = 1$, then $X_2 = 1$ (by $X_3 \Leftarrow \sim X_1 \vee X_2$). If $X_2 = 1$, then $X_4 = 0$ (by $X_4 \Leftarrow \sim X_2 \wedge X_3$).

[18]Calculation: $X_3 = 1$ (by extrapolation). $X_2 = 0$ (by assumption). If $X_3 = 1$ and $X_2 = 0$, then $X_1 = 0$ (by $X_3 \Leftarrow \sim X_1 \vee X_2$). If $X_2 = 0$ and $X_3 = 1$, then $X_4 = 1$ (by $X_4 \Leftarrow \sim X_2 \wedge X_3$).

[19]The term "relevant submodel," suggested by an anonymous reviewer, is from Hiddleston [10]. Also see Hiddleston ([10], 650ff.) for a related discussion.

[20]It is not necessary that the context always determines a unique submodel.

[21]According to the aforementioned formulation, intervention will always determine a unique submodel. Intervention, hence, is vacuously context-sensitive, namely, different contexts will give rise to the same (set of) relevant submodels. However, the context-insensitivity of intervention may have more to do with the way intervention is formulated here than with the general notion of intervention. For instance, we have limited our attention to intervention involved conjunctions, i.e., ($A_1 \wedge \ldots \wedge A_n$), since we only consider counterfactuals whose antecedents are of the form "$A_1 \wedge \ldots \wedge A_n$." Intervention of this specific sort determines a unique submodel. However, to intervene in a model with respect to a disjunction may fail to determine a unique submodel (cf. Briggs [3], 152ff.). Hence, the notion of relevant submodels will apply to intervention as well.

[22]Hiddleston [10] has proposed a causal modeling semantics of counterfactuals that bears some similarities to CM_{EX}. There are two main differences between them, though. First, while the causal modeling semantics presented above takes structural equations to specify deterministic laws between a variable Y and its parents X's (see Footnote 10), Hiddleston's semantics takes structural equations to be indeterministic laws formulated in probabilistic terms. Second, Hiddleston's semantics

Some remarks are in order.

First, let us unpack some terminology. On CM_{IN} and CM_{EX}, the truth-condition of counterfactuals is determined by two modes of counterfactualization—one is related to intervention and the other to extrapolation (as indicated by the subscripts). Call them "intervention-counterfactualization" ("counterfactualization$_{IN}$") and "extrapolation-counterfactualization" ("counterfactualization$_{EX}$") respectively. "A > C" can be true under counterfactualization$_{IN}$, but false under counterfactualization$_{EX}$, and vice versa. Hence, we distinguish counterfactuals being true by counterfactualization$_{IN}$ ("true$_{IN}$") from counterfactuals being true by counterfactualization$_{EX}$ ("true$_{EX}$").

Both CM_{IN} and CM_{EX} are context-sensitive. While issues related to context-sensitivity are important on their own, they are not the main concerns of this paper. So long as no confusion will arise, I will omit the term "relevant" when talking about submodels.

Second, while the causal modeling semantics has gradually gained its importance in recent literature, the distinction between CM_{IN} and CM_{EX} has not been widely

(Footnote 22 continued)
concerns only with positive causal influences, while CM_{EX} takes into account both positive and negative causal influences.

Let us call (X = x) has a *direct positive influence on* (Y = y) in a causal model M if the probability of (Y = y) is raised by (X = x) other things being equal. We call all the variables that have a direct positive influence on (Y = y) *the positive parents of Y*. Suppose that M′ is a submodel of M. If the value of Y in M′ is different from Y's value in M, while Y's positive parents' values in M′ and M are the same, then we call that M′ contain a *Causal Break*. If Y's values and Y's positive parents' values in M′ and M are the same, then we call that M′ contains a *Causal Intact*. According to Hiddleston's semantics, very roughly, "A > C" is true in M iff for all submodels M′ such that A is true in M′ and that M′ contains the maximal amount of Causal Intacts and the minimal amount of Causal Breaks, C is also true in M′. Let us call that "A > C" is true in M in Hiddleston's sense *"A > C" is true in the Maximal-Intact-and-Minimal-Break M′*.

For the present purposes, it is worth pointing out that if a causal model M contains no probabilistic equations (i.e., Y's parents raise the probability of Y getting the value y to 1), and if all Y's parents X's are positive parents, then being true in the Maximal-Intact-and-Minimal-Break M′ and being true$_{EX}$ in M converge. That is, in such limited cases, "A > C" is true in Maximal-Intact-and-Minimal-Break M′ iff "C" is true in M^A(i.e., iff "A > C" is true$_{EX}$ in M).

However, even in such cases, Hiddleston's semantics and CM_{EX} are still fundamentally different. First, Hiddleston's semantics is supposed to be a complete semantics on its own. It does not admit the ambiguity of counterfactuals indicated by *Ask* . In particular, it does not allow the same counterfactual to have a forward-tracking as well as a backtracking interpretation. Hence, Hiddleston's semantics faces the same problem as the possible-worlds semantics does.

Second, Hiddleston's semantics characterizes the truth-condition of counterfactuals in terms of the notion of being true in the Maximal-Intact-and-Minimal-Break M′. Now, we know that CM_{EX}cannot account for cases of forward-tracking counterfactuals, which are best suit for CM_{IN}. Given that Hiddleston's semantics basically is CM_{EX} when no probabilistic equations are involved, it follows that the only way for Hiddleston's semantics to explain forward-tracking counterfactuals, say, A > C, is to stipulate that A raises the probability of C to n, where n < 1. I think this approach will lead to some serious problems. But I will not pursue this line of thought here. What this shows is that Hiddleston's semantics and the present account handle the truth-condition of counterfactuals very differently from each other.

I would like to thank an anonymous reviewer for pushing me to elaborate this point.

recognized. The orthodox view is that the truth-condition of counterfactuals is captured by CM_{IN} (cf., e.g., Pearl [17]; Briggs [3]). As a result, the orthodox view, like the possible-worlds semantics, is unable to respect the distinction between forward-tracking and backtracking counterfactuals (see Sect. 5.6).

Still, the distinction between intervention and extrapolation is not unheard of. The distinction was first brought to my attention by David Galles and Judea Pearl's distinction between *doing* and *seeing* ([7], 159).[23] But they do not develop it as I do.

Third, intervention and extrapolation are different kinds of causal manipulation. Intervention is to a causal model as an event-changing action is to a scenario. As noted, to intervene in K with respect to (MAD = 0) is to disconnect the causal relationship between MAD and QUARREL and to set MAD to take on the value 0. Intervening in K with respect to (MAD = 0) is like an act of easing Jack's anger in *Ask*—we inject tranquilizer into Jack's body, we erase Jack's memory about the quarrel, etc. In that case, Jack will not be mad at Jim regardless of the yesterday's quarrel.

By contrast, extrapolation is to a causal model as a supposition is to a scenario. As noted, to extrapolate K with respect to (MAD = 0) is to make MAD to take on the value 0, while preserving its causal relations to other variables. Extrapolating K with respect to (MAD = 0) is like supposing that Jack is not mad at Jim in *Ask*. In that case, Jack must not have a quarrel with Jim yesterday since Jack will be mad at Jim if he had a quarrel with Jim.

Fourth, submodels generated by intervention contain all necessary information regarding the causal effect of a certain action (cf. Galles and Pearl [7], 159). For instance, suppose that we intervene in M with respect to ($C_i = c_i$), giving rise to the submodel $M_{Ci=ci}$. The primary difference between M and $M_{Ci=ci}$ is that $C_i = c_i$ obtains in $M_{Ci=ci}$ but not in M in such a way that only the values of C_i's children, but not its parents', are subject to change. In this way, C_i screens off its parents from its children. Intuitively, $M_{Ci=ci}$ gives us a clear picture of the *causal impact of* $C_i = c_i$ *in M*.

By contrast, submodels generated by extrapolation contain information regarding what the original model could have come about. For instance, suppose we extrapolate M with respect to ($C_i = c_i$), giving rise to the submodel $M^{Ci=ci}$. The primary difference between M and $M^{Ci=ci}$ is that $C_i = c_i$ obtains in $M^{Ci=ci}$ but not in M in such a way that both the values of C_i's parents and the values of its children are subject to change. In this way, both C_i's parents and its children have to adjust in order to cope with $C_i = c_i$. I think it is appropriate to say that $M^{Ci=ci}$ contains information regarding how M would have "evolved" (all things considered) if $C_i = c_i$ were to obtain in M. Intuitively, $M^{Ci=ci}$ tells us *what M would have come about if* $C_i = c_i$ *were to obtain in M*.

Fifth,[24] intervention and extrapolation converge when only exogenous variables are causally manipulated. That is, to intervene in M with respect to ($C_i = c_i$) is tantamount to extrapolating M with respect to ($C_i = c_i$), when C_i is exogenous. Informally

[23] For an elaboration, see Sloman ([21], Chap. 5).

[24] Thanks for an anonymous reviewer for urging me to elaborate this point.

speaking, intervening in M with respect to ($C_i = c_i$) is a two-step procedure: we first surgically remove the structural equation corresponding to C_i, and then stipulate C_i to take on the value c_i. Extrapolation, by contrast, consists only of the second step of intervention, namely, to extrapolate M with respect to ($C_i = c_i$) is to stipulate C_i to take on the value c_i, while C_i's structural equation remains intact. When C_i is exogenous, intervention and extrapolation converge, since the first step of intervention becomes vacuous.

That intervention and extrapolation may sometimes converge indicating that there is no clear-cut distinction between the two. This point is not implausible once we notice that intervening in M with respect to ($C_i = c_i$), where C_i is exogenous, not only gives us the information about C_i's causal impacts in M, but also the information about what would need to happen in order for C_i to take on the value c_i in M. In other words, these two offer the same kind of information when the variable in question is exogenous. Still the distinction between intervention and extrapolation is not undermined as they give rise to different kinds of information if the variables involved are endogenous.[25]

5.6 Backtracking Counterfactuals Revisited

The causal modeling semantics constructed in this paper has an edge over the possible-worlds semantics on two scores. First, unlike the possible-worlds semantics, the causal modeling semantics is immune to the counterexamples mentioned in Sect. 5.3. Second, the causal modeling semantics, but not the possible-worlds one, has resources enough for accounting for the distinction between forward-tracking and backtracking counterfactuals. This section is dedicated to the second point. The next section comes back to the first point.

According to CM_{IN} and CM_{EX}, "Ask > Help" is true under one mode of counterfactualization but false under the other. Let us first intervene in K with respect to (ASK = 1). $K_{(ASK=1)}$'s $S_{(ASK=1)}$ consists of the following:

$$\begin{aligned} &\text{MAD} \Leftarrow \text{QUARREL} \\ &\text{ASK} \Leftarrow 1 \\ &\text{HELP} \Leftarrow (\text{ASK} \vee \sim \text{MAD}) \end{aligned}$$

$A_{(ASK=1)}$ is that (Fig. 5.4):

[25] An anonymous reviewer also points out to me that the existence of $M^{Ci=ci}$ depends on $C_i = c_i$ being compatible with the set of structural equations S of M, while the existence of $M_{Ci=ci}$ is not so constrained. This feature is worth exploring, but I will not carry out the task here.

Fig. 5.4 DAG of $K_{(ASK=1)}$

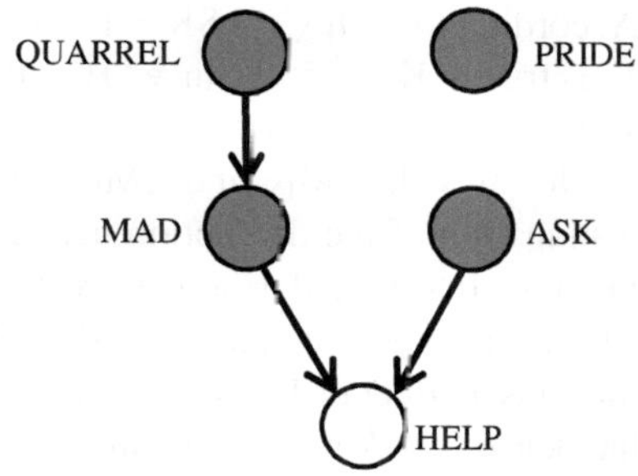

Fig. 5.5 DAG of $K^{(ASK=1)}$

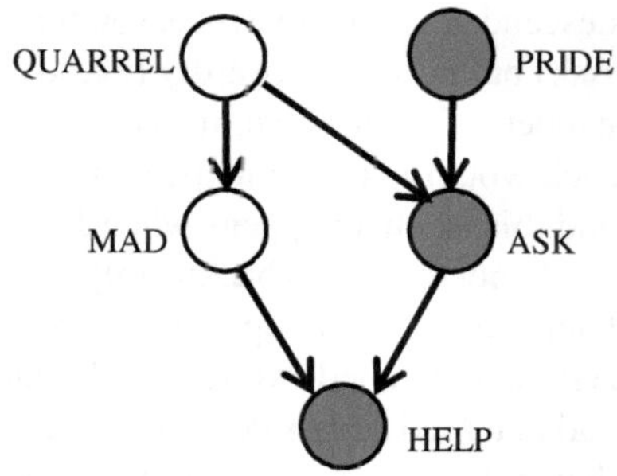

$A_{(ASK=1)}(HELP) = 0$, and
$A_{(ASK=1)}(QUARREL) = A_{(ASK=1)}(MAD) = A_{(ASK=1)}(PRIDE) = A_{(ASK=1)}(ASK) = 1$.[26]

According to CM_{IN}, "ASK = 1 > HELP = 1" is $true_{IN}$ in K if and only if "HELP = 1" is true in $K_{(ASK=1)}$. Since "HELP = 1" is false in $K_{(ASK=1)}$, "ASK = 1 > HELP = 1" is not $true_{IN}$ in K.

To extrapolate K with respect to (ASK = 1), on the other hand, gives rise to $K^{(ASK=1)}$. $K^{(ASK=1)}$ and K consist of the same set of structural equations. Moreover, $A^{(ASK=1)}$ is as follows (Fig. 5.5):

$A^{(ASK=1)}(QUARREL) = A^{(ASK=1)}(MAD) = 0$, and
$A^{(ASK=1)}(PRIDE) = A^{(ASK=1)}(ASK) = A^{(ASK=1)}(HELP) = 1$[27]

[26]Calculation: QUARREL = 1 and PRIDE = 1 (by assumption). ASK = 1 (by intervention). If QUARREL = 1, then MAD = 1 (by MAD ⇐ QUARREL). If MAD = 1, then HELP = 0 (HELP ⇐ (ASK ∧ ∼MAD)).

[27]Calculation: ASK = 1 (by extrapolation). PRIDE = 1 (by assumption). If ASK = 1 and PRIDE = 1, then QUARREL = 0 (by ASK ⇐ (∼PRIDE ∨ ∼QUARREL)). If QUAREL = 0, then MAD = 0 (by MAD ⇐ QUARREL). If MAD = 0 and ASK = 1, then HELP = 1 (by HELP ⇐ (ASK ∧ ∼MAD)). However, acute readers may notice that the calculation above has held (PRIDE = 1) fixed. It is by doing so that we deduce HELP = 1. Suppose that we hold (QUARREL = 1) fixed instead. We would then get the opposite result: if QUARREL = 1, then MAD = 1 (by MAD ⇐ QUARREL). If MAD = 1 and ASK = 1, then HELP = 0 (by HELP ⇐ (ASK ∧ ∼MAD)).

According to CM_{EX}, "ASK = 1 > HELP = 1" is $true_{EX}$ in K if and only if "HELP = 1" is true in $K^{(ASK=1)}$. Since "HELP = 1" is true in $K^{(ASK=1)}$, "ASK = 1 > HELP = 1" is $true_{EX}$ in K.

Not only do CM_{IN} and CM_{EX} give the correct predictions. They offer a natural explanation of the distinction between forward-tracking and backtracking counterfactuals. Interpreted as a forward-tracking counterfactual, "Ask > Help" is false in *Ask*. More precisely, on forward-tracking counterfactualization, we focus solely on the causal effect of Jim asking Jack for help (i.e., Ask), namely, on what would have happened if Ask were to obtain, while ignoring Ask's causal ancestors. In so doing, we appeal only to our knowledge of the causal relations between Ask and its causal descendants. We always reason forwardly (i.e., on what would follow causally from Ask) but never backwardly (i.e., on what would need to happen in order for Ask to happen). For instance, in *Ask*, we reason forwardly that if Ask had obtained, then Jack would not have helped Jim (i.e., ~Help) since Jack is mad at Jim (i.e., Mad), and this is what happens when Jack gets mad.

By not reasoning backwardly, we do not attempt to rationalize how Ask could have happened in the first place. For instance, when asking what would have happened if Ask had obtained, we ignore the fact that Jim being a prideful person (i.e., Pride), and that Mad and Pride prevent Help from happening. In a sense, we simply stipulate that Ask had somehow come about without a specific story. In many cases, filling in such stories would be inappropriate. Suppose that we try to rationalize Ask. We quickly encounter problems: how could the prideful Jim ask Jack for help after the two have had such a quarrel yesterday? This kind of questions cannot be answered unless we shift to the backtracking mode of reasoning. But doing so simply ruins the point of forward-tracking counterfactualization.

As should be obvious by now, forward-tracking reasoning is nicely captured by $counterfactualization_{IN}$. Intervention gives us everything we need to know about the causal impact of a certain action. To intervene in K with respect to (ASK = 1), for instance, is to disconnect ASK from its parents, to set ASK to take on the value 1, and to calculate the values of ASK's children accordingly. It thereby allows $K_{(ASK=1)}$ to contain just the information regarding the causal impact ASK has on its children in K.

"Ask > Help," by contrast, is true under the backtracking reading. That is, on backtracking counterfactualization, we focus on rationalizing how Ask could have

(Footnote 27 continued)

As noted, $counterfactualization_{EX}$ is *context-sensitive*; to extrapolate a causal model with respect to (C_i = c_i) needs to hold something fixed, and what should be held fixed is always a matter determined by the context.

The idea that extrapolation is context-sensitive is quite intuitive in this case, as $counterfactualization_{EX}$ is context-sensitive in a parallel way. For instance, there are two ways to $counterfactualize_{EX}$ what would have happened if Jim were to ask Jack for help. On the one hand, if Jim were to ask Jack for help, it must be that Jim had somehow swallowed his pride, since they had had a quarrel yesterday, and if Jim did not swallow his pride, he would not have asked Jack for help. On the other hand, if Jim were to ask Jack for help, it must be that Jim was not mad at him, since Jim was a prideful person, who would not ask Jack for help after quarreling with him. Both are legitimate counterfactualization $_{EX}$, and only the context could tell which one is to be adopted.

happened *all things considered.* We exploit our knowledge of the causal relations among Ask, its causal ancestors, and its causal descendants in order to determine under what condition Ask could have happened in *Ask*. We reason forwardly as well as backwardly, searching for the most plausible and still consistent story. For instance, in *Ask*, we reason, backwardly, that if Ask were to obtain, Jim must not be mad at Jack (i.e., ~Mad), since Pride prevents Ask from obtaining if Mad has obtained. To reason further still, we conclude that Jim and Jack must not have a quarrel yesterday (i.e., ~Quarrel), since if there were a quarrel, ~Mad could not have happened. Reasoning backwardly and (then) forwardly, we then conclude that Help must have obtained, for this is what should have happened if ~Mad and Ask both obtain. By reasoning backwardly, we attempt to provide the most plausible and still consistent story as to how Ask could have happened in the first place. In a sense, backtracking reasoning tells us what "really" would have happened in *Ask*, if Ask were to have had happened.

Likewise, it should be clear that backtracking counterfactualization is nicely captured by counterfactualization$_{\mathrm{EX}}$. Extrapolation tells us what a causal model would have been all things considered. To extrapolate M with respect to (ASK = 1), for instance, is first to set ASK to take on the value 1 and then to calculate the values of ASK's parents and children accordingly. $K^{(ASK=1)}$ thereby contains the information about what K would have "really" become were ASK to take on the value 1.

5.7 Troubling Cases Revisited

We have seen that the causal modeling semantics has resources enough for accounting for the distinction between forward-tracking and backtracking counterfactuals, which has eluded the possible-worlds semantics. In this section, I will further show that the causal modeling semantics is immune to cases like *Bomb* and *Bet*, which have caused serious problems for the possible-worlds semantics.

Let us construct a causal model B for *Bomb*. Intuitively, B consists of the following set of variables V:

PUSH represents whether or not the beetle pushes the button.
SIGNAL represents whether or not a signal runs along a wire and out of the box.
BOX represents whether or not the black box and all of its contents are destroyed after t.
DESTROY represents whether or not the universe is destroyed.

As stipulated, whether or not a signal runs along a wire and out of the box causally depends on whether or not the beetle pushes the button. The signal will run along a wire and out of the box if and only if the beetle pushes the button. Whether or not the universe is destroyed causally depends on whether or not a signal runs along a wire and out the box (if the signal runs out the box, the mega-bomb will be detonated). The universe will be destroyed if and only if a signal runs along a wire and out the box. Whether or not the black box and all of its contents are destroyed after t causally depends on whether or not a signal has run along a wire and out of the box. The black

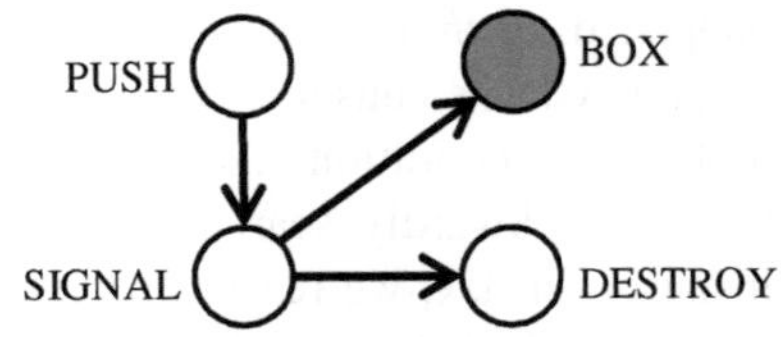

Fig. 5.6 DAG of B

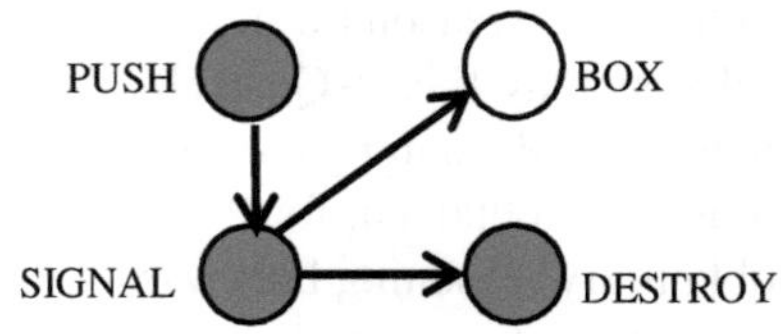

Fig. 5.7 DAG of $B_{(PUSH=1)}$

box and all of its contents will be destroyed after t if and only if no signal has run along a wire and out of the box. Hence, the set of structural equations of B is as follows:

$$\begin{aligned}&\text{SIGNAL} \Leftarrow \text{PUSH}\\&\text{BOX} \Leftarrow\sim \text{SIGNAL}\\&\text{DESTROY} \Leftarrow \text{SIGNAL}.\end{aligned}$$

B's value assignment *A* is (Fig. 5.6):

A(PUSH) = A(SIGNAL) = A(DESTROY) = 0, and
A(BOX) = 1.[28]

In words, in *Bomb*, the beetle does not push the button, there is no signal running along a wire and out of the box, the universe is not destroyed, and the black box and all of its contents are destroyed after t.

The causal modeling semantics is able to explain the intuition that "Push > Destroy" is true in *Bomb*. Suppose that we intervene in B with respect to (PUSH = 1). In this case, B and $B_{(PUSH=1)}$ consist of the same set structural equations, since PUSH is an exogenous variable, which does not have a corresponding structural equation.

$A_{(PUSH=1)}$ is as follows (Fig. 5.7):

$A_{(PUSH=1)}(BOX) = 0$, and
$A_{(PUSH=1)}(PUSH) = A_{(PUSH=1)}(SIGNAL) = A_{(PUSH=1)}(DESTROY) = 1$.[29]

---

[28]Calculation: PUSH = 0 (by assumption). If PUSH = 0, then SIGNAL = 0 (by SIGNAL ⇐ PUSH). If SIGNAL = 0, then BOX = 1 (by BOX ⇐∼SIGNAL). If SIGNAL = 0, then DESTROY = 0 (by DESTROY ⇐ SIGNAL).

[29]Calculation: PUSH = 1 (by intervention). If PUSH = 1, then SIGNAL = 1 (by SIGNAL ⇐ PUSH). If SIGNAL = 1, then BOX = 0 (by BOX ⇐∼SIGNAL). If SIGNAL = 1, then DESTROY = 1 (by DESTROY ⇐ SIGNAL).

Fig. 5.8 DAG of T

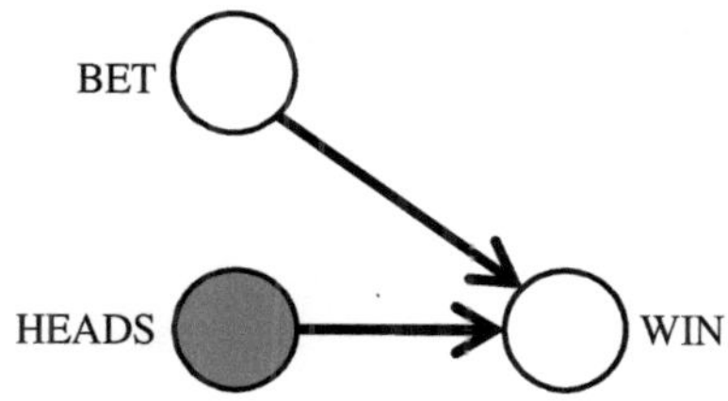

Since "DESTROY = 1" is true in $B_{(PUSH=1)}$, "PUSH = 1 > DESTROY = 1" is $true_{IN}$ in B, as desired.[30]

Let us construct a causal model T for *Bet*.[31] Intuitively, T consists of the following variables *V*:

HEADS represents whether or not the coin comes out heads.
BET represents whether or not the hearer bets (heads).
WIN represents whether or not the hearer wins the bet.

As stipulated, whether or not the hearer wins the bet causally depends on whether or not the coin comes out heads and whether or not the hearer bets (heads). The hearer will win the bet if and only if the coin lands heads and she bets (heads). Hence, the set of structural equations of T is:

$$\text{WIN} \Leftarrow (\text{HEADS} \wedge \text{BET}).$$

T's value assignment *A* is (Fig. 5.8):

A(BET) = A(WIN) = 0, and
A(HEADS) = 1.[32]

In words, in *Bet*, the coin does land heads. But the hearer does not bet (heads), and thus does not win the bet. Notice that the case also stipulates that whether or not the coin lands heads is indeterministic; it is not necessary that the coin would land heads should the hearer's friend flip it. But this indeterministic feature of HEADS has no direct bearing on the following discussion. For simplicity's sake, I take HEADS to be an exogenous variable.[33]

---

[30]Notice that given that PUSH is an exogenous variable, to intervene in B with respect to (PUSH = 1) is tantamount to extrapolating B with respect to (PUSH = 1). That is, $B_{(FUSH=1)}$ is identical to $B^{(PUSH=1)}$. It follows that "PUSH = 1 > DESTROY = 1" is also $true_{EX}$ in B.

That $B_{(PUSH=1)}$ is identical to $B^{(PUSH=1)}$ should not be surprising given that PUSH is an exogenous variable. The difference between intervention and extrapolation consists in that the latter, but not the former, allows the values of PUSH's parents be subject to change. Since PUSH has no parents, $B_{(PUSH=1)}$ and $B^{(PUSH=1)}$ naturally converge. Also see the end of Sect. 5.5.

[31]This part was omitted in the original draft. Thanks for an anonymous reviewer for urging me to put in it in the main text.

[32]Calculation: BET = 0 (by assumption). HEADS = 1 (by assumption). If BET = 0, then WIN = 0 (by WIN ⇐ (HEADS ∧ BET)).

[33]An explanation of *Bet* may not need to assign indeterministic (probabilistic) causal connections among variables. But one may wonder whether in some other cases the causal connections among

Fig. 5.9 DAG of $T_{(BET=1)}$

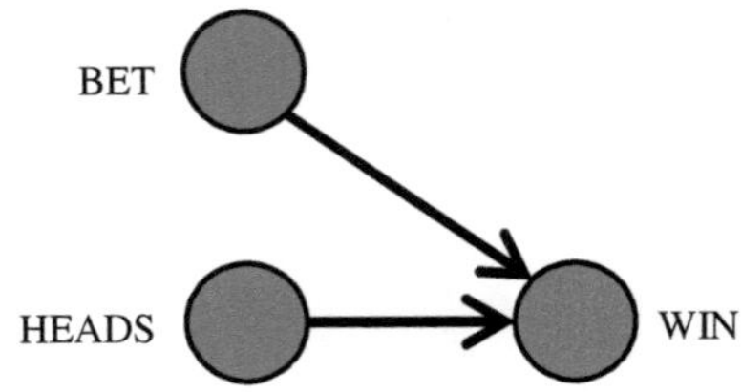

The causal modeling semantics is able to explain our intuitions that "Bet > Win" is true in *Bet*. Suppose that we intervene in T with respect to (BET = 1). T and $T_{(BET=1)}$ consist of the same set of structural equations, since BET is an exogenous variable, which does not have a corresponding structural equation.

$A_{(BET=1)}$ is as follows (Fig. 5.9):

$A_{(BET=1)}(BET) = A_{(BET=1)}(HEADS) = A_{(BET=1)}(WIN) = 1$.[34]

Since "WIN = 1" is true in $T_{(BET=1)}$, "BET = 1 > WIN = 1" is $true_{IN}$ in T, as desired.[35]

I conclude that the causal modeling semantics has an advantage over the possible-worlds semantics in that the former, but not the latter, is immune to the troubling cases discussed in Sect. 5.3.

5.8 Conclusion

The possible-worlds semantics has been the prominent account in the literature. Yet, despite its widespread acceptance, the possible-worlds semantics is theoretically less desirable than the causal modeling semantics. First, it suffers from a specific type of counterexamples, which indicates that the notion of similarity must be characterized in terms of causal dependence. If so, however, the possible-worlds semantics has devolved into a cumbersome causal modeling semantics. Second, the possible-worlds semantics is incomplete at best since it lacks the resources necessary for accounting for backtracking counterfactuals.

The causal modeling semantics, by contrast, faces none of these problems. First, the causal modeling semantics can explain cases that cause serious problems for the

(Footnote 33 continued)
variables should be characterized in probabilistic terms. The present account, however, does not allow such characterization, as we have implicitly assumed that what Galles and Pearl call "inhibiting" and "triggering abnormalities" do not hold (see Footnote 9). This line of thought assumes that indeterministic relationships between events are the result of our ignorance. While this assumption may not square well with quantum physics, it does fit well with our ordinary notion of causation (also see Pearl [17], 26–7).

[34]Calculation: HEADS = 1 (by assumption). BET = 1 (by intervention). If HEADS = 1 and BET = 1, then WIN = 1 (by WIN $\Leftarrow$ (HEADS $\wedge$ BET)).

[35]Since BET is an exogenous variable, being $true_{IN}$ in T is tantamount to being $true_{EX}$ in T. Also see the end of Sect. 5.5.

possible-worlds semantics. Second, the causal modeling semantics (with appropriate modifications) has resources enough for accounting for backtracking counterfactuals.

The causal modeling semantics constructed above features a distinction between intervention and extrapolation. While this framework has not been widely recognized, it is intuitively plausible, as it offers a natural explanation of the distinction between forward-tracking and backtracking counterfactuals. The present work is just a first step toward a full-fledged causal modeling semantics. I have mainly focused on the issues concerning the truth-condition of counterfactuals. Even so, some aspects (such as the context-sensitivity of submodels) are not fully explored. And I have left out questions about *validity* of inferences involving counterfactuals. A lot more needs to be said, but that will have to be left for another occasion.

Acknowledgments I am grateful to two anonymous reviewers for helpful comments. Specifically, one reviewer has given me invaluable suggestions and corrections, which greatly improve the original draft as well as inspire my thoughts on the issues. I also want to thank Daniel Marshall for helpful comments and proofreading of an earlier draft. I am also indebted to the participants of the Taiwan Philosophical Logic Colloquium in 2014 for comments and discussions. The present work has received funding from the Ministry of Science and Technology (MOST) of Taiwan (R.O.C.) (MOST 103-2410-H-194-125).

References

1. Bennett, J.: Counterfactuals and temporal direction. Philos. Rev. **93**(1), 57–91 (1984)
2. Bennett, J.: A Philosophical Guide to Conditionals. Clarendon Press, Oxford (2003)
3. Briggs, R.: Interventionist counterfactuals. Philos. Stud. **160**(1), 139–166 (2012)
4. Downing, P.B.: Subjunctive conditionals, time order, and causation. Proc. Aristotelian Soc. **59**(January), 125–140 (1958)
5. Edgington, D.: Counterfactuals and the benefit of hindsight. In: Dowe, P., Noordhof, P. (eds.) Cause and Chance: Causation in an Indeterministic World, pp. 12–27. Routledge, New York (2004)
6. Fine, K.: Critical notice to Lewis (1973). Mind **84**(1), 451–458 (1975)
7. Galles, D., Pearl, J.: An axiomatic characterization of causal counterfactuals. Found. Sci. **3**(1), 151–182 (1998)
8. Halpern, J.Y.: Axiomatizing causal reasoning. J. Artif. Intell. Res. **12**(1), 317–337 (2000)
9. Hawthorne, J.: Chance and counterfactuals. Philos. Phenomenol. Res. LXX **2**, 396–405 (2005)
10. Hiddleston, E.: A causal theory of counterfactuals. Noûs **39**(4), 632–657 (2005)
11. Hitchock, C.: The intransitivity of causation revealed in equations and graphs. J. Philos. **98**(6), 273–299 (2001)
12. Kahneman, D.: Thinking: Fast and Slow. Farrar, Straus and Giroux, New York (2011)
13. Lewis, D.: Counterfactuals. Blackwell, Malden (1973)
14. Lewis, D.: Counterfactual dependence and time's arrow. Noûs **13**(4), 455–476 (1979)
15. Lewis, D.: Postcripts to 'Counterfactual dependence and time's arrow'. In: Philosophical papers **II**, 52–66. Oxford University Press, Oxford (1986)
16. Northcott, R.: On Lewis, Schaffer and the non-reductive evaluation of counterfactuals. Theoria **75**(4), 336–343 (2009)
17. Pearl, J.: Causality: Models, Reasoning, and Inference. Cambridge University Press, Cambridge (2000)
18. Pearl, J.: Reasoning with cause and effect. AI Magazine **23**(1), 95–111 (2002)
19. Pruss, A.R.: David Lewis's counterfactual arrow of time. Noûs **37**(4), 606–637 (2003)

20. Schaffer, J.: Counterfactuals, causal independence and conceptual circularity. Analysis **64**(4), 299–309 (2004)
21. Sloman, S.A.: Causal Models: How People Think about the World and Its Alternatives. Oxford University Press, Oxford (2009)
22. Slote, M.A.: Time in counterfactuals. Philos. Rev. **87**(1), 3–27 (1978)
23. Stalnaker, R.: A theory of conditional. In: Harper, W.L., Stalnaker, R., Pearce, G. (eds.) Ifs: Conditionals, Belief, Decision, Chance, and Time, pp. 41–55. D. Reidel Publishing Company, Boston (1968)
24. Tooley, M.: Backward causation and the Stalnaker-Lewis approach to counterfactuals. Analysis **62**(3), 191–197 (2002)
25. Wasserman, R.: The future similarity objection tevisited. Synthese **150**(1), 57–67 (2006)
26. Woodward, J.: Causation and manipulability. In: Zalta, E.N. (ed.) The stanford encyclopedia of philosophy(Winter 2013 Edition). http://plato.stanford.edu/archives/win2013/entries/causation-mani/

Chapter 6
The Meaning of Epistemic Modality and the Absence of Truth

Hanti Lin

Abstract When one asserts the disjunction 'the keys might be in the drawer, or they might be in the car,' the speaker seems committed to both of the disjuncts, 'the keys might be in the drawer' and 'they might be in the car' (Kamp, Proc Aristotelian Soc N S 74:57–74 (1973), [12]). Namely, 'or' behaves like a conjunction 'and' when it meets epistemic modality 'might'. It has been noted that it is very difficult to explain this phenomenon in terms of conversational implicature (Zimmermann, Nat Lang Seman 8:255–290 (2000), [19]); a semantic explanation is worth pursuing. This paper proposes the first semantics that explains the conjunctive 'or' as a semantic phenomenon and still preserves classical logic when 'might' is absent, all done without *ad hoc* case distinctions. The truth-conditional approach to semantics has not been able to do that. Instead of truth conditions, the proposed semantics provides *acceptability conditions*. To be more specific, information states are modeled by sets of possible worlds, and each sentence is compositionally evaluated at each information state as: acceptable, deniable, or undecided. Working with acceptability conditions does not mean that we abandon truth conditions altogether. In fact, we can employ a sentence's acceptability condition to determine whether it has a truth condition. Epistemic modals turn out to *lack* truth conditions, while sentences like "snow is white" can have truth conditions if you wish. Although the above may appear to be a mere case study in linguistics, the result points to a new, general semantic framework for addressing a central issue in philosophical logic and meta-ethics: Which types of declarative sentences lack truth conditions, especially epistemic modals, indicative conditionals, and moral claims?

Keywords Epistemic modal · Disjunction · Conjunctive 'or' · Acceptability condition · Truth condition · Compositional semantics

H. Lin (✉)
Philosophy Department, 1240 Social Science and Humanities, University of California, One Shields Avenue, Davis, CA 95616, USA
e-mail: ika@ucdavis.edu

S.C.-M. Yang et al. (eds.), *Structural Analysis of Non-Classical Logics*,
Logic in Asia: Studia Logica Library, DOI 10.1007/978-3-662-48357-2_6

6.1 Introduction: A Puzzle About 'Or' and 'Might'

Looking at the cloudy sky, I assert 'it might rain today.' This describes or expresses features of my belief, knowledge, or perhaps evidence; namely, the word 'might' expresses *epistemic modality*. The present paper aims to give a novel semantics to explain a common—but mysterious—phenomenon about 'might'-assertions, known as the *free choice disjunction* or *conjunctive 'or'* [12]. Suppose that you are looking for your car keys and ask someone for help, who replies with a disjunction:

(1) The keys might be in the drawer, or they might be in the car.

Then the speaker seems committed to *both* of the disjuncts:

(2) The keys might be in the drawer.
(3) And they might be in the car.

That is, when 'or' meets 'might', it somehow becomes conjunctive, behaving like an 'and'. The conjunctive reading is not easy to explain. Everyone's first idea is to explain the conjunctive reading as a conversational implicature [10]. But that does not work, as pointed out by Zimmermann ([19]: 259).[1] A conversational implicature in general can be *canceled* by outright denial, but this is not the case for the conjunctive reading:

(4) # The keys might be in the drawer, or they might be in the car. Indeed, they cannot be in the car.

That sounds contradicting oneself rather than canceling an implicature. Furthermore, a conversational implicature in general can be *reinforced* by explicitly stating it *without redundancy*, but that is not the case for the conjunctive reading:

(5) # The keys might be in the drawer, or they might be in the car. Indeed, they might be in car.

The last remark sounds redundant. So the conjunctive reading seems not a conversational implicature. Perhaps we can explain it as a conventional implicature but, to prevent *ad hoc* postulations of conventions, it seems better to take that as the last resort.

Given that pragmatic explanations are difficult to find, it is interesting to see whether we can have a semantic explanation. This paper aims to explore that possibility. The goal is to work out a sufficiently simple semantics that satisfies the following features:

Feature (A) The semantics is to validate the inference from "might-ϕ or might-ψ" to each of the disjuncts.
Feature (B) The semantics is to save the 'or'-introduction rule of inference (from ϕ to $\phi \vee \psi$) when 'might' is absent.

[1]There are similar phenomena when 'or' meets the deontic 'may', but in that case, the conjunctive reading of 'or' can be easily canceled. See Zimmermann [19] for discussion.

Feature (C) The above two features are to be achieved with a uniform semantics of 'or' without ad-hoc case distinctions, so that it unifies the two apparently different uses of 'or'.

Every semantics in the existing literature violates at least one of the three features (see Sect. 6.2 for a literature review).

To satisfy those three features, I propose a new approach to natural language semantics. On the standard, truth-conditional approach, each sentence is compositionally evaluated at a world as true or false. I propose that each sentence be compositionally evaluated at (a formal model of) an information state as acceptable, deniable, or undecided. This idea will be developed into a formal semantics. Validity is defined to be preservation of acceptability.

Working with acceptability conditions does not mean that we abandon the concept of truth conditions altogether. In fact, we can employ a sentence's acceptability condition to determine whether it has a truth condition. According to the proposed semantics, it turns out that epistemic modals do not have truth conditions, while sentences such as "snow is white" can have truth conditions if you wish. As I will explain in Sect. 6.6, the result points to a new, general semantic framework for addressing a central issue in philosophical logic and meta-ethics: Which types of declarative sentences lack truth conditions, especially epistemic modals, indicative conditionals, and moral claims?

Perhaps epistemic modals in English do not really have the conjunctive 'or' inference valid. Whether this is the case is ultimately an empirical question; whether our semantics of English (or any other natural language) should satisfy features (A)–(C) is ultimately an empirical question. But the semantics to be constructed in this paper shows the following: if the format of natural language semantics is supposed to be so general that every possible language can be correctly described by a particular implementation of this format (as what Lewis [13] has in mind in his paper "General Semantics"), then the format of truth-conditional semantics is not general enough. It seems to me that there is a possible language in which, first, the conjunctive 'or' inference for epistemic modals is valid and, second, 'or'-introduction is valid when epistemic modals are absent. Such a possible language is best described by a semantics that satisfies features (A)–(C). It seems that such a semantics cannot be a truth-conditional one, but can be something like an acceptability-conditional one—as we will see in the following.

This paper is structured as follows. Section 6.2 presents a literature review. Section 6.3 provides an illustrated introduction to the proposed semantics, and explains the conjunctive 'or' by drawing Venn diagrams. Then the semantics is presented formally in Sect. 6.4, followed by an extension in Sect. 6.5. Section 6.6 discusses the philosophical work that has been done, and to be done, by the semantics.

6.2 Recent Explanations of the Conjunctive 'Or'

Almost all recent explanations of the conjunctive 'or' adopt the truth-conditional approach to semantics. They either continue from, or respond to, Zimmermann's [19] semantic explanation. To a first approximation, Zimmermann proposes that a disjunction is true only relative to a speaker, and that a disjunction is true for a speaker only if:[2]

(GENUINENESS) No disjunct is known by the speaker to be false.

In Zimmermann's words, each disjunct is a *genuine* epistemic possibility for the speaker. Zimmermann's semantics of epistemic modality $\Diamond$ is quite standard: $\Diamond\phi$ is true for a speaker iff ϕ is not known by the speaker to be false. Then Zimmermann is able to explain the conjunctive 'or' by proving the following: whenever $\Diamond\phi \vee \Diamond\psi$ is true for a speaker, both disjuncts are true for the speaker.[3] However, the (GENUINENESS) condition is so strong that it invalidates the classical 'or'-introduction rule of inference (6), for most logically consistent sentences ψ.

(6) ϕ; therefore, $\phi \vee \psi$.

The reason is simply that a logically consistent ψ may be known by the speaker to be false and, in that case, the (GENUINENESS) condition would preclude the truth of the conclusion $\phi \vee \psi$ for the speaker. So almost all uses of the 'or'-introduction rule in everyday life become invalid, which is intuitively wrong. In case you want to see an argument rather than a mere claim about intuition, please see appendix A.

So Zimmermann seems to face a difficulty: to explain the conjunctive 'or' semantically, the semantics of 'or' seems to have to be modified in such a way that no longer accommodates other cases of reasoning with 'or'. Namely, features (A) and (B) seem incompatible. Indeed, that is the difficulty for all earlier semantic accounts. For example, Geurts [7] only slightly modifies Zimmermann's approach; so, like Zimmermann, his treatment violates feature (B). Simons [15] proposes a novel semantics of 'or', which ultimately explains the conjunctive 'or' as a conversational implicature (Simons [15]: 300–302) and, hence, violates feature (A).[4]

There is a variant of the conjunctive 'or' phenomenon. When one asserts (7), the speaker seems to be committed to the conjunctive reading (8).

---

[2]Zimmermann adds a further condition to turn the necessary condition into a necessary and sufficient condition, but that is omitted because it has nothing to do with explaining the conjunctive 'or'.

[3]Proof. Suppose that $\Diamond\phi \vee \Diamond\psi$ is true for the speaker. Then, by (GENUINENESS), both $\Diamond\Diamond\phi$ and $\Diamond\Diamond\psi$ are true for the speaker. Now Zimmermann makes an assumption: knowing that p implies knowing that one knows that p. So both disjuncts, $\Diamond\phi$ and $\Diamond\psi$, are true for the speaker. The assumption that knowing implies knowing that one knows is very controversial in epistemology. But perhaps Zimmermann can replace knowledge by belief in his semantics and only assume that believing always implies believing that one beliefs, which is much less controversial.

[4]Aloni's [1] focus is on the deontic 'may' rather than the epistemic 'might'. She sketches how to extend her work to the epistemic 'might' in a footnote (Aloni [1]: 78, fn. 8).

(7) The keys might be in the drawer or in the car.
(8) The keys might be in the drawer, and they might be in the car.

To explain the conjunctive reading, every author just mentioned tries to develop a semantics that validates the following inference:

(9) $\Diamond(\phi \vee \psi)$; therefore, $\Diamond\phi$ and $\Diamond\psi$.

Note that the premise itself is an epistemic modal that embeds a disjunction; by contrast, what we have been discussing is $\Diamond\phi \vee \Diamond\psi$, the disjunction of two epistemic modals. Although those authors try to validate inference (9), that seems to me on the wrong track. The reason is that inference (9) is actually invalid. When one asserts an instance of $\Diamond(\phi \vee \psi)$ such as sentence (10), the speaker is not always committed to the conjunctive reading (8).

(10) It might be the case that the keys are in the drawer or in the car.

So, *pace* those authors, I propose the following:

Feature (D) The semantics is to invalidate the inference from $\Diamond(\phi \vee \psi)$ to $\Diamond\phi \wedge \Diamond\psi$.

So the real puzzle is: why sentences (7) and (10) look so similar in terms of syntactic structure but behave so differently in terms of semantic entailment: the former seems to always have the conjunctive reading, while the latter is not. That is a puzzle concerning both syntactic and semantic issues. The present paper will not address that puzzle because, for the time being, I want to focus on the semantic side.

6.3 New Explanation

This section provides the minimal elements of the new proposal that suffices for explaining the conjunctive 'or'.

6.3.1 Acceptability at a Information State

Let W be a set of possible worlds. *Information states* are subsets of W.[5] A information state I is understood to rule out all possibilities outside and leave open the possibilities inside (Fig. 6.1). So each information state is assumed to have a truth condition: it is true at all and only the worlds that it contains. The proposed semantics evaluates

[5]Following Hintakka [11] and Stalnaker [16].

Fig. 6.1 α is acceptable at I

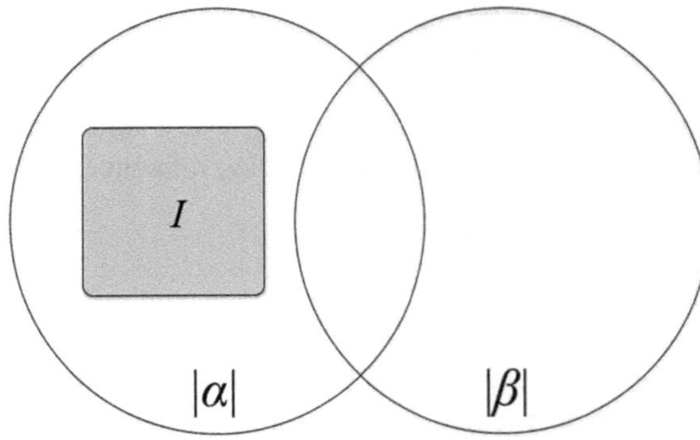

each sentence ϕ as *acceptable* or not at each information state I.[6] Just like the notion of truth employed in a truth-conditional semantics is not analyzed, I do not think I have to analyze the notion of acceptability to be employed. But I need to say what it is not and what it is like. It is not "warranted acceptability" in Dummett's [4] sense or any verificationist sense. That should be obvious: I talk about acceptability of a sentence at an *information state*, which is *doxastic* rather than evidential. Without trying to provide an analysis, we may understand acceptability as follows (if you find it helpful): "ϕ as a sentence in language $\mathcal{L}$ is acceptable at information state I" means that any competent speaker of language $\mathcal{L}$ with information state I can accept sentence ϕ while staying in information state I.

Assume, just in this section, that every atomic sentence α has a truth condition $|\alpha|$, which denotes the set of worlds at which α is true. This assumption is made only for the sake of pictorial illustration and will be relaxed in the formal presentation of the semantics (see next section). Then, α is acceptable at I just in case it is true at every world left open by I, i.e., $I \subseteq |\alpha|$ (Fig. 6.1). Similarly for atomic sentence β. Beyond the atomic level, there will be no reference to truth conditions any more.

6.3.2 Semantics of 'Might' and 'Or'

When one asserts an epistemic modal $\Diamond\alpha$, the speaker envisages a possible future in which she obtains new information that *strengthens* her current information state I into a consistent (i.e., nonempty) information state, say I' ($\subseteq I$), at which α comes to be acceptable (Fig. 6.2).

---

[6]Strictly speaking, the semantics to be developed evaluates each sentence as *acceptable*, *deniable*, or *undecided* in each information state, which will be presented in the next section. Deniability and undecidedness are ignored in the present section only because they are not essential for explaining the conjunctive 'or'; only acceptability is essential.

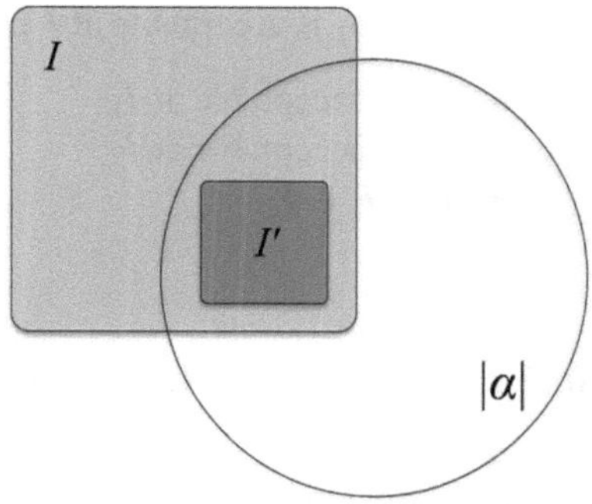

Fig. 6.2 $\Diamond\alpha$ is acceptable at I

So I propose that, in general:

(MIGHT) $\Diamond\phi$ is acceptable at information state I iff there exists an information state I' such that:

- $\varnothing \neq I' \subseteq I$,
- ϕ is acceptable at I'.

Note that this semantic rule does not presuppose that ϕ has a truth condition or not. Then we have:

Lemma 1 *Assume semantic rule* (MIGHT). *Assume, just for the sake of pictorial illustration, that atomic sentence α has its truth condition and, hence, α is acceptable at I iff $I \subseteq |\alpha|$. Then:*

$$\Diamond\alpha \textit{ is acceptable at } I \iff I \cap |\alpha| \neq \varnothing.$$

This result can be easily verified by drawing Venn diagrams (cf. Fig. 6.2).

The above is straightforward, while the crux lies in developing the right semantics of disjunctions. The following principle employs set-theoretic union as a way to construct information states that make a disjunction acceptable:

(UNION) Whenever ϕ_1 is acceptable at I_1 and ϕ_2 is acceptable at I_2, then the union $I_1 \cup I_2$ is an information state at which the disjunction $\phi_1 \vee \phi_2$ is acceptable.

Although the union operation $\cup$ is just one way to construct information states that make disjunction $\phi_1 \vee \phi_2$ acceptable, it seems general enough for constructing *all* such information states. To illustrate, let the disjuncts be atomic sentences α, β with truth conditions. The information states at which α is acceptable are exactly the subsets of $|\alpha|$ (Fig. 6.1); similarly for β. By taking the unions of subsets of $|\alpha|$ and subsets of $|\beta|$, we can construct *all and only* subsets of $|\alpha| \cup |\beta|$, which are exactly the information states at which disjunction $\alpha \vee \beta$ is supposed to be acceptable. In Fig. 6.1, for example, I is a subset of $|\alpha| \cup |\beta|$ and it can be constructed as the union of I (a subset of $|\alpha|$) and $\varnothing$ (a subset of $|\beta|$). Hence, the (UNION) principle generates all and only information states in which disjunction $\phi_1 \vee \phi_2$ is acceptable—whenever the disjuncts have truth conditions. I propose that the same applies to arbitrary disjuncts:

(OR) $\phi_1 \vee \phi_2$ is acceptable at I iff there exist information states I_1, I_2 such that:

- ϕ_1 is acceptable at I_1,
- ϕ_2 is acceptable at I_2,
- $I = I_1 \cup I_2$.

6.3.3 Conjunctive 'Or' Explained

Then the semantics predicts the conjunctive 'or' phenomenon:

Claim 1 *Assume semantic rules* (OR) *and* (MIGHT). *Then, for all sentences ϕ, ψ and all information states I,*

$$\Diamond\phi \vee \Diamond\psi \text{ is acceptable at } I \implies \text{both } \Diamond\phi \text{ and } \Diamond\psi \text{ are acceptable at } I.$$

This general claim is an immediate corollary of Proposition 1 below. Here let us prove the following special case, which is provable by drawing Venn diagrams—perhaps this is more explanatory than a set-theoretic proof.

Claim 2 (Special Case) *Assume semantic rules* (OR) *and* (MIGHT). *Assume, further, that atomic sentence α has its truth condition and, hence, α is acceptable at I iff $I \subseteq |\alpha|$; similarly for atomic sentence β. Then we have:*

$$\Diamond\alpha \vee \Diamond\beta \text{ is acceptable at } I \implies \text{both } \Diamond\alpha \text{ and } \Diamond\beta \text{ are acceptable at } I.$$

Proof Suppose that one of the disjuncts fails to be acceptable at I, say $\Diamond\alpha$. It suffices to show that disjunction $\Diamond\alpha \vee \Diamond\beta$ fails to be acceptable at I too. Since $\Diamond\alpha$ is not acceptable at I, it follows from Lemma 1 that I is disjoint from $|\alpha|$ (Fig. 6.3). So, no matter how we express I as a union $I_1 \cup I_2$, the first component I_1 is *still* disjoint from $|\alpha|$ and, hence, is an information state at which $\Diamond\alpha$ is not acceptable (by Lemma 1). In other words, disjunction $\Diamond\alpha \vee \Diamond\beta$ is not acceptable at I because there is no way

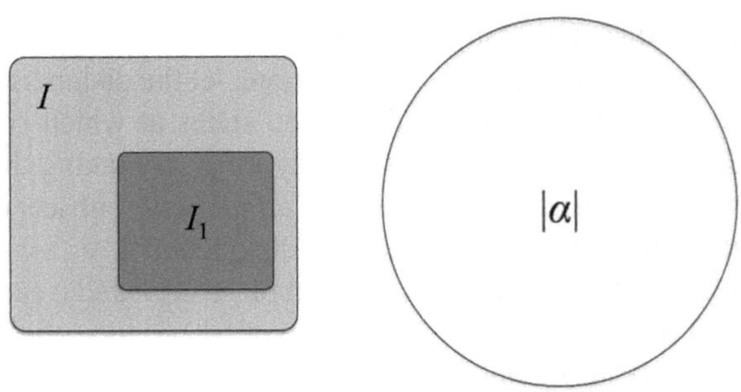

Fig. 6.3 Whenever $\Diamond\alpha$ is not acceptable at I, neither is $\Diamond\alpha \vee \Diamond\beta$

to satisfy the first clause of semantic rule (OR). That explains the conjunctive 'or' phenomenon. □

Although the above explanation assumes that atomic sentences have truth conditions, this assumption is made only for the sake of visualizing the explanation with Venn diagrams. The next section frees us from that assumption and presents the details of the acceptability-conditional semantics.

6.4 Acceptability-Conditional Semantics

Understand $[\![\phi]\!]^I = \textsf{Acceptable}$ as saying that sentence ϕ is acceptable at information state I. $\textsf{Acceptable}$ is only one of the totally three semantic values in use: we also have $\textsf{Deniable}$ and $\textsf{Undecided}$, standing for deniability and undecidedness, respectively. The formal semantics defines valuation function $[\![\cdot]\!]$ compositionally. Let the atomic case be given; i.e., for each atomic sentence α and for each information state I, let the value of $[\![\alpha]\!]^I$ be given. Only one constraint is imposed on the atomic case:

Semantic Rule 1 For each atomic sentence α that has truth condition $|\alpha|$:

$$[\![\alpha]\!]^I = \begin{cases} \textsf{Acceptable} & \text{iff } I \subseteq |\alpha|; \\ \textsf{Deniable} & \text{iff } I \cap |\alpha| = \varnothing \text{ and } I \neq \varnothing; \\ \textsf{Undecided} & \text{otherwise (i.e., iff } I \cap |\alpha| \neq \varnothing \text{ and } I \cap (W \setminus |\alpha|) \neq \varnothing) \end{cases}$$

The above raises an issue: which sentences have truth conditions? We will talk more about that in the concluding section. As for present purposes, it suffices to note that the formal semantics itself is neutral about that issue. As for negation, what it does is just to switch acceptability and deniability, except for the inconsistent information state $\varnothing$ as a limiting case:

Semantic Rule 2 (*Negation*) If I is the empty set, $[\![\neg\phi]\!]^I = [\![\phi]\!]^I$. If I is nonempty, then:

$$[\![\neg\phi]\!]^I = \begin{cases} \textsf{Acceptable} & \text{iff } [\![\phi]\!]^I = \textsf{Deniable}; \\ \textsf{Deniable} & \text{iff } [\![\phi]\!]^I = \textsf{Acceptable}; \\ \textsf{Undecided} & \text{iff } [\![\phi]\!]^I = \textsf{Undecided} \end{cases}$$

The acceptability condition of a conjunction is straightforward. Its deniability condition captures the following idea: deny the sentence if it is not acceptable to you right now nor acceptable at any possible future you can envisage.[7]

[7]It is inspired from the Beth–Krikpe semantics for negation in intuitionistic logic.

Semantic Rule 3 (*Conjunction*)

$$[\![\phi_1 \wedge \phi_2]\!]^I = \begin{cases} \textsf{Acceptable} & \text{iff } [\![\phi_i]\!]^I = \textsf{Acceptable} \text{ for each } i \in \{1,2\}; \\ \textsf{Deniable} & \text{iff } [\![\phi_1 \wedge \phi_2]\!]^I \neq \textsf{Acceptable} \text{ and} \\ & [\![\phi_1 \wedge \phi_2]\!]^{I'} \neq \textsf{Acceptable} \text{ for each nonempty } I' \subseteq I; \\ \textsf{Undecided} & \text{otherwise} \end{cases}$$

If you are worried that the deniability condition makes the semantics non-compositional because it refers to the conjunction itself rather than its conjuncts, just use the acceptability condition of a conjunction to unpack "$[\![\phi_1 \wedge \phi_2]\!]^I \neq$ Acceptable" into: "$[\![\phi_i]\!]^I \neq$ Acceptable for some $i \in \{1,2\}$." As for disjunctions, their acceptability conditions are as explained in the preceding section, while their deniability conditions are inspired by the same dynamic perspective as above[8,9]:

Semantic Rule 4 (*Disjunction*)

$$[\![\phi_1 \vee \phi_2]\!]^I = \begin{cases} \textsf{Acceptable} & \text{iff } I \text{ is the union of two sets } I_1, I_2 \text{ such that} \\ & [\![\phi_i]\!]^{I_i} = \textsf{Acceptable} \text{ for each } i \in \{1,2\}; \\ \textsf{Deniable} & \text{iff } [\![\phi_1 \vee \phi_2]\!]^I \neq \textsf{Acceptable} \text{ and} \\ & [\![\phi_1 \vee \phi_2]\!]^{I'} \neq \textsf{Acceptable} \text{ for each nonempty } I' \subseteq I; \\ \textsf{Undecided} & \text{otherwise.} \end{cases}$$

Here is the semantic rule for epistemic modals:

Semantic Rule 5 (*Epistemic Modal*)

$$[\![\Diamond\phi]\!]^I = \begin{cases} \textsf{Acceptable} & \text{iff } I \text{ has a nonempty subset } I' \text{ such that} \\ & [\![\phi]\!]^{I'} = \textsf{Acceptable}; \\ \textsf{Deniable} & \text{iff } I \text{ has no nonempty subset } I' \text{ such that} \\ & [\![\phi]\!]^{I'} = \textsf{Acceptable}; \\ \textsf{Undecided} & \text{otherwise (in fact, in no cases).} \end{cases}$$

Definition 1 (*Model*) An *acceptability model* is an ordered pair $(W, [\![\cdot]\!])$, where W is a nonempty set (of objects to be called possible worlds) and $[\![\cdot]\!]$ is a valuation function that satisfies the above five semantic rules.

---

[8] Note that in the standard, truth-table semantics for classical propositional logic, conjunction and disjunction have a duality: switching truth and falsity in the truth table for conjunction, we get the truth table for disjunction, and vice versa. But such duality is lost in the proposed semantics: switching Acceptable and Deniable, we cannot transform the rule for conjunction into the rule for disjunction. I thank Robert Stalnaker for bringing my attention to that. I suspect that it is a price we have to play if we want to explain the conjunctive 'or'. Indeed, the classical duality is broken not only by me, but all earlier semantic explanations of the conjunctive 'or'.

[9] I thank Alexander Worsnip for pointing to me that I made a mistake in an earlier version of the deniability condition for disjunction.

Definition 2 (*Validity*) Validity is defined to be preservation of acceptability. Namely, an argument is *valid* just in case: under any acceptability model, whenever the premises are all acceptable at a nonempty information state, the conclusion is also acceptable at the same information state.

In light of the discussion in the preceding section, it should not be surprising that the semantics predicts the conjunctive 'or':

Proposition 1 (Conjunctive 'Or') *For all sentences ϕ, ψ:*

$$[\![\Diamond\phi_1 \vee \Diamond\phi_2]\!]^I = \textsf{Acceptable} \iff [\![\Diamond\phi_1]\!]^I = [\![\Diamond\phi_2]\!]^I = \textsf{Acceptable}.$$

This result relies *solely* on the acceptability conditions of disjunctions and epistemic modals, independent of their deniability and undecidedness conditions. The left-to-right side is what feature (A) requires.

Classical logic can be shown to hold for what I call *classical sentences*, which are defined to be the sentences constructed from (i) atomic sentences that have truth conditions, (ii) connectives $\neg$, $\wedge$, $\vee$, and no more. Due to the way classical sentences are constructed, they can be assigned truth conditions in the standard way:

Definition 3 For all classical sentences ϕ and ψ:

$$\begin{aligned} |\neg\phi| &= W \setminus |\phi|, \\ |\phi \wedge \psi| &= |\phi| \cap |\psi|, \\ |\phi \vee \psi| &= |\phi| \cup |\psi| \end{aligned}$$

Then we have:

Proposition 2 *For each classical sentence ϕ,*

1. $[\![\phi]\!]^I = \textsf{Acceptable}$ *iff* $I \subseteq |\phi|$*;*
2. $[\![\phi]\!]^I = \textsf{Deniable}$ *iff* $I \cap |\phi| = \varnothing$ *and* $I \neq \varnothing$.

The above is what we expect for any sentence ϕ that has a truth condition. It follows immediately that the logic of classical sentences is exactly classical logic:

Corollary (Validity of Classical Inference) *For each classical sentence ϕ and each set Γ of classical sentences, the following three conditions are equivalent:*

1. *The inference from Γ to ϕ is valid with respect to classical logic.*
2. *Under any acceptability model,* $\bigcap_{\gamma\in\Gamma} |\gamma| \subseteq |\phi|$.
3. *Under any acceptability model, the inference from Γ to ϕ is valid with respect to the acceptability-conditional semantics; namely, for every information state I, if* $[\![\gamma]\!]^I = \textsf{Acceptable}$ *for all $\gamma \in \Gamma$, then* $[\![\phi]\!]^I = \textsf{Acceptable}$.

This covers what feature (B) requires.

Feature (C) asks us to provide a *uniform* semantics for 'or' without case disjunctions, which we have done. What feature (D) requires is accomplished in the following example:

Example (Invalidity of the Inference from $\Diamond(\phi \vee \psi)$ to $\Diamond\phi \wedge \Diamond\psi$) Let ϕ and ψ be classical sentences with disjoint, nonempty truth conditions $|\phi|$ and $|\psi|$, respectively. Consider information state $I = |\phi|$. Then $\Diamond(\phi \vee \psi)$ is acceptable at I, because I can be trivially strengthen into itself I, at which $\phi \vee \psi$ is acceptable by Proposition 2. But $\Diamond\psi$ is not acceptable at I, because I is disjoint from $|\psi|$ and, hence, cannot be strengthened into a nonempty information state included in $|\psi|$. Since $\Diamond\psi$ as the second conjunct is not acceptable at I, the conjunction $\Diamond\phi \wedge \Diamond\psi$ is not acceptable at I.

To finish the presentation of this new style of semantics, the concept of logical equivalence is defined as follows:

Definition 4 (*Logical Equivalence*) Logical equivalence is defined as necessary identity of semantic values. Namely, sentences ϕ, ψ are *logically equivalent* just in case $[\![\phi]\!]^I = [\![\psi]\!]^I$, for each acceptability model and for the valuation function $[\![\cdot]\!]$ and each nonempty information state I in that model.[10]

Note that the logical equivalence of two sentences requires something more than the validity of inferring from each one to the other, which concerns acceptability alone. Logical equivalence concerns identity in all the three semantic values: Acceptable, Deniable, and Undecided. For example, let α be an atomic sentence that has a truth condition $|\alpha|$. Then $\neg\alpha$ and $\neg\Diamond\alpha$ can be shown to have exactly the same acceptability conditions if we only consider nonempty information states I, i.e., that I is disjoint from $|\alpha|$. So the inference from each one to the other is valid. But clearly $\neg\alpha$ and $\neg\Diamond\alpha$ are not logically equivalent, at least for this intuitive reason: those two sentences are not intersubstitutable in a negated context $\neg(\ )$. In other words, those two sentences do not have the same deniability conditions, as correctly predicted by the proposed semantics.[11]

6.5 A New Puzzle and Its Solution

At a certain stage of a treasure hunt, the father (F) decides to provide some hint to the child (C):

(11) F: "The prize might be in the garden, or it might be in the attic."
(12) C: "So... it might be in the garden?"
(13) F: "Yes, it might be in the garden, and it might be in the attic."

In this case, the father's assertion of disjunction (11) seems to commit him to both of the disjuncts, as he admits in (13). But, assuming that he remembers where he puts

[10]I thank David Etlin for suggesting the this definition of logical equivalence, which explains the importance of deniability in my semantics better than I attempted in an earlier version of this paper.

[11]To see why, it suffices to let I be nonempty. $\neg(\neg\alpha)$ is acceptable at I iff $\neg\alpha$ is deniable in I iff α is acceptable at I iff $I \subseteq |\alpha|$. $\neg(\neg\Diamond\alpha)$ is acceptable at I iff $\neg\Diamond\alpha$ is deniable in I iff $\Diamond\alpha$ is acceptable at I iff I has a nonempty subset included in $|\alpha|$ iff $I \cap |\alpha| \neq \varnothing$.

the prize, his assertion (13) is insincere: if the prize is put in the garden, then for him it cannot be in the attic; if the prize is put in the attic, then for him it cannot be the garden. No matter which is the case, (13) is not acceptable at the father's information state. So, given the validity of conjunctive 'or', (11) is also not acceptable at the father's information state. With so much insincerity, how can the father's assertions be felicitous?[12]

The solution lies in understanding the semantics appropriately. The formula $[\![\phi]\!]^I = \textsf{Acceptable}$ has been understood as saying that ϕ is acceptable at information state I, which leaves open the question as to whose information state is involved. But note that I can be, for example, the informational common ground shared by the participants of a conversation; that is, I can represent what they commonly believe. When the father asserts disjunction (11), he proposes to modify the common ground I between his child and himself so that the disjunction is acceptable at I—following Stalnaker's [17] account of assertion. Since the semantics validates the conjunctive 'or' inference, the father's proposal carries with a commitment: the common ground I be modified so that both disjuncts are acceptable at I—this is exactly what the father makes explicit in assertion (13). The father's assertions are indeed insincere with respect to *his own* information state, but that is not important for the game. What is important for this game is to make the game fun, which requires the father to make appropriate sentences accepted at the common ground between him and his child.

After getting the hint, the child continues the treasure hunt and escapes the father's sight. Then the child's mother (M), who does not participate in the game, asks the father:

(14) M: "Where did you put the prize, seriously?"
(15) F: "It is in the garden."
(16) M: "So it cannot be in the attic."
(17) F: "No, it cannot be in attic."

In this case, the father's assertions are not only intended to be proposals to modify the common ground between his wife and himself, but also intended to be acceptable at his own information state.

6.6 Which Declarative Sentences Lack Truth Conditions?

Although the proposed semantics aims to provide acceptability conditions, it does not mean that we have to abandon the concept of truth conditions altogether. If a sentence ϕ has truth condition T (which is a set of possible worlds), then that sentence has

[12]This puzzle, together with the solution I propose below, is inspired by Justin Khoo's comments on an earlier version of this paper.

the following property: the information states at which ϕ is acceptable are exactly the subsets of T. This property, I propose, is not only necessary but also sufficient:

Definition 5 (*Having A Truth Condition*) A sentence ϕ is said to *have truth condition* T just in case, for each information state I, ϕ is acceptable at I iff $I \subseteq T$.

Namely, a sentence's acceptability condition determines whether it has a truth condition or not. Then, given the proposed acceptability-conditional semantics, it is routine to verify the following:

Claim 3 *If a sentence has a truth condition, it has a unique truth condition.*

Claim 4 *All classical sentences have truth conditions.*

Claim 5 *No epistemic modal has a truth condition.*

We have been talking about acceptability a *information states*, and we can generalize and talk about acceptability at *mental states*. Model a mental state S as an n-tuple $(I_{S,\ldots})$, where the first component I_S is the information state that underlies mental state S, and the other components may model what one desires, prefers, or approves. Then, to have a truth condition is to have the acceptability condition depend solely on information states in the way we have seen:

Definition 6 (*Having A Truth Condition: Generalized Version*) A sentence ϕ is said to *have truth condition* T just in case, for each mental state S, ϕ is acceptable at S iff the information state I_S that underlies S is a subset of T.

Allowing for the concept of truth conditions, the semantics is neutral about whether, for example, indicative conditionals or moral claims have truth conditions. It depends on how we develop the semantics in order to accommodate linguistic data. For example, we may insist that indicative conditional "if ϕ then ψ" has the same acceptability condition as material implication $\neg\phi \vee \psi$, so most indicative conditionals have truth conditions. Alternatively, we may follow Ramsey's test [14] for indicative conditionals, and construct a semantics that proceeds roughly as follows: "if ϕ then ψ" is acceptable at an information state I iff the consequent ψ is acceptable at the information state that results from I by supposing the antecedent ϕ. In that treatment, indicative conditionals are expected to lack truth conditions. For moral claims, we may build moral facts into possible worlds and make them objects of belief. Alternatively, we may follow non-cognitivists' idea that moral claims lack truth conditions, and extend the semantics so that the acceptability of a moral claim depends also on one's desire-like state.[13] So the style of the acceptability-conditional semantics I propose is very flexible.

Such flexibility suggests a new, general semantic framework for addressing the following question: Which types of declarative sentences lack truth conditions? For

[13] The details have to be left to another paper, because a complete treatment requires a thorough discussion of the so-called Frege-Geach Problem in meta-ethics, which has nothing to do with the main theme of this paper: conjunctive 'or'.

indicative conditionals, let us develop truth-conditional theories in the proposed semantic framework, and also let us develop anti-truth-conditional theories in the *same* framework. Then we can evaluate them in terms of how good they accommodate linguistic data. We may decide to be truth-conditionalists for one type of sentence, and yet be anti-truth-conditionalists for another type of sentence—both in the *same* framework of acceptability-conditional semantics. The present paper argues for an anti-truth-conditional theory about epistemic modals, and that in itself says nothing about whether we should be truth-conditionalists about other types of declarative sentences.[14]

For example, let me sketch how one may proceed to develop an anti-truth-conditional treatment of sentences like "you should do that," and explain how it pertains to the so-called expressivism in meta-ethics. According to expressivism, to assert that Bob should work hard is to express one's policy that requires Bob to work hard. I propose to rewrite that idea in terms of acceptability: "Bob should work hard" is acceptable at mental state S iff S is committed to such a policy. But what is it to be committed to such a policy? Suppose that, for each possible world w, if the agent had believed that w is the actual world, she would take all and only the worlds in $P(w)$ as permissible. Call $P(w)$ the agent's hyper-policy at world w. But the agent might not know which world is the actual world, so she is committed to a (possibly unspecific) policy if and only if that policy is required by the hyper-policy $P(w)$ at each world w in her information state. To be precise, let a mental state (that we are interested in for now) be an ordered pair (I, P), where I is an information state and P is a function from worlds to hyper-policies.

Semantic Rule 6

$$[\![\text{SHOULD}\,\phi]\!]^{I,P} = \text{Acceptable iff } [\![\phi]\!]^{I',P} = \text{Acceptable for every } I' \text{ in } \{P(w) : w \in I\}.$$

According to the proposed semantics, the acceptability conditions of 'should'-claims depend not only on one's information state but also on one's assignment P of hyper-policies. So, according to that semantics, 'should'-claims do not have truth conditions.

6.7 Concluding Remarks

The thesis that underlies the proposed semantics is that the semantic value of a declarative sentence should be characterized by the conditions in which the sentence is acceptable, deniable, and undecided, respectively. My ultimate argument for it is simply that it explains linguistic data better than the orthodox thesis that the semantic value of a declarative sentence is its truth condition. This paper does not examine a wide range of data, of course. What I intend to do here is only to examine a hard

[14]For anti-truth-conditionalism about moral claims, see, e.g., Gibbard [8] and Blackburn [2]; about indicative conditionals, see, e.g., Edgington [5]; about epistemic modals, see, e.g., Yalcin [18].

problem in linguistics—the conjunctive 'or'—and to make a first step toward a full-fledged explanatory semantics of natural languages. Let me sketch what the next few steps will be like.

The achievements of standard truth-conditional semantics include, for example, accounts of quantification, alethic modality, and propositional attitude attribution. The proposed semantics can easily inherit those achievements. To incorporate alethic modality, let each world w be associated with the set $R(w)$ of worlds that are metaphysically accessible from w, following standard Kripke semantics. Then 'it is metaphysically necessary that ϕ' is acceptable at an information state I iff of each world $w \in I$, ϕ is acceptable at $R(w)$ (taken as an information state). The same strategy applies to ascriptions of belief and knowledge, if it is agreed that a Kripke semantics of belief and knowledge ascriptions is appropriate [11]. Quantification can be incorporated by letting each possible world be a standard model of a first-order language. Then, since a formula may contain free variables, the acceptability of a formula should be evaluated at a mental state *plus* an assignment of objects to variables. To interpret identity, it is not a trivial task to provide the transworld identity relation between objects in different worlds, especially when the worlds are epistemically possible worlds (rather than metaphysically possible worlds). This is not my own problem—it is common to *all* semantic theories that employ epistemically possible worlds.[15] Quantifiers, names, belief ascriptions, and transworld identity will interact with one another, which requires careful treatments. In particular, Frege's puzzle about the morning star and the evening star [6] has to be taken care of, but it is every semanticist's problem.

The proposed semantics can work well with the standard pragmatics. We have seen how it works with Stalnaker's pragmatic account of assertion in Sect. 6.5. It can also work smoothly with the Gricean pragmatics. What is said in an utterance is represented by, not a truth condition, but an acceptability condition. What is meant is still to be (defeasibly) inferred from the Gricean maxims [10]. Only the maxim of quality has to be restated carefully: "assert only what you believe to be true" has to be replaced by "assert only what is acceptable to you." The Gricean pragmatics itself does presuppose some theory of contents, but it does not force contents to be truth conditions.

The idea of compositional acceptability-conditional semantics is not entirely new. The Beth–Kripke semantics of intuitionistic logic is a forerunner. What I have done is to propose a new style of compositional acceptability-conditional semantics that is plausible as a semantics for natural languages—or at least for a fragment of English that contains epistemic modals. It is expected to have the applications mentioned earlier: to linguistics, philosophical logic, and meta-ethics. Those applications would constitute a big project, and I hope the present case study about the conjunctive 'or' makes the project appear not so crazy.

[15] But if one insists on only using worlds that are metaphysically possible, she may nonetheless use metaphysically possible worlds to 'simulate' epistemically possible worlds, following Stalnaker [16].

Acknowledgments The author is indebted to Anders Schoubye, Mandy Simons, Maria Aloni, Jeroen Groenendijk, and Florian Steinberger for discussion. I am also indebted to the participants of the graduate conference at Yale University in 2012, especially Robert Stalnaker, Justin Khoo, and Alexander Worsnip. I am indebted to the participants of the graduate conference at the University of Western Ontario in 2012, especially Hartry Field. I am indebted to the participants of the Ninth Conference on Logic and Engineering of Natural Language Semantics (LENLS 9), especially David Etlin and Hans Kamp. I am also indebted to the participants of the Deontic Modality Workshop at the University of Southern California in 2013, and the participants of the Taiwan Philosophical Logic Colloquium in 2014.

Arguement for 'Or'-Introduction in Ordinary Cases

Consider the following conversation.

X: "Everyone in the party got drunk or overate."[16]
Y: "Really?!"
X: "Yeah. Alice, Bob, and Charles got drunk, and they have almost nothing to eat because Dorothy and I ate too much."

The general claim entails the truth of the instance "Alice got drunk or overate." But the speaker knows that Alice did not overate, as can be seen from the above conversation. So, if Zimmermann's GENUINENESS condition is correct, the instance "Alice got drunk or overate" is false and, hence, the general claim is false—but that is counterintuitive.[17] Furthermore, the speaker X uses his second claim to justify his first claim, and the justification is naturally understood as follows:

> Alice got drunk, so (by *'or'-introduction*) she got drunk or overate. Similarly, Bob, Charles, Dorothy, and I got drunk or overate. So everyone in the party got drunk or overate.

That is why I insist on the classical 'or'-introduction rule of inference, the very rule of inference that contradicts Zimmermann's GENUINENESS condition. In general, classical inferences should be preserved as much as possible—that is why I take (B) as a feature.

Proofs

Proof of Proposition 1 For ($\Longrightarrow$), suppose that $[\![\Diamond\phi_1 \vee \Diamond\phi_2]\!]^I = \mathsf{Acceptable}$. Then, by the acceptability conditions of disjunctions, I equals $I_1 \cup I_2$ for some sets I_1, I_2 such that $[\![\Diamond\phi_i]\!]^{I_i} = \mathsf{Acceptable}$ for $i = 1, 2$. Then, by the acceptability conditions of epistemic modals, I_i has a nonempty subset I_i' such that $[\![\phi]\!]^{I_i'} = \mathsf{Acceptable}$.

[16]This example is adapted from Simons [15], although she uses it for different purposes.

[17]Zimmermann does notice the present difficulty, but he only provides a sketchy response in a footnote (Zimmermann [19]: 276, fn.31).

It follows that I has a nonempty subset, namely I_i', such that $[\![\phi]\!]^{I_i'} = \textsf{Acceptable}$ (because $I_i' \subseteq I_i \subseteq I$). So, by the acceptability conditions of epistemic modals, $[\![\Diamond\phi_i]\!]^I = \textsf{Acceptable}$, for $i = 1, 2$.

For ($\Longleftarrow$), suppose that $[\![\Diamond\phi_1]\!]^I = [\![\Diamond\phi_2]\!]^I = \textsf{Acceptable}$. Then, since $I = I \cup I$, it follows from the acceptability conditions of disjunctions that $[\![\Diamond\phi \vee \Diamond\psi]\!]^I = \textsf{Acceptable}$. □

Proof of Proposition 2 Prove by induction on the complexity of ϕ as follows. Inductive basis: suppose that ϕ is an atomic sentence α that has truth condition $|\alpha|$. Then the proposition holds by the acceptability and deniability conditions of α. Inductive step for ($\neg$): suppose that ϕ is a negation $\neg\psi$. If I is empty, then the derivation is almost trivial:

$$\begin{aligned} [\![\neg\psi]\!]^I = \textsf{Acceptable} &\Leftrightarrow [\![\psi]\!]^I = \textsf{Acceptable} \\ &\Leftrightarrow I \subseteq |\psi| \\ &\Leftrightarrow I \subseteq |\neg\psi| \text{ (since the empty set } I \text{ is included in every set).} \end{aligned}$$

$$\begin{aligned} [\![\neg\psi]\!]^I = \textsf{Deniable} &\Leftrightarrow [\![\psi]\!]^I = \textsf{Deniable} \\ &\Leftrightarrow I \cap |\psi| = \varnothing \text{ and } I \neq \varnothing \text{ (which is impossible)} \\ &\Leftrightarrow I \cap |\neg\psi| = \varnothing \text{ and } I \neq \varnothing \text{ (which is impossible, too).} \end{aligned}$$

If I is nonempty, then:

$$\begin{aligned} [\![\neg\psi]\!]^I = \textsf{Acceptable} &\Leftrightarrow [\![\psi]\!]^I = \textsf{Deniable} \\ &\Leftrightarrow I \cap |\psi| = \varnothing \text{ and } I \neq \varnothing \\ &\Leftrightarrow I \cap |\psi| = \varnothing \\ &\Leftrightarrow I \subseteq |\neg\psi|. \end{aligned}$$

$$\begin{aligned} [\![\neg\psi]\!]^I = \textsf{Deniable} &\Leftrightarrow [\![\psi]\!]^I = \textsf{Acceptable} \\ &\Leftrightarrow I \subseteq |\psi| \\ &\Leftrightarrow I \cap |\neg\psi| = \varnothing \\ &\Leftrightarrow I \cap |\neg\psi| = \varnothing \text{ and } I \neq \varnothing. \end{aligned}$$

Inductive step for ($\wedge$): suppose that ϕ is a conjunction $\phi_1 \wedge \phi_2$. Then:

$$\begin{aligned} [\![\phi_1 \wedge \phi_2]\!]^I = \textsf{Acceptable} &\Leftrightarrow [\![\phi_1]\!]^I = \textsf{Acceptable and } [\![\phi_2]\!]^I = \textsf{Acceptable} \\ &\Leftrightarrow I \subseteq |\phi_1| \text{ and } I \subseteq |\phi_2| \\ &\Leftrightarrow I \subseteq |\phi_1| \cap |\phi_2| \\ &\Leftrightarrow I \subseteq |\phi_1 \wedge \phi_2|. \end{aligned}$$

$$
\begin{aligned}
[\![\phi_1 \wedge \phi_2]\!]^I = \textsf{Deniable} &\Leftrightarrow \text{for each } I', \text{ if } I' \text{is} I \text{or a nonempty subset of} I \\
&\quad \text{then } [\![\phi_1 \wedge \phi_2]\!]^{I'} \neq \textsf{Acceptable} \\
&\Leftrightarrow \text{for each } I', \text{ if } I' \text{ is } I \text{ itself or a nonempty subset of } I \\
&\quad \text{then } [\![\phi_1]\!]^{I'} \neq \textsf{Acceptable} \text{ or } [\![\phi_1]\!]^{I'} \neq \textsf{Acceptable} \\
&\Leftrightarrow \text{for each } I', \text{ if } I' \text{ is } I \text{ itself or a nonempty subset of } I \\
&\quad \text{then } I' \not\subseteq |\phi_1| \text{ or } I' \not\subseteq |\phi_2| \\
&\Leftrightarrow \text{for each } I', \text{ if } I' \text{ is } I \text{ itself or a nonempty subset of } I \\
&\quad \text{then it is not the case that } I' \subseteq |\phi_1| \text{ and } I' \subseteq |\phi_2| \\
&\Leftrightarrow \text{for each } I', \text{ if } I' \text{ is } I \text{ itself or a nonempty subset of } I \\
&\quad \text{then it is not the case that } I' \subseteq |\phi_1| \cap |\phi_2| \\
&\Leftrightarrow \text{for each } I', \text{ if } I' \text{ is } I \text{ itself or a nonempty subset of } I \\
&\quad \text{then it is not the case that } I' \subseteq |\phi_1 \wedge \phi_2| \\
&\Leftrightarrow I \neq \varnothing \text{ and } I \cap |\phi_1 \wedge \phi_2| = \varnothing
\end{aligned}
$$

Inductive step for ($\vee$): suppose that ϕ is a disjunction $\phi_1 \vee \phi_2$. Then:

$$
\begin{aligned}
[\![\phi_1 \vee \phi_2]\!]^I = \textsf{Acceptable} &\Leftrightarrow B \text{ equals the union of some sets } I_1, I_2 \text{ such that} \\
&\qquad [\![\phi_i]\!]^{I_i} = \textsf{Acceptable} \text{ for } i = 1, 2 \\
&\Leftrightarrow B \text{ equals the union of some sets } I_1, I_2 \text{ such that} \\
&\qquad I_i \subseteq |\phi_i| \text{ for } i = 1, 2 \\
&\overset{(a)}{\Leftrightarrow} I \subseteq |\phi_1| \cup |\phi_2| \\
&\Leftrightarrow I \subseteq |\phi_1 \vee \phi_2|.
\end{aligned}
$$

To establish the ($\Rightarrow$) side of (a), it suffices to note that, if $I_i \subseteq |\phi_i|$ for $i = 1, 2$, then $I_1 \cup I_2 \subseteq |\phi_1| \cup |\phi_2|$. To establish the ($\Leftarrow$) side of ($a$), it suffices to let $I_i = I \cap |\phi_i|$ for $i = 1, 2$.

$$
\begin{aligned}
[\![\phi_1 \vee \phi_2]\!]^I = \textsf{Deniable} &\Leftrightarrow \text{for each } I', \text{ if } I' \text{ is } I \text{ or a nonempty subset of } I \\
&\quad \text{then } [\![\phi_1 \vee \phi_2]\!]^{I'} \neq \textsf{Acceptable} \\
&\Leftrightarrow \text{for each } I', \text{ if } I' \text{ is } I \text{ or a nonempty subset of } I \\
&\quad \text{then } I' \text{ is not the union of some sets } I_1, I_2 \text{ such that} \\
&\qquad [\![\phi_i]\!]^{I_i} = \textsf{Acceptable} \text{ for } i = 1, 2 \\
&\Leftrightarrow \text{for each } I', \text{ if } I' \text{ is } I \text{ or a nonempty subset of } I \\
&\quad \text{then } I' \text{ is not the union of some sets } I_1, I_2 \text{ such that} \\
&\qquad I_i \subseteq |\phi_i| \text{ for } i = 1, 2 \\
&\overset{(b)}{\Leftrightarrow} I \neq \varnothing \text{ and } I \text{ is disjoint from both } |\phi_1| \text{ and } |\phi_2| \\
&\Leftrightarrow I \neq \varnothing \text{ and } I \text{ is disjoint from } |\phi_1| \cup |\phi_2| \\
&\Leftrightarrow I \neq \varnothing \text{ and } I \text{ is disjoint from } |\phi_1 \vee \phi_2|
\end{aligned}
$$

To establish the ($\Rightarrow$) side of (b), suppose that the left hand side is true. If $I = \varnothing$, then the left hand side has a counterexample: $I' = I_1 = I_2 = \varnothing$. So $I \neq \varnothing$. If I is not disjoint from $|\phi_1|$, then the left hand side has a counterexample: $I' = I_1 = I \cap |\phi_1|$, $I_2 = \varnothing$. So I is disjoint from $|\phi_1|$. By symmetry, I is disjoint from $|\phi_2|$. To establish the ($\Leftarrow$) side of (a), suppose (for reductio) that the right hand side is true and the left hand side is false. Since $I \neq \varnothing$, it follows from the falsity of the left hand side that there exist subsets I', I_1, I_2 of I such that $I' \neq \varnothing$, $I = I_1 \cup I_2$, and $I_i \subseteq |\phi_i|$ for $i = 1, 2$. Since I' is nonempty and $I' = I_1 \cup I_2$, I_j is nonempty for some $j \in \{1, 2\}$. Since I_j is a nonempty subset both of I and of $|\phi_j|$, I is not disjoint from $|\phi_j|$, which contradicts the right hand side. □

References

1. Aloni, M.: Free choice, modals, and imperatives. Nat. Lang. Seman. **15**(1), 65–94 (2007)
2. Blackburn, S.: Essays in Quasi-Realism. Oxford University Press, Oxford (1993)
3. Brandom, R.: Truth and assertibility. J. Philos. **73**(60), 137–149 (1976)
4. Dummett, M.: The philosophical basis of intuitionistic logic. His Truth and Other Enigmas, pp. 97–129. Harvard University Press, Cambridge (1978)
5. Edgington, D.: The mystery of the missing matter of fact. Proc. Aristotelian Soc. Supplementary **65**, 185–209 (1991)
6. Frege, G.: (1892/[1980]) On sense and reference. In: Geach, P. Black, M. (eds. and trans.) (1980) Translations from the Philosophical Writings of Gottlob Frege, Blackwell, Oxford (1980)
7. Geurts, B.: Entertaining alternatives: disjunctions as modals. Nat. Lang. Seman. **13**(4), 383–410 (2005)
8. Gibbard, A.: Two Recent Theories of Conditionals. In: Harper, WL., Stalnaker, R., Pearce, G., (eds.) (1981)
9. Gibbard, A.: Wise Choices, Apt Feelings. Harvard University Press, Cambridge (1990)
10. Grice, H.P.: Logic and conversation, Reprinted. In: Grice, H.P. (1989) (ed.) Studies in the Way of Words, pp. 22–40. Harvard University Press, Cambridge (1975)
11. Hintikka, J.: Knowledge and Belief: An Introduction to the Logic of the Two Notions. Cornell University Press, Cornell (1962)
12. Kamp, H.: Free choice permission. Proc. Aristotelian Soc. N.S. **74**, 57–74 (1973)
13. Lewis, D.: General semantics. Synthese **22**, 18–67 (1970)
14. Ramsey, F.P.: (1929) General propositions and causality. In: Mellor, H.A. (ed.) F. Ramsey, Philosophical Papers. Cambridge University Press, Cambridge (1990)
15. Simons, M.: Dividing things up: the semantics of or and the modal/or interaction. Nat. Lang. Seman. **13**(3), 271–316 (2005)
16. Stalnaker, R.: Inquiry. MIT Press, Cambridge (1984)
17. Stalnaker, R.: Context and Content. Oxford University Press, Oxford (1999)
18. Yalcin, S.: (2011) Nonfactualism about Epistemic Modality. In: Egan, A., Weatherson, B. (eds.) Epistemic Modality
19. Zimmermann, E.: Free choice disjunction and epistemic possibility. Nat. Lang. Seman. **8**, 255–290 (2000)

Chapter 7
Revising a Labelled Sequent Calculus for Public Announcement Logic

Shoshin Nomura, Katsuhiko Sano and Satoshi Tojo

Abstract We first show that a labelled sequent calculus **G3PAL** for Public Announcement Logic (PAL) by Maffezioli and Negri (2011) has been lacking rules for deriving an axiom of Hilbert-style axiomatization of PAL. Then, we provide our revised calculus **GPAL** to show that all the formulas provable in Hilbert-style axiomatization of PAL are also provable in **GPAL** together with the *cut* rule. We also establish that our calculus enjoys cut elimination theorem. Moreover, we show the soundness of our calculus for Kripke semantics with the notion of surviveness of possible worlds in a restricted domain. Finally, we provide a direct proof of the semantic completeness of **GPAL** for the link-cutting semantics of PAL.

7.1 Introduction

Public Announcement Logic (PAL) was first presented by Plaza [12], and it has been the basis of Dynamic Epistemic Logics. PAL is a logic for formally expressing changes of human knowledge. Specifically, when we obtain some information through communication with others, our state of knowledge may change. For example, if 'John does not know whether it will rain tomorrow or not' is true and he gets information from the weather forecast which says that 'it will not rain tomorrow,' then the state of John's knowledge changes and so 'John knows that it will not rain tomorrow' becomes true. While a Kripke model of the standard epistemic logic stands for the state of knowledge, the standard epistemic logic does not have any syntax for properly expressing changes of the state of knowledge. PAL was introduced for the purpose of dealing with flexibility of human knowledge; and Dynamic Epistemic

S. Nomura (✉) · K. Sano · S. Tojo
School of Information Science, Japan Advanced Institute of Science and Technology, Nomi, Japan
e-mail: nomura@jaist.ac.jp

K. Sano
e-mail: v-sano@jaist.ac.jp

S. Tojo
e-mail: tojo@jaist.ac.jp

S.C.-M. Yang et al. (eds.), *Structural Analysis of Non-Classical Logics*,
Logic in Asia: Studia Logica Library, DOI 10.1007/978-3-662-48357-2_7

Logics based on PAL contain many possibilities to be applied to various fields such as artificial intelligence, epistemology in philosophy, formalizing law, and so on.

A proof system for PAL has been provided in terms of Hilbert-style axiomatization (we call it **HPAL**) which is complete for Kripke semantics; however, an easier system to calculate theorems should be desirable, since Hilbert-style proof systems are, in general, hard to handle for proving theorems. One possible candidate for such a proof system is a celebrated Gentzen-style sequent calculus [4], where a basic unit of a derivation is the notion of a *sequent*

$$\Gamma \Rightarrow \Delta,$$

which consists of two lists (or multisets or sets) of formulas. How can we read $\Gamma \Rightarrow \Delta$ intuitively? There are at least two ways of reading it. First, we may read it as 'if all formulas in Γ hold, then some formula in Δ holds'. Second, we may also read it as 'it is not the case that all formulas in Γ hold and all formulas in Δ fail'. We may wonder if these two readings are equivalent, but in fact the equivalence depends on an underlying logic. For example, two readings are equivalent in the classical propositional logic, provided we understand that 'a formula A holds' by 'A is true in a given truth assignment' and 'A fails' by 'A is false under the assignment' (note that, under these readings, A does not holds if and only if A fails). One of the most uniform approaches for sequent calculus for modal logic is *labelled sequent calculus* (c.f., [9]), where each formula has a label corresponding to an element of a domain in Kripke semantics for modal logic. The proof system we are concerned with in this paper is one of variants of labelled sequent calculus. An existing labelled sequent calculus for PAL, named **G3PAL**, was devised by Maffezioli and Negri [7]; however, a deficiency of **G3PAL** has been pointed out by Balbiani et al. [1].[1] In this paper, we also suggest a different defect in it. In brief, because **G3PAL** does not have inference rules relating to accessibility relations, there exists a problem in case of proving one of axioms of **HPAL**. Therefore, we introduce a revised labelled sequent calculus **GPAL** (with the rule of cut, **GPAL**$^+$) to compensate for the deficiency by adding some rules for accessibility relations.

Moreover, we especially focus on the soundness theorem of **GPAL**, since there is a hidden factor behind the definition of validity of the sequent $\Gamma \Rightarrow \Delta$, of which the researchers of this field (e.g., [1, 7]) seemingly have not made a point. In particular, we notice that the above two readings of a sequent in our setting are not equivalent and that the notion of validity based on the first reading of a sequent is not sufficient to prove the soundness of our calculus for Kripke semantics; however, we employ the notion of validity based on the second reading of a sequent to establish **GPAL**'s soundness. One of the reasons why two notions of validity are not equivalent consists of deleting possible worlds by a (truthful) public announcement. In fact, we will show the completeness of our calculus for PAL's another semantics, a version of the

---

[1] They stated that there are some valid formulas such as $[A \wedge A]B \leftrightarrow [A]B$ which may be unprovable in **G3PAL**.

link-cutting semantics by van Benthem et al. [14] where only the accessibility relation is restricted in a model and two notions of validity become equivalent.

The outline of this paper is as follows: Sect. 7.2 provides definitions of syntax of PAL and Kripke semantics for it, then introduces one simple example of Kripke model that is used throughout the paper. Additionally, the existing Hilbert-style axiomatization **HPAL** of PAL and its semantic completeness are outlined. Section 7.3 reviews Maffezioli and Negri's labelled sequent calculus **G3PAL** and specifies which part of **G3PAL** is problematic. Section 7.4 introduces our calculus **GPAL**, a revised version of **G3PAL**, and we show that all the theorems of **HPAL** are provable in **GPAL**$^+$ (Theorem 1), and establish the cut elimination theorem of **GPAL**$^+$ (Theorem 2). Section 7.5 focuses on its soundness theorem (Theorem 3) in terms of two notions of validity based on the above two readings of a sequent. Section 7.6 introduces the link-cutting semantics of PAL to provide a direct proof of the completeness of **GPAL** for the link-cutting semantics (Theorem 4). Finally, Sect. 7.7 concludes the paper.

7.2 Kripke Semantics and Axiomatization of PAL

First of all, we will address the syntax of PAL. Let $\mathsf{Prop} = \{p, q, r, \ldots\}$ be a countably infinite set of propositional variables and $\mathsf{G} = \{a, b, c, \ldots\}$ a nonempty finite set with elements called agents. Then the set $\mathsf{Form} = \{A, B, C, \ldots\}$ of formulas of PAL is inductively defined as follows ($p \in \mathsf{Prop}$, $a \in \mathsf{G}$):

$$A ::= p \mid \neg A \mid (A \to A) \mid \mathsf{K}_a A \mid [A]A.$$

Other logical connectives ($\wedge$, $\vee$, etc.) are defined as usual. $\mathsf{K}_a A$ is read as 'agent a knows that A', and $[A]B$ is read as 'after public announcement of A, it holds that B'.

Example 1 Let us consider a propositional variable p to read 'it will rain tomorrow'. Then a formula $\neg(\mathsf{K}_a p \vee \mathsf{K}_a \neg p)$ means that a does not know whether it will rain tomorrow or not, and $[\neg p]\mathsf{K}_a \neg p$ means that after a public announcement (e.g., a weather report) of $\neg p$, a knows that it will not rain tomorrow.

7.2.1 Kripke Semantics of PAL

We should now consider the Kripke semantics of PAL. The sequent calculus introduced in the next section can be regarded as a formalized version of Kripke semantics of PAL. We mainly follow the semantics introduced in van Ditmarsch et al. [15]. We call $\mathfrak{M} = \langle W, (R_a)_{a \in \mathsf{G}}, V \rangle$ a *Kripke model* if W is a nonempty set of possible worlds, $R_a \subseteq W \times W$, and V is a valuation function which assigns an propositional variable to a subset of W. W is also called the *domain* of $\mathfrak{M}$, denoted by $\mathcal{D}(\mathfrak{M})$. Next, let us define the satisfaction relation.

Definition 1 Given a Kripke model $\mathfrak{M}$, $w \in \mathcal{D}(\mathfrak{M})$, and $A \in \mathsf{Form}$, we define $\mathfrak{M}, w \Vdash A$ as follows:

$$
\begin{array}{lll}
\mathfrak{M}, w \Vdash p & \text{iff} & w \in V(p), \\
\mathfrak{M}, w \Vdash \neg A & \text{iff} & \mathfrak{M}, w \nVdash A, \\
\mathfrak{M}, w \Vdash A \to B & \text{iff} & \mathfrak{M}, w \Vdash A \text{ implies } \mathfrak{M}, w \Vdash B, \\
\mathfrak{M}, w \Vdash \mathsf{K}_a A & \text{iff} & \text{for all } v \in W : w R_a v \text{ implies } \mathfrak{M}, v \Vdash A, \text{ and} \\
\mathfrak{M}, w \Vdash [A]B & \text{iff} & \mathfrak{M}, w \Vdash A \text{ implies } \mathfrak{M}^A, w \Vdash B,
\end{array}
$$

where the restriction $\mathfrak{M}^A$, at the definition of the announcement operator, is the restricted Kripke model to the truth set of A, defined as $\mathfrak{M}^A = \langle W^A, (R_a^A)_{a \in \mathsf{G}}, V^A \rangle$ with

$$
\begin{array}{ll}
W^A & := \{x \in W \mid \mathfrak{M}, x \Vdash A\}, \\
R_a^A & := R_a \cap (W^A \times W^A), \\
V^A(p) & := V(p) \cap W^A \ (p \in \mathsf{Prop}).
\end{array}
$$

As above, the restriction of a Kripke model is based on the restriction of the set of possible worlds, so that this can be said to be the *world-deletion* semantics of PAL, and this will be distinguished from the *link-cutting* semantics in Sect. 7.6. In the semantics above, we do not assume any requirement on the accessibility relations $(R_a)_{a \in \mathsf{G}}$, while it is usually assumed that R_a is an equivalent relation in Kripke semantics for the standard epistemic logic; however, since the previous works [1, 7] also start with a Kripke model with an arbitrary accessibility relation, we also follow them in this respect.

Definition 2 A formula A is *valid* in a Kripke model $\mathfrak{M}$ if $\mathfrak{M}, w \Vdash A$ for all $w \in \mathcal{D}(\mathfrak{M})$.

This is the definition of PAL's semantics, but readers who are not familiar with PAL may not easily see what it is, so the following example might help for understanding the heart of PAL.

Example 2 Example 1 can be semantically modeled as follows. Let us consider $\mathsf{G} = \{a\}$ and the following two models, such as $\mathfrak{M} = \langle \{w_1, w_2\}, \{w_1, w_2\}^2, V \rangle$ where $V(p) = \{w_1\}$, and $\mathfrak{M}^{\neg p} = \langle \{w_2\}, \{(w_2, w_2)\}, V^{\neg p} \rangle$ where $V^{\neg p}(p) = \emptyset$. These models can be shown in graphic forms as follows.

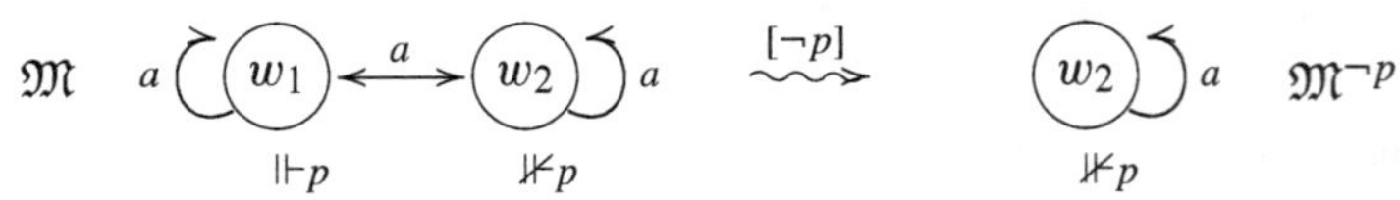

In $\mathfrak{M}$, agent a does not know whether p or $\neg p$ (i.e., $\neg(\mathsf{K}_a p \vee \mathsf{K}_a \neg p)$ is valid in $\mathfrak{M}$), but after announcement of $\neg p$, agent a comes to know $\neg p$ in the restricted model $\mathfrak{M}$ to $\neg p$.

7.2.2 Hilbert-Style Axiomatization of PAL

Hilbert-style axiomatization, **HPAL**, is defined in Table 7.1 below, and it includes some axioms with announcement operators as additional axioms to the axiomatization of **K**. These five additional axioms (from (RA1) to (RA5) are called *reduction axioms* (or sometimes, *recursion axioms*). They exist for reducing each of the theorems of **HPAL** into a theorem of modal logic **K**. The previous work [12] has shown the completeness theorem of **HPAL**.

Fact 1 (Completeness of PAL) *For any formula A, A is valid in all Kripke models iff A is provable in* **HPAL**.

Proof (Outline) In the case of the soundness theorem, it suffices to show validity of **HPAL**'s reduction axioms, which is straightforward. For the case of the completeness theorem, following [15, pp.186-7], the translation function t is defined as follows.

$$
\begin{array}{ll}
t(p) = p & t([A]p) = t(A \to p) \\
t(\neg p) = \neg t(p) & t([A]B \to C) = t([A]B \to [A]C) \\
t(A \to B) = t(A) \to t(B) & t([A]\mathsf{K}_a B) = t(A \to \mathsf{K}_a[A]B) \\
t(\mathsf{K}_a A) = \mathsf{K}_a t(A) & t([A][B]C) = t([A \wedge [A]B]C)
\end{array}
$$

Here the underlying idea of this translation is that, with the help of reduction axioms, we can push each of the outermost occurrences of the announcement operator to a propositional variable up to equivalence. Then, suppose that A is valid on all Kripke models. Since $t(A) \leftrightarrow A$ is valid on all models, we obtain $t(A)$ is valid on all models. Since the Hilbert-style axiomatization of **K** is complete with respect to all Kripke models, $t(A)$ is provable in the Hilbert-style axiomatization **K**, hence also in **HPAL**. Note that $t(A) \leftrightarrow A$ is provable in **HPAL**, we conclude that A is provable in **HPAL**. □

Table 7.1 Hilbert-style axiomatization of PAL: **HPAL**

Modal axioms
All instantiations of propositional tautologies
(K) $\mathsf{K}_a(A \to B) \to (\mathsf{K}_a A \to \mathsf{K}_a B)$
Reduction axioms
(RA1) $[A]p \leftrightarrow (A \to p)$
(RA2) $[A](B \to C) \leftrightarrow ([A]B \to [A]C)$
(RA3) $[A]\neg B \leftrightarrow (A \to \neg[A]B)$
(RA4) $[A]\mathsf{K}_a B \leftrightarrow (A \to \mathsf{K}_a[A]B)$
(RA5) $[A][B]C \leftrightarrow [A \wedge [A]B]C$
Inference rules
(MP) From A and $A \to B$, infer B
(Nec) From A, infer $\mathsf{K}_a A$

7.3 Sequent Calculus for PAL

As we have mentioned in the introduction, a labelled sequent calculus called **G3PAL** has been provided by [7] based on G3-style sequent calculus (or simply, G3-style) for modal logic **K**.[2]

7.3.1 G3PAL

In order to introduce **G3PAL**, as in [7], it is better to explicitly confirm the satisfaction relation with a list of formulas, that restricts a Kripke model, since the following inference rules of **G3PAL** are all obtained from those satisfaction relations. We denote finite lists $(A_1, A_2, \ldots, A_n)$ of formulas by α, β, etc., and do the empty list by ϵ from here and after. As an abbreviation, for any list $\alpha = (A_1, A_2, \ldots, A_n)$ of formulas, we define $\mathfrak{M}^\alpha$ inductively as: $\mathfrak{M}^\alpha := \mathfrak{M}$ (if $\alpha = \epsilon$), and $\mathfrak{M}^\alpha := (\mathfrak{M}^\beta)^{A_n} = \langle W^{\beta, A_n}, (R_a^{\beta, A_n})_{a \in \mathsf{G}}, V^{\beta, A_n} \rangle$ (if $\alpha = \beta, A_n$). We may also denote $(\mathfrak{M}^\beta)^{A_n}$ by $\mathfrak{M}^{\beta, A_n}$ for simplicity. The satisfaction relation with restricting formulas is shown explicitly as follows:

$$
\begin{array}{ll}
\mathfrak{M}^{\alpha, A}, w \Vdash p & \text{iff } \mathfrak{M}^\alpha, w \Vdash A \text{ and } \mathfrak{M}^\alpha, w \Vdash p, \\
\mathfrak{M}^\alpha, w \Vdash \neg A & \text{iff } \mathfrak{M}^\alpha, w \nVdash A, \\
\mathfrak{M}^\alpha, w \Vdash A \to B & \text{iff } \mathfrak{M}^\alpha, w \Vdash A \text{ implies } \mathfrak{M}^\alpha, w \Vdash B, \\
\mathfrak{M}^\alpha, w \Vdash \mathsf{K}_a A & \text{iff for all } v \in W : w R_a^\alpha v \text{ implies } \mathfrak{M}^\alpha, v \Vdash A, \text{ and} \\
\mathfrak{M}^\alpha, w \Vdash [A]B & \text{iff } \mathfrak{M}^\alpha, w \Vdash A \text{ implies } \mathfrak{M}^{\alpha, A}, w \Vdash B,
\end{array}
$$

where $p \in \mathsf{Prop}$, $A, B \in \mathsf{Form}$, $\mathfrak{M}$ is any Kripke model, $w \in \mathcal{D}(\mathfrak{M})$, and α is any list of formulas. According to the Kripke semantics defined in Sect. 7.2, $\langle w, v \rangle \in R_a^{\alpha, A}$ is equivalent to the following conjunction:

$$\langle w, v \rangle \in R_a^{\alpha, A} \textit{ iff } \langle w, v \rangle \in R_a^\alpha \text{ and } \mathfrak{M}^\alpha, w \Vdash A \text{ and } \mathfrak{M}^\alpha, v \Vdash A.$$

A point to notice here is that from an accessibility relation with restricting formulas, we may obtain three conjuncts.

Now we will introduce **G3PAL**. Let $\mathsf{Var} = \{x, y, z, \ldots\}$ be a countably infinite set of variables. Then, given any $x, y \in \mathsf{Var}$, any list of formulas α and any formula A, we say $x{:}^\alpha A$ is a *labelled formula*, and that, for any agent $a \in \mathsf{G}$, $x \mathsf{R}_a^\alpha y$ is a *relational atom*. Intuitively, the labelled formula $x{:}^\alpha A$ corresponds to '$\mathfrak{M}^\alpha, x \Vdash A$'

---

[2] G3-style sequent calculus for modal logic **K** named **G3K** has been introduced in Negri [8]. And G3-style sequent calculus is a calculus that does not have any structural rules and the most outstanding feature of this calculus is that the contraction rules are admissible. The specific introduction of G3-style sequent calculus (or G3-system) itself can be found in Negri and Plato [9] and Troelstra and Schwichtenberg [13].

Table 7.2 G3PAL

(Initial Sequent)

$$x{:}^{\epsilon} p, \Gamma \Rightarrow \Delta, x{:}^{\epsilon} p$$

(Rules for propositional connectives)

$$\frac{}{x{:}^{\alpha}\bot, \Gamma \Rightarrow \Delta}\ (L\bot)$$

$$\frac{\Gamma \Rightarrow \Delta, x{:}^{\alpha} A}{x{:}^{\alpha}\neg A, \Gamma \Rightarrow \Delta}\ (L\neg) \qquad \frac{x{:}^{\alpha} A, \Gamma \Rightarrow \Delta}{\Gamma \Rightarrow \Delta, x{:}^{\alpha}\neg A}\ (R\neg)$$

$$\frac{\Gamma \Rightarrow \Delta, x{:}^{\alpha} A \quad x{:}^{\alpha} B, \Gamma \Rightarrow \Delta}{x{:}^{\alpha} A \to B, \Gamma \Rightarrow \Delta}\ (L\to) \qquad \frac{x{:}^{\alpha} A, \Gamma \Rightarrow \Delta, x{:}^{\alpha} B}{\Gamma \Rightarrow \Delta, x{:}^{\alpha} A \to B}\ (R\to)$$

(Rules for knowledge operators)

$$\frac{y{:}^{\alpha} A, x{:}^{\alpha}\mathsf{K}_a A, x\mathsf{R}_a^{\alpha} y, \Gamma \Rightarrow \Delta}{x{:}^{\alpha}\mathsf{K}_a A, x\mathsf{R}_a^{\alpha} y, \Gamma \Rightarrow \Delta}\ (L\mathsf{K}_a) \qquad \frac{x\mathsf{R}_a^{\alpha} y, \Gamma \Rightarrow \Delta, y{:}^{\alpha} A}{\Gamma \Rightarrow \Delta, x{:}^{\alpha}\mathsf{K}_a A}\ (R\mathsf{K}_a)\dagger$$

† y does not appear in the lower sequent.

(Rules for PAL)

$$\frac{x{:}^{\alpha} A, x{:}^{\alpha} p, \Gamma \Rightarrow \Delta}{x{:}^{\alpha,A} p, \Gamma \Rightarrow \Delta}\ (Lat) \qquad \frac{\Gamma \Rightarrow \Delta, x{:}^{\alpha} A \quad \Gamma \Rightarrow \Delta, x{:}^{\alpha} p}{\Gamma \Rightarrow \Delta, x{:}^{\alpha,A} p}\ (Rat)$$

$$\frac{x{:}^{\alpha,A} B, x{:}^{\alpha}[A]B, x{:}^{\alpha} A, \Gamma \Rightarrow \Delta}{x{:}^{\alpha}[A]B, x{:}^{\alpha} A, \Gamma \Rightarrow \Delta}\ (L[.]) \qquad \frac{x{:}^{\alpha} A, \Gamma \Rightarrow \Delta, x{:}^{\alpha,A} B}{\Gamma \Rightarrow \Delta, x{:}^{\alpha}[A]B}\ (R[.])$$

$$\frac{x{:}^{\alpha,A,B} C, \Gamma \Rightarrow \Delta}{x{:}^{\alpha,A\wedge[A]B} C, \Gamma \Rightarrow \Delta}\ (L_{cmp}) \qquad \frac{\Gamma \Rightarrow \Delta, x{:}^{\alpha,A,B} C}{\Gamma \Rightarrow \Delta, x{:}^{\alpha,A\wedge[A]B} C}\ (R_{cmp})$$

and is to read 'after a sequence α of public announcements, x still survives[3] and A holds at x', and the relational atom $x\mathsf{R}_a^{\alpha} y$ is to read 'after a sequence α of public announcements both x and y survive and we can still access from x to y'. We also use the term, *labelled expressions* to indicate that they are either labelled formulas or relational atoms, and we denote them by $\mathfrak{A}$, $\mathfrak{B}$, etc. A *sequent* $\Gamma \Rightarrow \Delta$ is a pair of finite multisets of labelled expressions. The set of inference rules of **G3PAL** is given in Table 7.2. Hereinafter, for any sequent $\Gamma \Rightarrow \Delta$, if $\Gamma \Rightarrow \Delta$ is provable in **G3PAL**, we write **G3PAL** $\vdash \Gamma \Rightarrow \Delta$. The rules of (Lat) and (Rat) are obtained from the above satisfaction relation, hence if there is an announcement A and a propositional variable p, we get p with the restricting formula A. In the case of $(L[.])$ and $(R[.])$, although the satisfaction relation of the announcement operator is the

[3]The notion of *surviveness* will be referred in Sect. 7.5 more specifically.

same as that of implication only with the exception of restricting formulas, the rules, $(L[.])$ and $(R[.])$, are (probably) modified for G3-style. The last two rules (L_{cmp}) and (R_{cmp}) are for dealing with the proof of (RA5) of **HPAL** (we will discuss them shortly afterwards). Other inference rules result naturally from the semantics. As we have referred in the previous paragraph, while we could have sound inference rules corresponding to restricted relational atoms, there is, actually, no rule of relational atoms in **G3PAL**, and due to this fact, **G3PAL** may not have an ability to prove one of the reduction axioms, (RA4).

7.3.2 Problems of G3PAL

Maffezioli and Negri stated, in Sect. 7.5 of [7], that **G3PAL** may prove all inference rules and axioms of **HPAL**, namely if **HPAL** $\vdash A$, then **G3PAL** $\vdash \Rightarrow x{:}^{\epsilon} A$ (for any A and x). Nevertheless, there are, in fact, some problems in proving (RA4):

$$[A]\mathsf{K}_a B \leftrightarrow (A \rightarrow \mathsf{K}_a[A]B).$$

This axiom seemingly cannot be proven in **G3PAL**. Let us look at possible but plausible attempts to derive both directions of (RA4). First, a possible attempt of deriving the direction from right to left is given as follows:

$$
\dfrac{\dfrac{\dfrac{\begin{array}{c}\vdots\ \mathcal{D}_1\\ x{:}^{\epsilon}A \Rightarrow x{:}^{\epsilon}A, x{:}^{A}\mathsf{K}_a B\end{array}\qquad \dfrac{\begin{array}{c}\vdots\ ?\\ x{:}^{\epsilon}A, x{:}^{\epsilon}\mathsf{K}_a[A]B, x\mathsf{R}^{A}_{a}y \Rightarrow y{:}^{A}B\end{array}}{x{:}^{\epsilon}A, x{:}^{\epsilon}\mathsf{K}_a[A]B \Rightarrow x{:}^{A}\mathsf{K}_a B}\,(R\mathsf{K}_a)}{x{:}^{\epsilon}A, x{:}^{\epsilon}A \rightarrow \mathsf{K}_a[A]B \Rightarrow x{:}^{A}\mathsf{K}_a B}\,(L\rightarrow)}{x{:}^{\epsilon}A \rightarrow \mathsf{K}_a[A]B \Rightarrow x{:}^{\epsilon}[A]\mathsf{K}_a B}\,(R[.])}{\Rightarrow x{:}^{\epsilon}(A \rightarrow \mathsf{K}_a[A]B) \rightarrow [A]\mathsf{K}_a B}\,(R\rightarrow) \qquad (*)
$$

Starting from the bottom sequent, the bottom sequent of $\mathcal{D}_1$ is clearly derivable, but it is difficult to find the way to go step forward from the right uppermost sequent of the derivation. The problem here is that A in $x\mathsf{R}^{A}_{a}y$ and ϵ in $x{:}^{\epsilon}\mathsf{K}_a[A]B$ on the left side of the sequent do not match, and therefore we cannot apply the rule $(L\mathsf{K}_a)$.

Second, the other direction of (RA4) also seemingly cannot be proven by **G3PAL**. A possible attempt to derive it may be as follows:

$$\dfrac{\dfrac{\dfrac{\dfrac{\dfrac{\dfrac{\vdots\ ?}{y{:}^{\epsilon}A, x\mathsf{R}^{\epsilon}_{a}y, x{:}^{A}\mathsf{K}_{a}B, x{:}^{\epsilon}A, x{:}^{\epsilon}[A]\mathsf{K}_{a}B \Rightarrow y{:}^{A}B}}{x\mathsf{R}^{\epsilon}_{a}y, x{:}^{A}\mathsf{K}_{a}B, x{:}^{\epsilon}A, x{:}^{\epsilon}[A]\mathsf{K}_{a}B \Rightarrow y{:}^{\epsilon}[A]B}\,(R[.])}{x{:}^{A}\mathsf{K}_{a}B, x{:}^{\epsilon}A, x{:}^{\epsilon}[A]\mathsf{K}_{a}B \Rightarrow x{:}^{\epsilon}\mathsf{K}_{a}[A]B}\,(R\mathsf{K}_{a})}{x{:}^{\epsilon}A, x{:}^{\epsilon}[A]\mathsf{K}_{a}B \Rightarrow x{:}^{\epsilon}\mathsf{K}_{a}[A]B}\,(L[.])}{x{:}^{\epsilon}[A]\mathsf{K}_{a}B \Rightarrow x{:}^{\epsilon}A \to \mathsf{K}_{a}[A]B}\,(R\to)}{\Rightarrow x{:}^{\epsilon}[A]\mathsf{K}_{a}B \to (A \to \mathsf{K}_{a}[A]B)}\,(R\to) \qquad (**)$$

The derivation also comes to a dead end (in fact, the rule $(L[.])$ is applicable infinitely many times, but no new labelled expression is obtained by the application). The problem here is also that ϵ in $x\mathsf{R}^{\epsilon}_{a}y$ and A in $x{:}^{A}\mathsf{K}_{a}B$ on the left side of the left uppermost sequent do not match, and again the rule $(L\mathsf{K}_{a})$ cannot be applied.

In brief, for applying the rule $(L\mathsf{K}_{a})$, α in $x\mathsf{R}^{\alpha}_{a}y$, and β in $x{:}^{\beta}\mathsf{K}_{a}B$ must be the same and $(L\mathsf{K}_{a})$ is indispensable for proving both directions of (RA4); however, there seems no way to make them equal in **G3PAL**. To settle the problems, we introduce rules for relational atoms for decomposing $x\mathsf{R}^{A}_{a}y$ into $x\mathsf{R}^{\epsilon}_{a}y$ and related labelled formulas.

7.4 Revising G3PAL

In this section, we revise **G3PAL** to make it possible to cope with (RA4) of **HPAL**. Let us examine the problem of $(*)$ first. To overcome the dead end of the derivation, we introduce rules of the relational atom with a list of formulas, i.e., $(Lrel_{a}1)$, $(Lrel_{a}2)$, $(Lrel_{a}3)$ and $(Rrel_{a})$, and it is not trivial if these rules are derivable in **G3PAL**. Here are our additional rules:

$$\dfrac{x{:}^{\alpha}A, \Gamma \Rightarrow \Delta}{x\mathsf{R}^{\alpha,A}_{a}y, \Gamma \Rightarrow \Delta}\,(Lrel_{a}1) \qquad \dfrac{y{:}^{\alpha}A, \Gamma \Rightarrow \Delta}{x\mathsf{R}^{\alpha,A}_{a}y, \Gamma \Rightarrow \Delta}\,(Lrel_{a}2) \qquad \dfrac{x\mathsf{R}^{\alpha}_{a}y, \Gamma \Rightarrow \Delta}{x\mathsf{R}^{\alpha,A}_{a}y, \Gamma \Rightarrow \Delta}\,(Lrel_{a}3)$$

$$\dfrac{\Gamma \Rightarrow \Delta, x{:}^{\alpha}A \quad \Gamma \Rightarrow \Delta, y{:}^{\alpha}A \quad \Gamma \Rightarrow \Delta, x\mathsf{R}^{\alpha}_{a}y}{\Gamma \Rightarrow \Delta, x\mathsf{R}^{\alpha,A}_{a}y}\,(Rrel_{a})$$

These inference rules are obtained in PAL's Kripke semantics. Namely, as we have already seen in Sect. 7.3.1, any restricted accessibility relation $wR^{\alpha,A}_{a}v$ is equivalent to the conjunction of the following three conjuncts such as: $wR^{\alpha}_{a}v$, $\mathfrak{M}^{\alpha}, w \Vdash A$ and $\mathfrak{M}^{\alpha}, v \Vdash A$. These three conjuncts correspond to three $(Lrel_{a}i)$ rules and three uppersequents of $(Rrel_{a})$. If we use $(Lrel_{a}3)$ to the dead end of $(*)$, $x\mathsf{R}^{\epsilon}_{a}y$ which we desire is obtained and it is obvious that the new emerged sequent is provable.

However, in the case of (∗∗), the additional inference rules are not sufficient to make the branch reach initial sequent(s). This is because the new rules could not be applied to $x\mathsf{R}^{\epsilon}y$ and they will not change the situation. To settle the problem, we reformulate the rule of $(L\mathsf{K}_a)$ in a semantically natural way. Our reformulated rule $(L\mathsf{K}'_a)$ is then defined as follows.

$$\frac{\Gamma \Rightarrow \Delta, x\mathsf{R}^{\alpha}_{a}y \quad y{:}^{\alpha}A, \Gamma \Rightarrow \Delta}{x{:}^{\alpha}\mathsf{K}_a A, \Gamma \Rightarrow \Delta}\ (L\mathsf{K}'_a)$$

It is necessary to note that, by this change of the rule, we need to depart from G3-style.[4] Although a solution with keeping G3-style might be a better solution than ours, we choose the semantically natural way to reformulate the rule $(L\mathsf{K}_a)$ first, and at the same time we reformulate the rule $(L[.])$ in a natural form.

7.4.1 Revised Sequent Calculus GPAL

Now, we introduce our revised calculus, **GPAL**. The definition of **GPAL** is presented in Table 7.3. For drawing simpler derivations, we prepare the following lemma.

Lemma 1 *For any labelled expression* $\mathfrak{A}$ *and multisets of labelled expressions* Γ *and* Δ, **GPAL** $\vdash \mathfrak{A}, \Gamma \Rightarrow \Delta, \mathfrak{A}$.

Proof It is obvious by applying (Rw) and/or (Lw) a finite number of times. □

Let us now show the derivations of (RA4) of **HPAL**.

Proposition 1 **GPAL** $\vdash \Rightarrow x{:}^{\epsilon}[A]\mathsf{K}_a B \leftrightarrow (A \rightarrow \mathsf{K}_a[A]B)$

Proof We may find a derivation of $x{:}^{\epsilon}[A]\mathsf{K}_a B \rightarrow (A \rightarrow \mathsf{K}_a[A]B)$ in **GPAL** as follows:

$$\dfrac{\dfrac{\text{Lemma 1}}{x{:}^{\epsilon}A \Rightarrow x{:}^{\epsilon}A, x{:}^{\epsilon}\mathsf{K}_a[A]B} \quad \dfrac{\dfrac{\dfrac{\begin{array}{c}\vdots\ \mathcal{D}_1\\ x{:}^{\epsilon}A, y{:}^{\epsilon}A, x\mathsf{R}^{\epsilon}_{a}y \Rightarrow y{:}^{A}B, x\mathsf{R}^{A}_{a}y\end{array} \quad \dfrac{\text{Lemma 1}}{y{:}^{A}B, x{:}^{\epsilon}A, y{:}^{\epsilon}A, x\mathsf{R}^{\epsilon}_{a}y \Rightarrow y{:}^{A}B}}{x{:}^{\epsilon}A, y{:}^{\epsilon}A, x{:}^{A}\mathsf{K}_a B, x\mathsf{R}^{\epsilon}_{a}y \Rightarrow y{:}^{A}B}\,(L\mathsf{K}'_a)}{x{:}^{\epsilon}A, x{:}^{A}\mathsf{K}_a B, x\mathsf{R}^{\epsilon}_{a}y \Rightarrow y{:}^{\epsilon}[A]B}\,(R[.])}{x{:}^{\epsilon}A, x{:}^{A}\mathsf{K}_a B \Rightarrow x{:}^{\epsilon}\mathsf{K}_a[A]B}\,(R\mathsf{K}_a)}{\dfrac{\dfrac{x{:}^{\epsilon}A, x{:}^{\epsilon}[A]\mathsf{K}_a B \Rightarrow x{:}^{\epsilon}\mathsf{K}_a[A]B}{x{:}^{\epsilon}[A]\mathsf{K}_a B \Rightarrow x{:}^{\epsilon}A \rightarrow \mathsf{K}_a[A]B}\,(R\rightarrow)}{\Rightarrow x{:}^{\epsilon}[A]\mathsf{K}_a B \rightarrow (A \rightarrow \mathsf{K}_a[A]B)}\,(R\rightarrow),}\,(L[.]')$$

---

[4] Of course, there might still exist a possibility to keep G3-style with the additional rules for relational atoms.

Table 7.3 Gentzen-style Sequent Calculus **GPAL**

(Initial Sequents)

$$x{:}^{\alpha} A \Rightarrow x{:}^{\alpha} A \quad x\mathsf{R}_a^{\alpha} v \Rightarrow x\mathsf{R}_a^{\alpha} v$$

(Structural Rules)

$$\frac{\Gamma \Rightarrow \Delta}{\mathfrak{A}, \Gamma \Rightarrow \Delta}\ (Lw) \quad \frac{\Gamma \Rightarrow \Delta}{\Gamma \Rightarrow \Delta, \mathfrak{A}}\ (Rw)$$

$$\frac{\mathfrak{A}, \mathfrak{A}, \Gamma \Rightarrow \Delta}{\mathfrak{A}, \Gamma \Rightarrow \Delta}\ (Lc) \quad \frac{\Gamma \Rightarrow \Delta, \mathfrak{A}, \mathfrak{A}}{\Gamma \Rightarrow \Delta, \mathfrak{A}}\ (Rc)$$

(Rules for propositional connectives)

$$\frac{\Gamma \Rightarrow \Delta, x{:}^{\alpha} A}{x{:}^{\alpha} \neg A, \Gamma \Rightarrow \Delta}\ (L\neg) \quad \frac{x{:}^{\alpha} A, \Gamma \Rightarrow \Delta}{\Gamma \Rightarrow \Delta, x{:}^{\alpha} \neg A}\ (R\neg)$$

$$\frac{\Gamma \Rightarrow \Delta, x{:}^{\alpha} A \quad x{:}^{\alpha} B, \Gamma \Rightarrow \Delta}{x{:}^{\alpha} A \to B, \Gamma \Rightarrow \Delta}\ (L\to) \quad \frac{x{:}^{\alpha} A, \Gamma \Rightarrow \Delta, x{:}^{\alpha} B}{\Gamma \Rightarrow \Delta, x{:}^{\alpha} A \to B}\ (R\to)$$

(Rules for knowledge operators)

$$\frac{\Gamma \Rightarrow \Delta, x\mathsf{R}_a^{\alpha} y \quad y{:}^{\alpha} A, \Gamma \Rightarrow \Delta}{x{:}^{\alpha} \mathsf{K}_a A, \Gamma \Rightarrow \Delta}\ (L\mathsf{K}_a') \quad \frac{x\mathsf{R}_a^{\alpha} y, \Gamma \Rightarrow \Delta, y{:}^{\alpha} A}{\Gamma \Rightarrow \Delta, x{:}^{\alpha} \mathsf{K}_a A}\ (R\mathsf{K}_a)\dagger$$

† y does not appear in the lower sequent.

(Rules for PAL)

$$\frac{x{:}^{\alpha} p, \Gamma \Rightarrow \Delta}{x{:}^{\alpha, A} p, \Gamma \Rightarrow \Delta}\ (Lat') \quad \frac{\Gamma \Rightarrow \Delta, x{:}^{\alpha} p}{\Gamma \Rightarrow \Delta, x{:}^{\alpha, A} p}\ (Rat')$$

$$\frac{\Gamma \Rightarrow \Delta, x{:}^{\alpha} A \quad x{:}^{\alpha, A} B, \Gamma \Rightarrow \Delta}{x{:}^{\alpha} [A]B, \Gamma \Rightarrow \Delta}\ (L[.]') \quad \frac{x{:}^{\alpha} A, \Gamma \Rightarrow \Delta, x{:}^{\alpha, A} B}{\Gamma \Rightarrow \Delta, x{:}^{\alpha} [A]B}\ (R[.])$$

$$\frac{x{:}^{\alpha} A, \Gamma \Rightarrow \Delta}{x\mathsf{R}_a^{\alpha, A} y, \Gamma \Rightarrow \Delta}\ (Lrel_a1) \quad \frac{y{:}^{\alpha} A, \Gamma \Rightarrow \Delta}{x\mathsf{R}_a^{\alpha, A} y, \Gamma \Rightarrow \Delta}\ (Lrel_a2) \quad \frac{x\mathsf{R}_a^{\alpha} y, \Gamma \Rightarrow \Delta}{x\mathsf{R}_a^{\alpha, A} y, \Gamma \Rightarrow \Delta}\ (Lrel_a3)$$

$$\frac{\Gamma \Rightarrow \Delta, x{:}^{\alpha} A \quad \Gamma \Rightarrow \Delta, y{:}^{\alpha} A \quad \Gamma \Rightarrow \Delta, x\mathsf{R}_a^{\alpha} y}{\Gamma \Rightarrow \Delta, x\mathsf{R}_a^{\alpha, A} y}\ (Rrel_a)$$

where the derivation $\mathcal{D}_1$ is given as follows:

$$\frac{\dfrac{\text{Lemma 1}}{x{:}^{\epsilon} A, y{:}^{\epsilon} A, x\mathsf{R}_a^{\epsilon} y \Rightarrow y{:}^{A} B, x{:}^{\epsilon} A} \quad \dfrac{\text{Lemma 1}}{x{:}^{\epsilon} A, y{:}^{\epsilon} A, x\mathsf{R}_a^{\epsilon} y \Rightarrow y{:}^{A} B, y{:}^{\epsilon} A} \quad \dfrac{\text{Lemma 1}}{x{:}^{\epsilon} A, y{:}^{\epsilon} A, x\mathsf{R}_a^{\epsilon} y \Rightarrow y{:}^{A} B, x\mathsf{R}_a^{\epsilon} y}}{x{:}^{\epsilon} A, y{:}^{\epsilon} A, x\mathsf{R}_a^{\epsilon} y \Rightarrow y{:}^{A} B, x\mathsf{R}_a^{A} y}\ (Rrel).$$

We may also find a derivation of $x{:}^{\epsilon} (A \to \mathsf{K}_a [A]B) \to [A]\mathsf{K}_a B$ in **GPAL** as follows:

$$
\dfrac{
\dfrac{\text{Lemma 1}}{x{:}^{\epsilon} A \Rightarrow x{:}^{A}\mathsf{K}_a B, x{:}^{\epsilon} A}
\quad
\dfrac{\dfrac{\dfrac{
\dfrac{\dfrac{\text{Lemma 1}}{x\mathsf{R}^{\epsilon}_a y \Rightarrow y{:}^{A} B, x\mathsf{R}^{\epsilon}_a y}}{x\mathsf{R}^{A}_a y \Rightarrow y{:}^{A} B, x\mathsf{R}^{\epsilon}_a y}\,(Lrel_a3)
\quad
\dfrac{\dfrac{\dfrac{\text{Lemma 1}}{y{:}^{\epsilon} A \Rightarrow y{:}^{A} B, y{:}^{\epsilon} A}}{x\mathsf{R}^{A}_a y \Rightarrow y{:}^{A} B, y{:}^{\epsilon} A}\,(Lrel_a2)
\quad
\dfrac{\text{Lemma 1}}{y{:}^{A} B, x\mathsf{R}^{A}_a y \Rightarrow y{:}^{A} B}}{y{:}^{\epsilon}[A]B, x\mathsf{R}^{A}_a y \Rightarrow y{:}^{A} B}\,(L[.]')
}{x\mathsf{R}^{A}_a y, x{:}^{\epsilon}\mathsf{K}_a[A]B \Rightarrow y{:}^{A} B}\,(L\mathsf{K}'_a)
}{x{:}^{\epsilon}\mathsf{K}_a[A]B \Rightarrow x{:}^{A}\mathsf{K}_a B}\,(R\mathsf{K}_a)
}{x{:}^{\epsilon}\mathsf{K}_a[A]B, x{:}^{\epsilon} A \Rightarrow x{:}^{A}\mathsf{K}_a B}\,(Lw)
}{
\dfrac{\dfrac{x{:}^{\epsilon} A, x{:}^{\epsilon} A \to \mathsf{K}_a[A]B \Rightarrow x{:}^{A}\mathsf{K}_a B}{x{:}^{\epsilon} A \to \mathsf{K}_a[A]B \Rightarrow x{:}^{\epsilon}[A]\mathsf{K}_a B}\,(R[.])}{\Rightarrow (x{:}^{\epsilon} A \to \mathsf{K}_a[A]B) \to [A]\mathsf{K}_a B}\,(R\to)
}\,(L\to)
$$

□

As we can see above, the proof of (RA4) in **GPAL** can be done thanks to the rules of relational atoms.

Moreover, **GPAL**$^+$ is defined to be **GPAL** with the following rule (Cut),

$$
\dfrac{\Gamma \Rightarrow \Delta, \mathfrak{A} \quad \mathfrak{A}, \Gamma' \Rightarrow \Delta'}{\Gamma, \Gamma' \Rightarrow \Delta, \Delta'}\ (Cut).
$$

$\mathfrak{A}$ in (Cut) is called a *cut expression*, and we say that a labelled expression $\mathfrak{A}$ is a *principal expression* of an inference rule of **GPAL**$^+$ if $\mathfrak{A}$ is newly introduced on the left uppersequent or the right uppersequent by the rule of **GPAL**$^+$.

Let us briefly summarize our revised calculus in order. **GPAL** is different from **G3PAL** in respect to the following features:

1. **GPAL** is based on Gentzen's standard sequent calculus [4] but not in G3-style, and so it contains structural rules.
2. **GPAL** includes rules for relational atoms which **G3PAL** lacks.
3. $(L[.])$ and $(L\mathsf{K}_a)$ are redefined in a semantically natural way, and each of them is denoted by $(L[.]')$ and $(L\mathsf{K}'_a)$ in **GPAL**.
4. **GPAL** does not contain (L_{cmp}) and (R_{cmp}) of **G3PAL**, but without them it can prove (RA5). These rules are also derivable in **GPAL**$^+$ (see Proposition 2).
5. (Lat) and (Rat) are redefined taking into account of the notion of surviveness, and each of them is denoted by (Lat') and (Rat') in **GPAL**.

The last two features have not been mentioned so far, and the last feature of **GPAL** will be considered at the beginning of Sect. 7.6. In this paragraph, we focus on feature 4. According to [7], the following rules

$$
\dfrac{x{:}^{\alpha, A, B} C, \Gamma \Rightarrow \Delta}{x{:}^{\alpha, A \wedge [A]B} C, \Gamma \Rightarrow \Delta}\ (L_{cmp}) \qquad \dfrac{\Gamma \Rightarrow \Delta, x{:}^{\alpha, A, B} C}{\Gamma \Rightarrow \Delta, x{:}^{\alpha, A \wedge [A]B} C}\ (R_{cmp})
$$

are required to prove (RA5) of **HPAL**:

$$
[A][B]C \leftrightarrow [A \wedge [A]B]C.
$$

In what follows, however, we reveal that these rules of (L_{cmp}) and (R_{cmp}) are not necessary in the set of inference rules of **GPAL**. Let us see the details. First, let us define the length of a labelled expression $\mathfrak{A}$.

Definition 3 For any formula A, $\mathsf{len}(A)$ is equal to the number of the propositional variables and the logical connectives in A.

$$\mathsf{len}(\alpha) = \begin{cases} 0 & \text{if } \alpha = \epsilon \\ \mathsf{len}(\beta) + \mathsf{len}(A) & \text{if } \alpha = \beta, A \end{cases}$$

$$\mathsf{len}(\mathfrak{A}) = \begin{cases} \mathsf{len}(\alpha) + \mathsf{len}(A) & \text{if } \mathfrak{A} = x{:}^{\alpha} A \\ \mathsf{len}(\alpha) + 1 & \text{if } \mathfrak{A} = x\mathsf{R}_a^{\alpha} y \end{cases}$$

Then, let us show the following lemma.

Lemma 2 *For any $A, B \in \mathsf{Form}$, $x, y \in \mathsf{Var}$ and for any list α, β of formulas,*

(i) **GPAL** $\vdash x{:}^{\alpha, A, B, \beta} C \Rightarrow x{:}^{\alpha, A \wedge [A]B, \beta} C$,
(ii) **GPAL** $\vdash x{:}^{\alpha, A \wedge [A]B, \beta} C \Rightarrow x{:}^{\alpha, A, B, \beta} C$,
(iii) **GPAL** $\vdash x\mathsf{R}_a^{\alpha, A, B, \beta} y \Rightarrow x\mathsf{R}_a^{\alpha, (A \wedge [A]B), \beta} y$,
(iv) **GPAL** $\vdash x\mathsf{R}_a^{\alpha, (A \wedge [A]B), \beta} y \Rightarrow x\mathsf{R}_a^{\alpha, A, B, \beta} y$.

Proof The proofs of (i), (ii), (iii), and (iv) are done simultaneously by double induction on C and β. We only see the case where C is of the form $\mathsf{K}_a D$ and the case where C is of the form $[D]E$, because the provability of the other sequents (ii), (iii) and (iv) can also be shown similarly. First, let us consider the case where C is of the form $\mathsf{K}_a D$. Let γ be (α, A, B, β) and θ be $(\alpha, A \wedge [A]B, \beta)$.

$$\dfrac{\dfrac{\dfrac{\vdots\ \mathcal{D}_1}{x\mathsf{R}_a^{\theta} y \Rightarrow x\mathsf{R}_a^{\gamma} y}}{x\mathsf{R}_a^{\theta} y \Rightarrow y{:}^{\theta} D, x\mathsf{R}_a^{\gamma} y}\,(Rw) \quad \dfrac{\dfrac{\vdots\ \mathcal{D}_2}{y{:}^{\gamma} D \Rightarrow y{:}^{\theta} D}}{y{:}^{\gamma} D, x\mathsf{R}_a^{\theta} y \Rightarrow y{:}^{\theta} D}\,(Lw)}{\dfrac{x{:}^{\gamma}\mathsf{K}_a D, x\mathsf{R}_a^{\theta} y \Rightarrow y{:}^{\theta} D}{x{:}^{\gamma}\mathsf{K}_a D \Rightarrow x{:}^{\theta}\mathsf{K}_a D}\,(R\mathsf{K}_a)}\,(L\mathsf{K}_a)$$

Both $\mathcal{D}_1$ and $\mathcal{D}_2$ are obtained by induction hypothesis, since the length of the labelled expressions is reduced. We may need to pay attention to the length of the labelled expression at the bottom sequent of $\mathcal{D}_1$, but according to Definition 3, $\mathsf{len}(x{:}^{\gamma}\mathsf{K}_a D) > \mathsf{len}(x\mathsf{R}_a^{\gamma} y)$ (for any γ).

Second, let us consider the case where C is of the form $[D]E$. Let γ be (α, A, B, β) and θ be $(\alpha, A \wedge [A]B, \beta)$.

$$\dfrac{\dfrac{\dfrac{\vdots\ \mathcal{D}_3}{x{:}^{\theta} D \Rightarrow x{:}^{\gamma} D}}{x{:}^{\theta} D \Rightarrow x{:}^{\gamma} D, x{:}^{\theta,D} E}\,(Rw) \quad \dfrac{\dfrac{\vdots\ \mathcal{D}_4}{x{:}^{\gamma,D} E \Rightarrow x{:}^{\theta,D} E}}{x{:}^{\gamma,D} E, x{:}^{\theta} D \Rightarrow x{:}^{\theta,D} E}\,(Lw)}{\dfrac{x{:}^{\gamma}[D]E, x{:}^{\theta} D \Rightarrow x{:}^{\theta,D} E}{x{:}^{\gamma}[D]E \Rightarrow x{:}^{\theta}[D]E}\,(R[.])}\,(L[.]')$$

The derivations $\mathcal{D}_3$ and $\mathcal{D}_4$ are obtained by induction hypotheses. □

Now with the help of the rule (Cut), we can also show the derivability of more general rules than (L_{cmp}) and (R_{cmp}) of **G3PAL** as follows:

Proposition 2 *The following rules* (L'_{cmp}) *and* (R'_{cmp}) *are derivable in* **GPAL**$^+$.

$$\dfrac{x{:}^{\alpha,A,B,\beta} C, \Gamma \Rightarrow \Delta}{x{:}^{\alpha,A\wedge[A]B,\beta} C, \Gamma \Rightarrow \Delta}\,(L'_{cmp}) \qquad \dfrac{\Gamma \Rightarrow \Delta, x{:}^{\alpha,A,B,\beta} C}{\Gamma \Rightarrow \Delta, x{:}^{\alpha,A\wedge[A]B,\beta} C}\,(R'_{cmp})$$

where $a \in \mathsf{G}$, $A, B, C \in \mathsf{Form}$ *and* α *and* β *are arbitrary lists of formulas.*

Proof It is shown immediately from Lemma 2 and (Cut).[5] □

7.4.2 *All Theorems of* HPAL *are provable in* GPAL$^+$

We first define the substitution of variables in labelled expressions.

Definition 4 Let $\mathfrak{A}$ be any labelled expression. Then the substitution of x for y in $\mathfrak{A}$, denoted by $\mathfrak{A}[x/y]$, is defined by

$$\begin{array}{lll} z[x/y] & := z & (\text{if } y \neq z) \\ z[x/y] & := x & (\text{if } y = z) \\ (z{:}^{\alpha} A)[x/y] & := (z[x/y]){:}^{\alpha} A & \\ (z\mathsf{R}_a^{\alpha} w)[x/y] & := (z[x/y])\mathsf{R}_a^{\alpha}(w[x/y]). & \end{array}$$

Substitution $[x/y]$ to a multiset Γ of labelled expressions is defined as

$$\Gamma[x/y] := \{\mathfrak{A}[x/y] \mid \mathfrak{A} \in \Gamma\}.$$

Next, for a preparation of Theorem 1, we show the next lemma.

[5]The following rules are also derivable in **GPAL**$^+$.

$$\dfrac{x\mathsf{R}_a^{\alpha,A,B,\beta} y, \Gamma \Rightarrow \Delta}{x\mathsf{R}_a^{\alpha,(A\wedge[A]B),\beta} y, \Gamma \Rightarrow \Delta}\,(L_{cmpr}) \qquad \dfrac{\Gamma \Rightarrow \Delta, x\mathsf{R}_a^{\alpha,A,B,\beta} y}{\Gamma \Rightarrow \Delta, x\mathsf{R}_a^{\alpha,(A\wedge[A]B),\beta} y}\,(R_{cmpr})$$

Lemma 3

(i) **GPAL** $\vdash \Gamma \Rightarrow \Delta$ *implies* **GPAL** $\vdash \Gamma[x/y] \Rightarrow \Delta[x/y]$ *for any* $x, y \in \mathsf{Var}$.
(ii) **GPAL**$^+ \vdash \Gamma \Rightarrow \Delta$ *implies* **GPAL**$^+ \vdash \Gamma[x/y] \Rightarrow \Delta[x/y]$ *for any* $x, y \in \mathsf{Var}$.

Proof By induction on the height of the derivation, we go through almost the same procedure in the proof in Negri and von Plato [10, p. 194]. □

Finally, let us show the following theorem:

Theorem 1 *For any formula A, if* **HPAL** $\vdash A$, *then* **GPAL**$^+ \vdash \Rightarrow x{:}^{\epsilon} A$ *(for any x).*

Proof The proof is carried out by the height of the derivation in **HPAL**. Since the case of reduction axiom (RA4) has been shown in Proposition 1, let us prove one direction of (RA5) $[A][B]C \leftrightarrow [A \wedge [A]B]C$ of **HPAL** for one of the base cases (the derivation height of **HPAL** is equal to 0).

$$\dfrac{\dfrac{\dfrac{\dfrac{\dfrac{\dfrac{\text{Lemma 1}}{x{:}^{\epsilon} A, x{:}^{A} B \Rightarrow x{:}^{\epsilon} A, x{:}^{A,B} C} \quad \dfrac{\dfrac{\text{Lemma 1}}{x{:}^{\epsilon} A, x{:}^{A} B \Rightarrow x{:}^{A} B, x{:}^{A,B} C}}{x{:}^{\epsilon} A, x{:}^{A} B \Rightarrow x{:}^{\epsilon} [A]B, x{:}^{A,B} C}\,(R[.])}{x{:}^{\epsilon} A, x{:}^{A} B \Rightarrow x{:}^{\epsilon} A \wedge [A]B, x{:}^{A,B} C}\,(R\wedge) \quad \dfrac{\dfrac{\text{Lemma 2}}{x{:}^{A\wedge[A]B} C \Rightarrow x{:}^{A,B} C}}{x{:}^{\epsilon} A, x{:}^{A} B, x{:}^{A\wedge[A]B} C \Rightarrow x{:}^{A,B} C}\,(Lw)}{x{:}^{\epsilon} A, x{:}^{\epsilon} [A \wedge [A]B]C, x{:}^{A} B \Rightarrow x{:}^{A,B} C}\,(L[.]')}{x{:}^{\epsilon} A, x{:}^{\epsilon} [A \wedge [A]B]C \Rightarrow x{:}^{A} [B]C}\,(R[.])}{x{:}^{\epsilon} [A \wedge [A]B]C \Rightarrow x{:}^{\epsilon} [A][B]C}\,(R[.])}{\Rightarrow x{:}^{\epsilon} [A \wedge [A]B]C \rightarrow [A][B]C}\,(R\rightarrow)$$

In the inductive step, we show the inference rules, (MP) and (Nec), by **GPAL**. The former is shown with (Cut).

$$\dfrac{\dfrac{\text{Assumption}}{\Rightarrow x{:}^{\epsilon} A} \quad \dfrac{\dfrac{\text{Assumption}}{\Rightarrow x{:}^{\epsilon} A \rightarrow B} \quad \dfrac{\dfrac{\text{Lemma 1}}{x{:}^{\epsilon} A \Rightarrow x{:}^{\epsilon} B, x{:}^{\epsilon} A} \quad \dfrac{\text{Lemma 1}}{x{:}^{\epsilon} B, x{:}^{\epsilon} A \Rightarrow x{:}^{\epsilon} B}}{x{:}^{\epsilon} A \rightarrow B, x{:}^{\epsilon} A \Rightarrow x{:}^{\epsilon} B}\,(L\rightarrow)}{x{:}^{\epsilon} A \Rightarrow x{:}^{\epsilon} B}\,(Cut)}{\Rightarrow x{:}^{\epsilon} B}\,(Cut)$$

The latter is shown by $(R\mathsf{K}_a)$, (Lw) and Lemma 3. □

7.4.3 *Cut Elimination of* GPAL$^+$

Here we prove an important theorem of the paper, the (syntactic) cut elimination theorem of **GPAL**$^+$.

Theorem 2 (Cut elimination theorem of **GPAL**$^+$) *For any sequent* $\Gamma \Rightarrow \Delta$, *if* **GPAL**$^+ \vdash \Gamma \Rightarrow \Delta$, *then* **GPAL** $\vdash \Gamma \Rightarrow \Delta$.

Proof The proof is carried out using Ono and Komori's method [11] introduced in the reference [6] by Kashima where we employ the following rule $(Ecut)$ instead of the usual method of 'mix cut'. We denote the n-copies of the same labelled expression $\mathfrak{A}$ by $\mathfrak{A}^n$, and $(Ecut)$ is defined as follows:

$$\frac{\Gamma \Rightarrow \Delta, \mathfrak{A}^n \quad \mathfrak{A}^m, \Gamma' \Rightarrow \Delta'}{\Gamma, \Gamma' \Rightarrow \Delta, \Delta'} \; (Ecut)$$

where $n, m \geq 0$. The theorem is proven by double induction on the height of the derivation and the length of cut expression $\mathfrak{A}$ of $(Ecut)$. The proof is divided into four cases. In brief, (1) at least one of uppersequents of $(Ecut)$ is an initial sequent; (2) the last inference rule of either uppersequents of $(Ecut)$ is a structural rule; (3) the last inference rule of either uppersequents of $(Ecut)$ is a nonstructural rule, and the principal expression introduced by the rule is not the cut expression; and (4) the last inference rules of two uppersequents of $(Ecut)$ are both nonstructural rules, and the principal expressions introduced by the rules used on the uppersequents of $(Ecut)$ are both cut expressions. We look at one of significant subcases of (4) in which principal expressions introduced by nonstructural rules are both cut expressions.

Let us consider one of the cases (4) where both sides of $\mathfrak{A}$ are $x\mathsf{R}_a^{\alpha,A}y$ and principal expressions. When we obtain the following derivation:

$$\dfrac{\dfrac{\begin{matrix}\vdots\ \mathcal{D}_1 \\ \Gamma \Rightarrow \Delta, (x\mathsf{R}_a^{\alpha,A}y)^{n\text{-}1}, x{:}^{\alpha} A\end{matrix} \quad \begin{matrix}\vdots\ \mathcal{D}_2 \\ \Gamma \Rightarrow \Delta, (x\mathsf{R}_a^{\alpha,A}y)^{n\text{-}1}, y{:}^{\alpha} A\end{matrix} \quad \begin{matrix}\vdots\ \mathcal{D}_3 \\ \Gamma \Rightarrow \Delta, (x\mathsf{R}_a^{\alpha,A}y)^{n\text{-}1}, x\mathsf{R}_a^{\alpha}y\end{matrix}}{\Gamma \Rightarrow \Delta, (x\mathsf{R}_a^{\alpha,A}y)^{n}} (Rrel_a) \quad \dfrac{\begin{matrix}\vdots\ \mathcal{D}_4 \\ x{:}^{\alpha} A, (x\mathsf{R}_a^{\alpha,A}y)^{m\text{-}1}, \Gamma' \Rightarrow \Delta'\end{matrix}}{(x\mathsf{R}_a^{\alpha,A}y)^{m}, \Gamma' \Rightarrow \Delta'} (Lrel_a 3)}{\Gamma, \Gamma' \Rightarrow \Delta, \Delta'} (Ecut),$$

it is transformed into the following derivation:

$$\dfrac{\dfrac{\dfrac{\begin{matrix}\vdots\ \mathcal{D}_1 \\ \Gamma \Rightarrow \Delta, (x\mathsf{R}_a^{\alpha,A}y)^{n\text{-}1}, x{:}^{\alpha} A\end{matrix} \quad \begin{matrix}\vdots\ \mathcal{D}'_4 \\ (x\mathsf{R}_a^{\alpha,A}y)^{m}, \Gamma' \Rightarrow \Delta'\end{matrix}}{\Gamma, \Gamma' \Rightarrow \Delta, \Delta', x{:}^{\alpha} A} (Ecut) \quad \dfrac{\begin{matrix}\vdots\ \mathcal{D}'_{123} \\ \Gamma \Rightarrow \Delta, (x\mathsf{R}_a^{\alpha,A}y)^{n}\end{matrix} \quad \begin{matrix}\vdots\ \mathcal{D}_4 \\ x{:}^{\alpha} A, (x\mathsf{R}_a^{\alpha,A}y)^{m\text{-}1}, \Gamma' \Rightarrow \Delta'\end{matrix}}{x{:}^{\alpha} A, \Gamma, \Gamma' \Rightarrow \Delta, \Delta'} (Ecut)\ \text{height-1}}{\Gamma, \Gamma, \Gamma', \Gamma' \Rightarrow \Delta, \Delta, \Delta', \Delta'} (Ecut)\ \text{length-1}}{\Gamma, \Gamma' \Rightarrow \Delta, \Delta'} (Rc/Lc),$$

where $(Ecut)$ to the two uppersequents is applicable by induction hypothesis, since the derivation height of $(Ecut)$ is reduced by comparison with the original derivation. Additionally, the application of $(Ecut)$ to the lowersequents is also allowed by induction hypothesis, since the length of the cut expression is reduced, namely $\mathsf{len}(x{:}^{\alpha} A) < \mathsf{len}(x\mathsf{R}_a^{\alpha,A}y)$. □

As a corollary of Theorem 2, the consistency of **GPAL**$^+$ is shown.

Corollary 1 (Consistency of **GPAL**) *The empty sequent* $\Rightarrow$ *cannot be proven in* **GPAL**$^+$.

Proof Suppose for contradiction that $\Rightarrow$ is derivable in **GPAL**$^+$. By Theorem 2, $\Rightarrow$ is derivable in **GPAL**; however, there is no inference rule in **GPAL** which can derive the empty sequent. This is a contradiction. □

7.5 Soundness of GPAL

Now, we switch the subject to the soundness theorem of **GPAL**. For the theorem, we extend Kripke semantics of PAL to cover the labelled expressions. Given any Kripke model $\mathfrak{M}$, we say that $f : \mathsf{Var} \to \mathcal{D}(\mathfrak{M})$ is an *assignment*.

Definition 5 Let $\mathfrak{M}$ be a Kripke model and $f : \mathsf{Var} \to \mathcal{D}(\mathfrak{M})$ an assignment.

$$\begin{array}{lll} \mathfrak{M}, f \Vdash x{:}^{\alpha} A & \text{iff} & \mathfrak{M}^{\alpha}, f(x) \Vdash A \text{ and } \mathrm{f(x)} \in \mathcal{D}(\mathfrak{M}^{\alpha}) \\ \mathfrak{M}, f \Vdash x\mathsf{R}_a^{\epsilon} y & \text{iff} & \langle f(x), f(y)\rangle \in R_a \\ \mathfrak{M}, f \Vdash x\mathsf{R}_a^{\alpha,A} y & \text{iff} & \langle f(x), f(y)\rangle \in R_a^{\alpha} \text{ and } \mathfrak{M}^{\alpha}, \mathrm{f(x)} \Vdash \mathrm{A} \text{ and } \mathfrak{M}^{\alpha}, \mathrm{f(y)} \Vdash \mathrm{A} \end{array}$$

Here we have to be careful of the fact that $f(x)$ and $f(y)$ above must be defined in $\mathcal{D}(\mathfrak{M}^{\alpha})$. In the clause $\mathfrak{M}, f \Vdash x{:}^{\alpha} A$, for example, $f(x)$ should survive (well defined) in the restricted Kripke model $\mathfrak{M}^{\alpha}$. Taking into account of this fact, it is essential that we pay attention to the negation of $\mathfrak{M}, f \Vdash x{:}^{\alpha} A$.

Proposition 3 $\mathfrak{M}, f \nVdash x{:}^{\alpha} A$ *iff* $f(x) \notin \mathcal{D}(\mathfrak{M}^{\alpha})$ *or* $(f(x) \in \mathcal{D}(\mathfrak{M}^{\alpha})$ *and* $\mathfrak{M}^{\alpha}, f(x) \nVdash A)$.

As far as the authors know, this point has not been suggested in previous works (e.g., [1, 7]). Then, the reader may wonder if the following 'natural' definition of the validity for sequents (which we call *s-valid*) also works. The following notion can be regarded as an implementation of the reading of a sequent $\Gamma \Rightarrow \Delta$ as 'if all of the antecedent Γ hold, then some of the consequents Δ hold'.

Definition 6 (*s-validity*) $\Gamma \Rightarrow \Delta$ is *s-valid* in $\mathfrak{M}$ if, for all assignments $f : \mathsf{Var} \to \mathcal{D}(\mathfrak{M})$ such that $\mathfrak{M}, f \Vdash \mathfrak{A}$ for all $\mathfrak{A} \in \Gamma$, there exists $\mathfrak{B} \in \Delta$ such that $\mathfrak{M}, f \Vdash \mathfrak{B}$.

However, following this natural definition of validity of sequents, we come to a deadlock on the way to prove the soundness theorem, especially in the case of rules for logical negation, as we can see the following proposition.

Proposition 4 *There is a Kripke model* $\mathfrak{M}$ *such that* $(R\neg)$ *of* **GPAL** *does not preserve s-validity in* $\mathfrak{M}$.

Proof Let $\mathsf{G} = \{a\}$ for simplicity. We use the same model as in Example 2, that is, we consider a Kripke model $\mathfrak{M} = \langle \{w_1, w_2\}, \{w_1, w_2\}^2, V\rangle$ where $V(p) = \{w_1\}$.

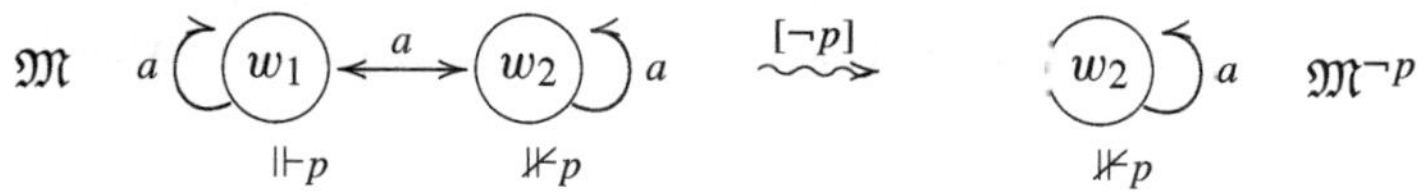

And the particular instance of the application of $(R\neg)$ is as follows:

$$\frac{x{:}^{\neg p} p \Rightarrow}{\Rightarrow x{:}^{\neg p} \neg p}\ (R\neg)$$

We show that the uppersequent is s-valid in $\mathfrak{M}$ but the lowersequent is not s-valid in $\mathfrak{M}$, and so $(R\neg)$ does not preserve s-validity in this case. Note that w_1 does not survive after $\neg p$, i.e., $w_1 \notin \mathcal{D}(\mathfrak{M}^{\neg p}) = \{w_2\}$.

First, we show that $x{:}^{\neg p} p \Rightarrow$ is s-valid in $\mathfrak{M}$, i.e., $\mathfrak{M}, f \not\Vdash x{:}^{\neg p} p$ for any assignment $f : \mathsf{Var} \to \mathcal{D}(\mathfrak{M})$. So, we fix any $f : \mathsf{Var} \to \mathcal{D}(\mathfrak{M})$. We divide our argument into: $f(x) = w_1$ or $f(x) = w_2$. If $f(x) = w_1$, $f(x)$ does not survive after $\neg p$, and so $\mathfrak{M}, f \not\Vdash x{:}^{\neg p} p$ by Proposition 3. If $f(x) = w_2$, $f(x)$ survives after $\neg p$ but $f(x) \notin \emptyset = V(p) \cap \mathcal{D}(\mathfrak{M}^{\neg p})$, which implies $\mathfrak{M}^{\neg p}, f(x) \not\Vdash p$ hence $\mathfrak{M}, f \not\Vdash x{:}^{\neg p} p$ by Proposition 3.

Second, we show that $\Rightarrow x{:}^{\neg p} \neg p$ is not s-valid in $\mathfrak{M}$, i.e., $\mathfrak{M}, f \not\Vdash x{:}^{\neg p} \neg p$ for some assignment $f : \mathsf{Var} \to W$. We fix some $f : \mathsf{Var} \to W$ such that $f(x) = w_1$. Since $f(x) \notin \mathcal{D}(\mathfrak{M}^{\neg p})$ ($f(x)$ does not survive after $\neg p$), $\mathfrak{M}, f \not\Vdash x{:}^{\neg p} \neg p$ by Proposition 3, as desired. □

Proposition 4 forces us to abandon the notion of s-validity and have an alternative notion of validity. Here we recall the second intuitive reading (in the introduction) of sequent $\Gamma \Rightarrow \Delta$ as 'it is not the case that all of the antecedents Γ hold and all of the consequents fail.' In order to realize the idea of 'failure', we first introduce the syntactic notion of the negated form $\overline{\mathfrak{A}}$ of a labelled expression $\mathfrak{A}$ and then provide the semantics $\mathfrak{M}, f \Vdash \overline{x{:}^{\alpha} A}$ with such negated forms, where we may read $\mathfrak{M}, f \Vdash \overline{x{:}^{\alpha} A}$ as '$\mathfrak{A}$ fails in $\mathfrak{M}$ under f.' Moreover, with this definition, our second notion of validity of a sequent, which we call *t-valid*,[6] is defined.

Definition 7 (*t-validity*) Let $\mathfrak{M}$ be a Kripke model and $f : \mathsf{Var} \to D(\mathfrak{M})$ an assignment. Then,

$$\begin{array}{lll}
\mathfrak{M}, f \Vdash \overline{x{:}^{\alpha} A} & \text{iff} & \mathfrak{M}^{\alpha}, f(x) \Vdash \neg A \text{ and } f(x) \in \mathcal{D}(\mathfrak{M}^{\alpha}), \\
\mathfrak{M}, f \Vdash \overline{x \mathsf{R}_a^{\epsilon} y} & \text{iff} & \langle f(x), f(y)\rangle \notin R_a, \\
\mathfrak{M}, f \Vdash \overline{x \mathsf{R}_a^{\alpha, A} y} & \text{iff} & \mathfrak{M}, f \Vdash \overline{x \mathsf{R}_a^{\alpha} y} \text{ or } \mathfrak{M}, f \Vdash \overline{x{:}^{\alpha} A} \text{ or } \mathfrak{M}, f \Vdash \overline{y{:}^{\alpha} A}.
\end{array}$$

We say that $\Gamma \Rightarrow \Delta$ is *t-valid* in $\mathfrak{M}$ if there is no assignment $f : \mathsf{Var} \to \mathcal{D}(\mathfrak{M})$ such that $\mathfrak{M}, f \Vdash \mathfrak{A}$ for all $\mathfrak{A} \in \Gamma$, and $\mathfrak{M}, f \Vdash \overline{\mathfrak{B}}$ for all $\mathfrak{B} \in \Delta$.

In this definition, we explicitly gave a condition of surviveness that $f(x) \in \mathcal{D}(\mathfrak{M}^{\alpha})$, e.g., in $\mathfrak{M}, f \Vdash \overline{x{:}^{\alpha} A}$. Therefore, '$x :^{\alpha} A$ fails in $\mathfrak{M}$ under f' means that $f(x)$ survives after α but A is false at $f(x)$ in $\mathfrak{M}^{\alpha}$. The following proposition shows that the clauses for relational atoms and their negated forms characterize what they intend to capture.

[6] We note that t-validity is close to the validity in the tableaux method of PAL [2].

Proposition 5 *For any Kripke model* $\mathfrak{M}$, *assignment* f, $a \in \mathsf{G}$ *and* $x, y \in \mathsf{Var}$,

(i) $\mathfrak{M}, f \Vdash x\mathsf{R}_a^\alpha y$ *iff* $\langle f(x), f(y)\rangle \in R_a^\alpha$,
(ii) $\mathfrak{M}, f \Vdash \overline{x\mathsf{R}_a^\alpha y}$ *iff* $\langle f(x), f(y)\rangle \notin R_a^\alpha$.

Proof Both are easily shown by induction of α. Let us consider the case of $\alpha = \alpha', A$ in the proof of (ii).

We show $\mathfrak{M}, f \nVdash \overline{x\mathsf{R}_a^{\alpha,A} y}$ iff $\langle f(x), f(y)\rangle \in R_a^{\alpha,A}$. $\mathfrak{M}, f \nVdash \overline{x\mathsf{R}_a^{\alpha,A} y}$ is, by Definition 7 and the induction hypothesis, equivalent to $\langle f(x), f(y)\rangle \in R_a^\alpha$ and $\mathfrak{M}^\alpha, f(x) \Vdash A$ and $\mathfrak{M}^\alpha, f(y) \Vdash A$. That is also equivalent to $\langle f(x), f(y)\rangle \in R_a^{\alpha,A}$. □

Following this, we may prove the soundness of **GPAL** properly.

Theorem 3 (Soundness of **GPAL**) *Given any sequent* $\Gamma \Rightarrow \Delta$ *in* **GPAL**, *if* **GPAL** $\vdash \Gamma \Rightarrow \Delta$, *then* $\Gamma \Rightarrow \Delta$ *is t-valid in every Kripke model* $\mathfrak{M}$.

Proof The proof is carried out by induction of the height of the derivation of $\Gamma \Rightarrow \Delta$ in **GPAL**. We only confirm one of base cases of relational atoms and some cases in the inductive step.

Base case: we show that $x\mathsf{R}_a^\alpha v \Rightarrow x\mathsf{R}_a^\alpha v$ is t-valid. Suppose for contradiction that $\mathfrak{M}, f \Vdash x\mathsf{R}_a^\alpha v$ and $\mathfrak{M}, f \Vdash \overline{x\mathsf{R}_a^\alpha v}$. By Proposition 5, this is impossible.

The case where the last applied rule is $(R\neg)$: We show the contraposition. Suppose that there is some $f : \mathsf{Var} \to W$ such that, $\mathfrak{M}, f \Vdash \mathfrak{A}$ for all $\mathfrak{A} \in \Gamma$, and $\mathfrak{M}, f \Vdash \overline{\mathfrak{B}}$ for all $\mathfrak{B} \in \Delta$, and $\mathfrak{M}, f \Vdash \overline{x{:}^\alpha \neg A}$. Fix such f. It suffices to show $\mathfrak{M}, f \Vdash x{:}^\alpha A$. Then, $\mathfrak{M}, f \Vdash \overline{x{:}^\alpha \neg A}$ iff $\mathfrak{M}^\alpha, f(x) \nVdash \neg A$ and $f(x) \in \mathcal{D}(\mathfrak{M}^\alpha)$, which is equivalent to: $\mathfrak{M}^\alpha, f(x) \Vdash A$ and $f(x) \in \mathcal{D}(\mathfrak{M}^\alpha)$. By Definition 5, $\mathfrak{M}, f \Vdash x{:}^\alpha A$. So, the contraposition has been shown.

The case where the last applied rule is $(L\mathsf{K}')$: We show the contraposition. Suppose that there is some $f : \mathsf{Var} \to W$ such that $\mathfrak{M}, f \Vdash \mathfrak{A}$ for all $\mathfrak{A} \in \Gamma$ and $\mathfrak{M}, f \Vdash x^\alpha{:}\mathsf{K}_a A$ and $\mathfrak{M}, f \Vdash \overline{\mathfrak{B}}$ for all $\mathfrak{B} \in \Delta$. Fix such f. It suffices to show $\mathfrak{M}, f \Vdash \overline{x\mathsf{R}_a^\alpha y}$ or $\mathfrak{M}, f \Vdash y{:}^\alpha A$. Then, from $\mathfrak{M}, f \Vdash x{:}^\alpha \mathsf{K}_a A$, we obtain $\langle f(x), f(y)\rangle \notin R_a^\alpha$ or $\mathfrak{M}^\alpha, f(y) \Vdash A$. Suppose the former disjunct, i.e., $\langle f(x), f(y)\rangle \notin R_a^\alpha$, which is, by Proposition 5, $\mathfrak{M}, f \Vdash \overline{x\mathsf{R}_a^\alpha y}$. Then, suppose the latter disjunct $\mathfrak{M}^\alpha, f(y) \Vdash A$. By definition, this is equivalent to $\mathfrak{M}, f \Vdash y :^\alpha A$. Then, the contraposition has been shown.

The case where the last applied rule is (Rat'): Similar to the above, we show the contraposition. Suppose there is some $f : \mathsf{Var} \to W$ such that, $\mathfrak{M}, f \Vdash \mathfrak{A}$ for all $\mathfrak{A} \in \Gamma$, and $\mathfrak{M}, f \Vdash \overline{\mathfrak{B}}$ for all $\mathfrak{B} \in \Delta$, and $\mathfrak{M}, f \Vdash \overline{x{:}^{\alpha,A} p}$. Fix such f. It suffices to show $\mathfrak{M}, f \Vdash \overline{x{:}^\alpha p}$. By Definition 7, $\mathfrak{M}, f \Vdash \overline{x{:}^{\alpha,A} p}$ is equivalent to $\mathfrak{M}^{\alpha,A}, f(x) \Vdash \neg p$ and $f(x) \in \mathcal{D}(\mathfrak{M}^{\alpha,A})$. By $f(x) \in \mathcal{D}(\mathfrak{M}^{\alpha,A})$, we obtain $f(x) \in \mathcal{D}(\mathfrak{M}^\alpha)$ and $\mathfrak{M}^\alpha, f(x) \Vdash A$. It follows from $\mathfrak{M}^\alpha, f(x) \Vdash A$ and $\mathfrak{M}^{\alpha,A}, f(x) \Vdash \neg p$ that $f(x) \notin V^\alpha(p)$. This is equivalent to $\mathfrak{M}, f \Vdash \overline{x{:}^\alpha p}$. Then, the contraposition has been shown.

The case where the last applied rule is $(Rrel)$: As before, we show the contraposition. Suppose there is some $f : \mathsf{Var} \to W$ such that, $\mathfrak{M}, f \Vdash \mathfrak{A}$ for all $\mathfrak{A} \in \Gamma$, and $\mathfrak{M}, f \Vdash \overline{\mathfrak{B}}$ for all $\mathfrak{B} \in \Delta$, and $\mathfrak{M}, f \Vdash \overline{x\mathsf{R}_a^{\alpha,A}y}$. Fix such f. By Definition 7, $\overline{x\mathsf{R}_a^{\alpha,A}y}$ is equivalent to $\mathfrak{M}, f \Vdash \overline{x\mathsf{R}_a^{\alpha}y}$ or $\mathfrak{M}, f \Vdash \overline{x{:}^{\alpha}A}$ or $\mathfrak{M}, f \Vdash \overline{y{:}^{\alpha}A}$. This is what we want to show. □

For the following corollary, we prepare the next proposition.

Proposition 6 *If* $\Rightarrow x{:}^{\epsilon}A$ *is* t*-valid in a Kripke model* $\mathfrak{M}$*, then* A *is valid in* $\mathfrak{M}$*.*

Proof Suppose that $\Rightarrow x{:}^{\epsilon}A$ is t-valid in $\mathfrak{M}$. So, it is not the case that there exists some assignment f such that $\mathfrak{M}, f \Vdash \overline{x{:}^{\epsilon}A}$. Equivalently, for all assignments f, $\mathfrak{M}, f \nVdash \overline{x{:}^{\epsilon}A}$. For any assignment f, $\mathfrak{M}, f \nVdash \overline{x{:}^{\epsilon}A}$ is equivalent to $\mathfrak{M}, f(x) \Vdash A$ because $f(x) \in \mathcal{D}(\mathfrak{M})$. So, it follows that $\mathfrak{M}, f(x) \Vdash A$ for all assignments f. Then, it is immediate to see that A is valid in $\mathfrak{M}$, as required. □

Then an indirect proof of completeness of **GPAL** can be provided as follows:

Corollary 2 *Given any formula* A *and label* $x \in \mathsf{Var}$*, the following are equivalent.*

(i) *A is valid on all Kripke models.*
(ii) **HPAL** $\vdash A$
(iii) **GPAL**$^{+}$ $\vdash \Rightarrow x{:}^{\epsilon}A$
(iv) **GPAL** $\vdash \Rightarrow x{:}^{\epsilon}A$

Proof The direction from (i) to (ii) is established by Fact 1 and the direction from (ii) to (iii) is shown in Theorem 1. Then, the direction from (iii) to (iv) is established by the admissibility of (Cut) (Theorem 2). Finally, the direction from (iv) to (i) is shown by Theorem 3 and Proposition 6. □

7.6 Completeness of GPAL for Link-Cutting Semantics

Let us denote by **GPALw** as the resulting sequent calculus of replacing (Lat') and (Rat') of **GPAL** with the following modified version of (Lat) and (Rat) in **G3PAL**:

$$\frac{x{:}^{\alpha}A, \Gamma \Rightarrow \Delta}{x{:}^{\alpha,A}p, \Gamma \Rightarrow \Delta}\ (Lat1) \quad \frac{x{:}^{\alpha}p, \Gamma \Rightarrow \Delta}{x{:}^{\alpha,A}p, \Gamma \Rightarrow \Delta}\ (Lat2) \quad \frac{\Gamma \Rightarrow \Delta, x{:}^{\alpha}A \quad \Gamma \Rightarrow \Delta, x{:}^{\alpha}p}{\Gamma \Rightarrow \Delta, x{:}^{\alpha,A}p}\ (Rat).$$

We checked that all results needed to show Corollary 2 hold also for **GPALw**, and so we can establish the similar result to Corollary 2 also for **GPALw**. While (Rat) *does* preserve t-validity in a Kripke model $\mathfrak{M}$ by the similar argument to the proof of Theorem 3, we remark that one premise $\Gamma \Rightarrow \Delta, x{:}^{\alpha}A$ of (Rat) becomes redundant when we prove that (Rat) preserves t-validity in a Kripke model. This is because, for any assignment f, $\mathfrak{M}, f \Vdash \overline{x{:}^{\alpha,A}p}$ already implies that A holds at $f(x)$ after

α, i.e., $\mathfrak{M}, f \Vdash x{:}^{\alpha} A$. We realize that this difference between **GPALw** and **GPAL** comes from the difference between the (standard) world-deletion semantics of PAL and the link-cutting semantics of PAL (see also Remark 1). In this section, we introduce our version of link-cutting semantics of PAL and provide a direct proof of completeness of **GPAL** for link-cutting semantics.[7] The specific definition of the link-cutting version of PAL's semantics is given as follows, where we keep the symbol $\Vdash$ for the previous world-deletion semantics of PAL and use the new symbol '$\models$' for the satisfaction relation for the link-cutting semantics.

Definition 8 (*Link-cutting semantics of PAL*) Given a Kripke model $\mathfrak{M}$, $w \in \mathcal{D}(\mathfrak{M})$ and a formula A, $\mathfrak{M}, w \models A$ is defined by

$$\begin{array}{lll}
\mathfrak{M}, w \models p & \text{iff} & w \in V(p), \\
\mathfrak{M}, w \models \neg A & \text{iff} & \mathfrak{M}, w \not\models A, \\
\mathfrak{M}, w \models A \to B & \text{iff} & \mathfrak{M}, w \models A \text{ implies } \mathfrak{M}, w \models B, \\
\mathfrak{M}, w \models \mathsf{K}_a A & \text{iff} & \text{for all } v \in W : w R_a v \text{ implies } \mathfrak{M}, v \models A, \text{ and} \\
\mathfrak{M}, w \models [A]B & \text{iff} & \mathfrak{M}, w \models A \text{ implies } \mathfrak{M}^{A!}, w \models B,
\end{array}$$

where the restriction $\mathfrak{M}^{A!}$ is defined by triple $\langle W, (R_a^{A!})_{a \in \mathsf{G}}, V \rangle$ with

$$R_a^{A!} := R_a \cap ([\![A]\!]_{\mathfrak{M}} \times [\![A]\!]_{\mathfrak{M}}), \quad \text{where } [\![A]\!]_{\mathfrak{M}} := \{x \in W \mid \mathfrak{M}, x \models A\}.$$

Remark 1 As far as the authors know, van Benthem et al. [14, p. 166] first provides an idea of link-cutting semantics of PAL. Their underlying idea is: cutting the links (pairs in an accessibility relation) between A-zone and $\neg A$-zone. Then, they state that all valid formulas in the resulting semantics are also the same as those in the world-deletion semantics [14, Fact1]. Their semantics is similar but different to our semantics above. Hansen [5, p. 145] touches on the same link-cutting semantics as ours in the public announcement extension of hybrid logic (an extended modal logic), but he does not investigate the semantics in detail there. A variant of our link-cutting semantics is also explained for logic of belief in [15], though the notion of public announcement there is not truthful and this is why the announcement there is called the 'introspective announcement.'

According to this definition, only the accessibility relation is restricted to A in $\mathfrak{M}^{A!}$, and the set of possible worlds and valuation stay as they were. Similar to the world-deletion semantics, we can also define the notion of validity in a Kripke model. The following soundness of **HPAL** for the link-cutting semantics is straightforward.

Proposition 7 *If A is a theorem of* **HPAL**, *A is valid in every Kripke model $\mathfrak{M}$ for the link-cutting semantics.*

As before, for any list $\alpha = (A_1, A_2, \ldots, A_n)$ of formulas , we define $\mathfrak{M}^{\alpha!}$ inductively as: $\mathfrak{M}^{\alpha!} := \mathfrak{M}$ (if $\alpha = \epsilon$), and $\mathfrak{M}^{\alpha!} := (\mathfrak{M}^{\beta!})^{A_n!} = \langle W, (R_a^{\beta!, A_n!})_{a \in \mathsf{G}}, V \rangle$ (if

[7]Thanks to a comment from Makoto Kanazawa in the annual meeting of MLG2014, we noticed that link-cutting semantics may be suitable for our labelled sequent calculus of PAL.

$\alpha = \beta, A_n$). Now we can show that the corresponding notions to s- and t-validity become equivalent under our link-cutting semantics.

Definition 9 Let $\mathfrak{M}$ be a Kripke model and $f : \mathsf{Var} \to \mathcal{D}(\mathfrak{M})$ an assignment.

$$\begin{array}{lll} \mathfrak{M}, f \models x{:}^{\alpha} A & \text{iff} & \mathfrak{M}^{\alpha !}, f(x) \models A \\ \mathfrak{M}, f \models x \mathsf{R}_a^{\epsilon} y & \text{iff} & \langle f(x), f(y) \rangle \in R_a \\ \mathfrak{M}, f \models x \mathsf{R}_a^{\alpha, A} y & \text{iff} & \langle f(x), f(y) \rangle \in R_a^{\alpha !} \text{ and } \mathfrak{M}^{\alpha !}, f(x) \models A \text{ and } \mathfrak{M}^{\alpha !}, f(y) \models A \end{array}$$

By this definition, the next proposition immediately follows.

Proposition 8 *For any Kripke model $\mathfrak{M}$, assignment f, $a \in \mathsf{G}$ and $x, y \in \mathsf{Var}$,*

$$\mathfrak{M}, f \models x \mathsf{R}_a^{\alpha} y \quad \textit{iff} \quad \langle f(x), f(y) \rangle \in R_a^{\alpha !}$$

The semantics of the negated form of a labelled expression $\overline{\mathfrak{A}}$ is also defined as before.

Definition 10 Let $\mathfrak{M}$ be a Kripke model and $f : \mathsf{Var} \to D(\mathfrak{M})$ an assignment. Then,

$$\begin{array}{lll} \mathfrak{M}, f \models \overline{x{:}^{\alpha} A} & \text{iff} & \mathfrak{M}^{\alpha !}, f(x) \not\models A, \\ \mathfrak{M}, f \models \overline{x \mathsf{R}_a^{\epsilon} y} & \text{iff} & \langle f(x), f(y) \rangle \notin R_a, \\ \mathfrak{M}, f \models \overline{x \mathsf{R}_a^{\alpha, A} y} & \text{iff} & \mathfrak{M}, f \Vdash \overline{x \mathsf{R}_a^{\alpha} y} \text{ or } \mathfrak{M}, f \not\models x{:}^{\alpha} A \text{ or } \mathfrak{M}, f \not\models y{:}^{\alpha} A \end{array}$$

Now we may confirm that, based on the semantics, t-validity and s-validity are equivalent since $\mathfrak{M}, f \not\models \overline{\mathfrak{B}}$ is equivalent to $\mathfrak{M}, f \models \mathfrak{B}$ in this semantics.

Proposition 9 *Under the link-cutting semantics, a sequent $\Gamma \Rightarrow \Delta$ is s-valid in a Kripke model $\mathfrak{M}$ iff it is t-valid in $\mathfrak{M}$.*

Proof Suppose $\Gamma \Rightarrow \Delta$ is t-valid in $\mathfrak{M}$. In other words, if there is no assignment $f : \mathsf{Var} \to \mathcal{D}(\mathfrak{M})$ such that $\mathfrak{M}, f \models \mathfrak{A}$ for all $\mathfrak{A} \in \Gamma$, and $\mathfrak{M}, f \models \overline{\mathfrak{B}}$ for all $\mathfrak{B} \in \Delta$. Equivalently, for all assignments $f : \mathsf{Var} \to \mathcal{D}(\mathfrak{M})$, $\mathfrak{M}, f \models \mathfrak{A}$ for all $\mathfrak{A} \in \Gamma$, there exists $\mathfrak{B} \in \Delta$ such that $\mathfrak{M}, f \models \mathfrak{B}$. □

Because the notion of surviveness is expelled, the definition of the satisfaction of labelled expressions becomes wholly natural. Thus, we do not need to worry about the notion of surviveness of possible worlds in this link-cutting semantics.

Hereafter, in this section we consider possibly infinite multisets of labelled expressions. That is, we call $\Gamma \Rightarrow \Delta$ an infinite sequent if Γ or Δ are infinite multisets. We use the notation **GPAL** $\vdash \Gamma \Rightarrow \Delta$ to mean that there are finite multisets Γ' and Δ' of labelled expressions such that **GPAL** $\vdash \Gamma' \Rightarrow \Delta'$ in the ordinary sense and $\Gamma' \subseteq \Gamma$ and $\Delta' \subseteq \Delta$. To establish the completeness result of **GPAL** for the link-cutting semantics, we first introduce the notion of saturation as follows.

Definition 11 A possibly infinite sequent $\Gamma \Rightarrow \Delta$ is *saturated* if it satisfies the following:

$(unprov)$ $\Gamma \Rightarrow \Delta$ is not derivable in **GPAL**,
$(\to l)$ if $x{:}^{\alpha} A \to B \in \Gamma$, then $x{:}^{\alpha} A \in \Delta$ or $x{:}^{\alpha} B \in \Gamma$,
$(\to r)$ if $x{:}^{\alpha} A \to B \in \Delta$, then $x{:}^{\alpha} A \in \Gamma$ and $x{:}^{\alpha} B \in \Delta$,
$(\neg l)$ if $x{:}^{\alpha} \neg A \in \Gamma$, then $x{:}^{\alpha} A \in \Delta$,
$(\neg r)$ if $x{:}^{\alpha} \neg A \in \Delta$, then $x{:}^{\alpha} A \in \Gamma$,
$(\mathsf{K}_a l)$ if $x{:}^{\alpha} \mathsf{K}_a A \in \Gamma$, then $x \mathsf{R}_a^{\alpha} y \in \Delta$ or $y{:}^{\alpha} A \in \Gamma$ for any label y,
$(\mathsf{K}_a r)$ if $x{:}^{\alpha} \mathsf{K}_a A \in \Delta$, then $x \mathsf{R}_a^{\alpha} y \in \Gamma$ and $y{:}^{\alpha} A \in \Delta$ for some label y,
$([.]l)$ if $x{:}^{\alpha}[A]B \in \Gamma$, then $x{:}^{\alpha} A \in \Delta$ or $x{:}^{\alpha,A} B \in \Gamma$,
$([.]r)$ if $x{:}^{\alpha}[A]B \in \Delta$, then $x{:}^{\alpha} A \in \Gamma$ and $x{:}^{\alpha,A} B \in \Delta$,
(atl) if $x{:}^{\alpha,A} p \in \Gamma$, then $x{:}^{\alpha} p \in \Gamma$,
(atr) if $x{:}^{\alpha,A} p \in \Delta$, then $x{:}^{\alpha} p \in \Delta$,
$(rell)$ if $x \mathsf{R}_a^{\alpha,A} y \in \Gamma$, then $x{:}^{\alpha} A \in \Gamma$ and $y{:}^{\alpha} A \in \Gamma$, and $x \mathsf{R}_a^{\alpha} y \in \Gamma$, and
$(relr)$ if $x \mathsf{R}_a^{\alpha,A} y \in \Delta$, then $x{:}^{\alpha} A \in \Delta$ or $y{:}^{\alpha} A \in \Delta$, or $x \mathsf{R}_a^{\alpha} y \in \Delta$.

We show the next lemma which states that any unprovable sequent in **GPAL** can be extended to a (possibility infinite) saturated sequent.

Lemma 4 *Let $\Gamma \Rightarrow \Delta$ be a finite sequent. If* **GPAL** $\nvdash \Gamma \Rightarrow \Delta$*, then there exists a possibility infinite saturated sequent $\Gamma^+ \Rightarrow \Delta^+$ where $\Gamma \subseteq \Gamma^+$ and $\Delta \subseteq \Delta^+$.*

Proof Fix any finite sequent $\Gamma \Rightarrow \Delta$ such that **GPAL** $\nvdash \Gamma \Rightarrow \Delta$. Let $\mathfrak{A}_1, \mathfrak{A}_2, \ldots$ be an enumeration of all labelled expressions such that each labelled expression appears infinitely many times. We inductively construct an infinite sequence $(\Gamma_i \Rightarrow \Delta_i)_{i \in \mathbb{N}}$ of finite sequents such that **GPAL** $\nvdash \Gamma_i \Rightarrow \Delta_i$ at each $i \in \mathbb{N}$ as follows and define $\Gamma^+ \Rightarrow \Delta^+$ as the 'limit' of such sequence.

Let $\Gamma_0 \Rightarrow \Delta_0$ be $\Gamma \Rightarrow \Delta$ as the basis of $\Gamma_i \Rightarrow \Delta_i$, and by the supposition **GPAL** $\nvdash \Gamma_0 \Rightarrow \Delta_0$. The $i+1$-th step consists of the procedures to define an underivable $\Gamma_{i+1} \Rightarrow \Delta_{i+1}$ from $\Gamma_i \Rightarrow \Delta_i$ depending on the shape of the labelled expression $\mathfrak{A}_i$. In the $i+1$-th step, one of the following operations is executed.

The case where $\mathfrak{A}_i$ is of the form $x{:}^{\alpha} A \to B$ and $\mathfrak{A}_i \in \Gamma_i$: Because $\Gamma_i \Rightarrow \Delta_i$ is unprovable, either $\Gamma_i \Rightarrow \Delta_i, x{:}^{\alpha} A$ or $x{:}^{\alpha} B, \Gamma_i \Rightarrow \Delta_i$ is also unprovable by $(L \to)$. Then we choose one unprovable sequent as $\Gamma_{i+1} \Rightarrow \Delta_{i+1}$.

The case where $\mathfrak{A}_i$ is of the form $x{:}^{\alpha} A \to B$ and $\mathfrak{A}_i \in \Delta_i$: We define $\Gamma_{i+1} \Rightarrow \Delta_{i+1} := x{:}^{\alpha} A, \Gamma_i \Rightarrow \Delta_i, x{:}^{\alpha} B$. By $(R \to)$ and **GPAL** $\nvdash \Gamma_i \Rightarrow \Delta_i$, the sequent $\Gamma_{i+1} \Rightarrow \Delta_{i+1}$ is also unprovable.

The case where $\mathfrak{A}_i$ is of the form $x{:}^{\alpha} \neg A$ and $\mathfrak{A}_i \in \Gamma_i$: We define $\Gamma_{i+1} \Rightarrow \Delta_{i+1} := \Gamma_i \Rightarrow \Delta_i, x{:}^{\alpha} A$. Because of $(L\neg)$ and **GPAL** $\nvdash \Gamma_i \Rightarrow \Delta_i$, the sequent $\Gamma_{i+1} \Rightarrow \Delta_{i+1}$ is also unprovable.

The case where $\mathfrak{A}_i$ is of the form $x{:}^{\alpha} \neg A$ and $\mathfrak{A}_i \in \Delta_i$: We define $\Gamma_{i+1} \Rightarrow \Delta_{i+1} := x{:}^{\alpha} A, \Gamma_i \Rightarrow \Delta_i$. Because of $(R\neg)$ and **GPAL** $\nvdash \Gamma_i \Rightarrow \Delta_i$, the sequent $\Gamma_{i+1} \Rightarrow \Delta_{i+1}$ is also unprovable.

The case where $\mathfrak{A}_i$ is of the form $x :^{\alpha} [A]B$ and $\mathfrak{A}_i \in \Gamma_i$: We define $\Gamma_{i+1} \Rightarrow \Delta_{i+1}$ as either $\Gamma_i \Rightarrow \Delta_i, x{:}^{\alpha} A$ or $x{:}^{\alpha,A} B, \Gamma_i \Rightarrow \Delta_i$. Because of $(L[.])$ and **GPAL** $\nvdash \Gamma_i \Rightarrow \Delta_i$, the sequent $\Gamma_{i+1} \Rightarrow \Delta_{i+1}$ is also unprovable.

The case where $\mathfrak{A}_i$ is of the form $x :^{\alpha} [A]B$ and $\mathfrak{A}_i \in \Delta_i$: We define $\Gamma_{i+1} \Rightarrow \Delta_{i+1} := x{:}^{\alpha} A, \Gamma_i \Rightarrow \Delta_i, x{:}^{\alpha,A} B$. Because of $(R[.])$ and **GPAL** $\nvdash \Gamma_i \Rightarrow \Delta_i$, the sequent $\Gamma_{i+1} \Rightarrow \Delta_{i+1}$ is also unprovable.

The case where $\mathfrak{A}_i$ is of the form $x{:}^{\alpha,A} p$ and $\mathfrak{A}_i \in \Gamma_i$: We define $\Gamma_{i+1} \Rightarrow \Delta_{i+1} := x{:}^{\alpha} p, \Gamma_i \Rightarrow \Delta_i$. Because of (Lat') and **GPAL** $\nvdash \Gamma_i \Rightarrow \Delta_i$, the sequent $\Gamma_{i+1} \Rightarrow \Delta_{i+1}$ is also unprovable.

The case where $\mathfrak{A}_i$ is of the form $x{:}^{\alpha,A} p$ and $\mathfrak{A}_i \in \Delta_i$: We define $\Gamma_{i+1} \Rightarrow \Delta_{i+1} := \Gamma_i \Rightarrow \Delta_i, x{:}^{\alpha} p$. Because of (Rat') and **GPAL** $\nvdash \Gamma_i \Rightarrow \Delta_i$, the sequent $\Gamma_{i+1} \Rightarrow \Delta_{i+1}$ is also unprovable.

The case where $\mathfrak{A}_i$ is of the form $x{:}^{\alpha}\mathsf{K}_a A$ and $\mathfrak{A}_i \in \Gamma_i$: Let $\{y_1, ..., y_n\}$ be the set of all labels appearing in $\Gamma_i \Rightarrow \Delta_i$. Suppose we have constructed $(\Gamma_i^{(k)} \Rightarrow \Delta_i^{(k)})_{1 \le k \le \ell}$ such that $(\Gamma_i^{(k)} \Rightarrow \Delta_i^{(k)})$ is unprovable, $\Gamma_i^{(k)} \subseteq \Gamma_i^{(k+1)}$, and $\Delta_i^{(k)} \subseteq \Delta_i^{(k+1)}$. Because of $(L\mathsf{K}_a)$ and **GPAL** $\nvdash \Gamma_i^{(l)} \Rightarrow \Delta_i^{(l)}$, either $\Gamma_i^{(l)} \Rightarrow \Delta_i^{(l)}, x\mathsf{R}_a^{\alpha} y_{\ell+1}$ or $y_{\ell+1}{:}A, \Gamma_i^{(l)} \Rightarrow \Delta_i^{(l)}$ is unprovable, and we choose one unprovable sequent as $\Gamma_i^{(l+1)} \Rightarrow \Delta_i^{(l+1)}$. Then we define $\Gamma_{i+1} \Rightarrow \Delta_{i+1} := \Gamma_i^{(n)} \Rightarrow \Delta_i^{(n)}$, and $\Gamma_{i+1} \Rightarrow \Delta_{i+1}$ is unprovable by construction.

The case where $\mathfrak{A}_i$ is of the form $x{:}^{\alpha}\mathsf{K}_a A$ and $\mathfrak{A}_i \in \Delta_i$: We define $\Gamma_{i+1} \Rightarrow \Delta_{i+1} := x\mathsf{R}_a^{\alpha} y, \Gamma_i \Rightarrow \Delta_i, y{:}^{\alpha} A$, where y is a fresh variable that does not appear in $\Gamma_i \Rightarrow \Delta_i$. Because of $(R\mathsf{K}_a)$ and **GPAL** $\nvdash \Gamma_i \Rightarrow \Delta_i$, the sequent $\Gamma_{i+1} \Rightarrow \Delta_{i+1}$ is also unprovable.

The case where $\mathfrak{A}_i$ is of the form $x\mathsf{R}_a^{\alpha,A} y$ and $\mathfrak{A}_i \in \Gamma_i$: We define $\Gamma_{i+1} \Rightarrow \Delta_{i+1} := x{:}^{\alpha} A, y{:}^{\alpha} A, x\mathsf{R}_a^{\alpha} y, \Gamma_i \Rightarrow \Delta_i$. Because of $(Lrel)$ and **GPAL** $\nvdash \Gamma_i \Rightarrow \Delta_i$, the sequent $\Gamma_{i+1} \Rightarrow \Delta_{i+1}$ is also unprovable.

The case where $\mathfrak{A}_i$ is of the form $x\mathsf{R}_a^{\alpha,A} y$ and $\mathfrak{A}_i \in \Delta_i$: We define $\Gamma_{i+1} \Rightarrow \Delta_{i+1}$ as either $\Gamma_i \Rightarrow \Delta_i, x{:}^{\alpha} A$ or $\Gamma_i \Rightarrow \Delta_i, y{:}^{\alpha} A$ or $\Gamma_i \Rightarrow \Delta_i, x\mathsf{R}_a^{\alpha} y$. Because of $(Rrel)$ and **GPAL** $\nvdash \Gamma_i \Rightarrow \Delta_i$, the sequent $\Gamma_{i+1} \Rightarrow \Delta_{i+1}$ is also unprovable.

Otherwise: We define $\Gamma_{i+1} \Rightarrow \Delta_{i+1} := \Gamma_i \Rightarrow \Delta_i$.

Finally, let $\Gamma^+ \Rightarrow \Delta^+$ be the union $\bigcup_{i \in \mathbb{N}} \Gamma_i \Rightarrow \bigcup_{i \in \mathbb{N}} \Delta_i$. Then, it is routine to check that $\Gamma^+ \Rightarrow \Delta^+$ is saturated and $\Gamma \subseteq \Gamma^+$ and $\Delta \subseteq \Delta^+$. □

We now prove the completeness of **GPAL** for the link-cutting semantics.

Theorem 4 *If a formula A is valid in every Kripke model $\mathfrak{M}$ for the link-cutting semantics, then* **GPAL** $\vdash \Rightarrow x{:}^{\epsilon} A$.

Proof We show its contraposition, and so suppose **GPAL** $\nvdash \Rightarrow x{:}^{\epsilon} A$. By Lemma 4, there exists a saturated sequent $\Gamma^+ \Rightarrow \Delta^+$ such that $\{x{:}^{\epsilon} A\} \subseteq \Delta^+$. Using the saturated sequent, we construct the derived Kripke model $\mathfrak{M} = \langle W, (R_a)_{a \in \mathsf{G}}, V \rangle$ from the saturated sequent $\Gamma^+ \Rightarrow \Delta^+$.

- W is a set of all labels appearing in $\Gamma^+ \Rightarrow \Delta^+$,
- $xR_a^\epsilon y$ iff $x\mathsf{R}_a^\epsilon y \in \Gamma^+$,
- $x \in V(p)$ iff $x{:}p \in \Gamma^+$.

In addition to this, let $f : \mathsf{Var} \to W$ be an arbitrary assignment such that $f(x) = x$ (if x is in W). Then, we can establish the following two items:

(i) $\mathfrak{A} \in \Gamma^+$ implies $\mathfrak{M}, f \models \mathfrak{A}$,
(ii) $\mathfrak{A} \in \Delta^+$ implies $\mathfrak{M}, f \not\models \mathfrak{A}$.

The second item implies that $\mathfrak{M}, f(x) \not\models A$ hence A is not valid in the derived model $\mathfrak{M}$. The proof for these two items is conducted by simultaneous induction on the length of $\mathfrak{A}$. Here we only look at the cases where $\mathfrak{A}$ is $x{:}^{\alpha,A}p$ or $x{:}^{\alpha}\mathsf{K}_aA$.

The case where $\mathfrak{A}$ is $x{:}^{\alpha,A}p$: (i) If $x{:}^{\alpha,A}p \in \Gamma^+$, then by saturatedness, we have $x{:}^{\alpha}p \in \Gamma^+$. Then by induction hypothesis, $\mathfrak{M}, f \models x{:}^{\alpha}p$ is obtained. This is equivalent to $\mathfrak{M}^{\alpha}, f(x) \models p$, i.e., $f(x) \in V(p)$. Hence $\mathfrak{M}, f \models x{:}^{\alpha,A}p$.
(ii) If $x{:}^{\alpha,A}p \in \Delta^+$, then by the saturatedness, we have $x{:}^{\alpha}p \in \Delta^+$. Then by induction hypothesis, $\mathfrak{M}, f \not\models x{:}^{\alpha}p$ is obtained. This is equivalent to $f(x) \notin V(p)$, and so $\mathfrak{M}, f \not\models x{:}^{\alpha,A}p$.

The case where $\mathfrak{A}$ is $x{:}^{\alpha}\mathsf{K}_aA$: (i) Suppose $x{:}^{\alpha}\mathsf{K}_aA \in \Gamma^+$. What we show is $\mathfrak{M}, f \models x{:}^{\alpha}\mathsf{K}_aA$, i.e., for all $y \in \mathcal{D}(\mathfrak{M})$, $xR_a^{\alpha!}y$ implies $\mathfrak{M}^{\alpha!}, y \models A$. So, fix any $y \in \mathcal{D}(\mathfrak{M})$ such that $xR_a^{\alpha!}y$. Now it suffices to show $\mathfrak{M}^{\alpha!}, y \models A$. By Proposition 8, we have $\mathfrak{M}, f \models x\mathsf{R}_a^{\alpha}y$. Suppose for contradiction that $x\mathsf{R}_a^{\alpha}y \in \Delta^+$. By induction hypothesis, $\mathfrak{M}, f \not\models x\mathsf{R}_a^{\alpha}y$. A contradiction. Therefore, $x\mathsf{R}_a^{\alpha}y \notin \Delta^+$. Since $\Gamma^+ \Rightarrow \Delta^+$ is saturated and $x{:}^{\alpha}\mathsf{K}_aA \in \Gamma^+$, we have $x\mathsf{R}_a^{\alpha}y \in \Delta^+$ or $y{:}^{\alpha}A \in \Gamma^+$. It follows that $y{:}^{\alpha}A \in \Gamma^+$, hence $\mathfrak{M}^{\alpha!}, y \models A$ by induction hypothesis.
(ii) Suppose $x{:}^{\alpha}\mathsf{K}_aA \in \Delta^+$. By Definition 11, $x\mathsf{R}_a^{\alpha}y \in \Gamma^+$ and $y{:}^{\alpha}A \in \Delta^+$, for some y. By induction hypothesis, $\mathfrak{M}, f \models x\mathsf{R}_a^{\alpha}y$ and $\mathfrak{M}, f \not\models y{:}^{\alpha}A$, for some y. By Proposition 8, the definition of f and Definition 5, $\langle x, f(y)\rangle \in R_a^{\alpha!}$ and $\mathfrak{M}^{\alpha!}, f(y) \not\models A$, for some y. Then, we get the goal: $\mathfrak{M}, f \not\models x{:}^{\alpha}\mathsf{K}_aA$. □

Corollary 3 *Given any formula A and label $x \in \mathsf{Var}$, the following are equivalent.*

(i) *A is valid on all Kripke models for the world-deletion semantics.*
(ii) **HPAL** $\vdash A$
(iii) **GPAL**$^+ \vdash \Rightarrow x{:}^{\epsilon}A$
(iv) **GPAL** $\vdash \Rightarrow x{:}^{\epsilon}A$
(v) *A is valid on all Kripke models for the link-cutting semantics.*

Proof The direction from (v) to (iv) is established by Theorem 4 and the direction from (ii) to (v) is shown by Propostion 7. Then, Corollary 2 implies the equivalence between five items. □

7.7 Conclusion

We found that inference rules for accessibility relations were missing in the existing labelled sequent calculus of **G3PAL**, and that (RA4), one of axioms in **HPAL**, was not provable by the system, although it should be if it is complete for Kripke semantics. Therefore, we have revised **G3PAL** by reformulating and adding some rules to it and named our calculus **GPAL**. During this revision, we also make the notion of *surviveness* explicit. According to this revision, we can show that **GPAL** is sound for Kripke semantics. Moreover, by carefully considering the notion of surviveness, we found the link-cutting version of PAL's semantics is more applicable to our labelled sequent calculus than the standard semantics i.e., the world-deletion semantics, and then we have shown **GPAL** is complete for the link-cutting semantics. Lastly, we would like to stress that the consideration of surviveness in the the restricted domain may be significant not only to PAL but also to other dynamic epistemic logics, such as Action Model Logic (cf. [3, 15]), in general where we need a restriction on possible worlds.

Acknowledgments We would like to thank an anonymous reviewer for his/her constructive comments to our manuscript. We also would like to thank the audiences in the Second Taiwan Philosophical Logic Colloquium (TPLC 2014) in Taiwan and the 49th MLG meeting at Kaga, Japan, particularly Makoto Kanazawa for a helpful comment on the link-cutting semantics at the MLG meeting. The second author would like to thank Didier Galmiche for a discussion on the topic of this paper. Finally, we are grateful to Sean Arn for his proofreading of the final version of the paper. This work of the first author was supported by Grant-in-Aid for JSPS Fellows, and that of the second author was supported by JSPS KAKENHI, Grant-in-Aid for Young Scientists (B) 24700146 and 15K21025. This work was conducted also by JSPS Core-to-Core Program (A. Advanced Research Networks).

References

1. Balbiani, P., Demange, V., Galmiche, D.: A sequent calculus with labels for PAL. Presented in Advances in Modal Logic, 2014
2. Balbiani, P., van Ditmarsch, H., Herzig, A., de Lima, T.: Taleaux for public announcement logic. J. Logic Comput. **20**, 55–76 (2010)
3. Baltag, A., Moss, L., Solecki, S.: The logic of public announcements, common knowledge and private suspicions. In: Proceedings of TARK, pp. 43–56. Morgan Kaufmann Publishers, Los Altos (1989)
4. Gentzen, G.: Untersuchungen Über das logische Schließen. I. Mathematische Zeitschrift **39**, (1934)
5. Hansen, J.U.: A logic toolbox for modeling knowledge and information in multi-agent systems and social epistemology. PhD thesis, Roskilde University (2011)
6. Kashima, R.: Mathematical Logic. Asakura Publishing Co., Ltd (2009). (in Japanese)
7. Maffezioli, P., Negri, S.: A Gentzen-style analysis of public announcement logic. In: Proceedings of the International Workshop on Logic and Philosophy of Knowledge, Communication and Action, pp. 293–313 (2010)
8. Negri, S.: Proof analysis in modal logic. J. Philos. Logic **34**, 507–544 (2005)
9. Negri, S., von Plato, J.: Structural Proof Theory. Cambridge University Press (2001)

10. Negri, S., von Plato, J.: Proof Analysis. Cambridge University Press (2011)
11. Ono, H., Komori, Y.: Logics without contraction rule. J. Symbolic Logic **50**(1), 169–201 (1985)
12. Plaza, J.: Logic of public communications. In: Proceedings of the 4th International Symposium on Methodologies for Intellingent Systems: Poster Session Program, pp. 201–216 (1989)
13. Troelstra, A.S., Schwichtenberg, H.: Basic Proof Theory. Cambridge University Press, 2 edn (2000)
14. van Benthem, J., Liu, F.: Dynamic logic of preference upgrade. J. Appl. Non-Classical Logics **17**, 157–182 (2007)
15. van Ditmarsch, H., Hoek, W., Kooi, B.: Dynamic Epistemic Logic. Springer Verlag Gmbh (2008)

Chapter 8
Logics for Dynamic Epistemic Behavioral Strategies

Joshua Sack

Abstract This paper shows how the probabilistic logic of communication and change can be used to reason about finite extensive-form games with incomplete or imperfect information and with probabilistic nature moves. The results of probabilistic behavioral strategies can be expressed, as well as the results of strategies that are sensitive not also just to the history of the game, but also to the beliefs of agents. Using this logic, game-theoretic concepts, such as best response, Nash equilibrium, and rationality can be expressed with respect to a finite set of possible strategies. Extensions to the logic are also proposed to allow for the comparison between one strategy and infinitely many others, thus providing less restricted expressions for best response, Nash equilibrium, and rationality.

Keywords Dynamic epistemic logic · Valuation change · Behavioural strategies · Imperfect information games

8.1 Introduction

In imperfect or incomplete information games with nature moves, hints about the structure of the game can be revealed by the moves of both chance (nature) and agent players. One example of such a game is the Urn Game. In this game, people line up to enter a room they all know contains either MW, the "majority white" urn with two white balls and one black ball, or MB, the "majority black" urn with two black balls and one white ball (but no one knows which one of these urns is in the room). Each player enters the room one by one to (1) draw a ball, observe its color, and replace it to the urn, and then (2) write down for everyone to see either MW or MB, typically

---

The research by this author has been made possible by VIDI grant 639.072.904 of the Netherlands Organization of Scientific Research (NWO).

J. Sack (✉)
Department of Mathematics and Statistics, California State University Long Beach,
1250 Bellflower Blvd, Long Beach CA 90840, USA
e-mail: joshua.sack@gmail.com

S.C.-M. Yang et al. (eds.), *Structural Analysis of Non-Classical Logics*,
Logic in Asia: Studia Logica Library, DOI 10.1007/978-3-662-48357-2_8

reflecting a guess the player has as to which urn the player thinks is in the room (the majority white or the majority black). A natural choice for payoffs may be to reward each player who guesses correctly. When forming the guess, the player puts into consideration both the color drawn (nature's move) as well as all the choices (or guesses) of either MW or MB made by the players who moved earlier (agent player moves), and in considering the actions of others, a player is likely to make assumptions about the beliefs of others as well (higher order beliefs). If a certain point is reached where the consideration of the previously made choices outweighs the direct signal from nature (the color drawn), we say an "informational cascade" has formed. Without payoffs, this scenario is called the Urn Example, and to make it a fully defined game, there are many possibilities for how players can be rewarded for various patterns of choices (see [1] for a number of such possibilities).

The Urn Game and the Urn Example are among a number of scenarios, called *Social Proof*, that illustrate group behavior. Examples of social proof include informational cascades (where agents act in sequence, which the Urn Example illustrates), conformity (where behavior is orchestrated by a common sense of obligation), and herd behavior (where agents act together), and being able to analyze such games using logic will shed light on such social phenomena. A goal of this paper is to show how logic can be used for reasoning about imperfect or incomplete information games with nature moves, thus promising further to help us reason about informational cascades and bring us closer to reasoning formally about other types of social proof. Well-known logics for games, such as Alternating-time Temporal Logic [4], Strategy Logic [9], and Game Logic [14], fall short of this goal as they are all in a perfect information game setting, and even Alternating-time Temporal Epistemic Logic, as in [12], which does allow one to reason about qualitative uncertainty of agents, does not involve probabilities, such as subjective probabilities or behavioral strategies. Game Logic and Alternating-time Temporal Logic do well to express the powers groups of agents have over time in concurrent game settings, with strategy logic being more explicit about the effects of strategies over time, but they do not capture the epistemic uncertainty players may have about the game and each other; even Probabilistic Alternating-time Temporal Logic, as in [10], only provides probabilistic uncertainty about the outcomes of actions, not uncertainty about the current state or the other players' thoughts. Another probabilistic game logic is the Modal Logic for Mixed Strategies in [15], which was the first modal logic to reason about certain game concepts such as mixed Nash equilibria, but this logic is static in time and only addresses normal form games. Epistemic approaches for reasoning about extensive games include variants of dynamic epistemic logic such as those in [6] and [5], which focus on qualitative epistemic aspects of extensive-form games. (See also [7] for more information about logics in games.) Although these works touch on probability, to the best of my knowledge a fully developed probabilistic extension of this line of work has not yet been developed.

The main focus of this paper is to show how the Probabilistic Logic of Communication and Change (PLCC), which was developed in [1] to reason about probabilistic dynamic multi-agent settings such as the Urn Example without explicit mention of strategies or payoffs, can also be used for reasoning about extensive-form games with both probabilistic and epistemic structure. This logic combines epistemics (to reason about the beliefs agents have about each other), common knowledge (of, for example, the structure of the scenario), probability (for Bayesian reasoning), and dynamic updates (to reflect how everyone's beliefs change after an action is made). Although it lacks explicit components for reasoning about preferences, we will see how, given a fixed game, we can express preferences of one game node over another, or even one strategy over another, by quantifying over finitely many propositional valuations, each of which represent a node of a game. Although strategies are not explicitly expressed as they are in, say strategy logic, we will see that they are implicit in many instances of the primary dynamic semantic structure: the event model. The update semantics involves an event model that, when satisfying appropriate conditions, effectively encodes behavioral strategies of each player (what action would be performed given certain preconditions). While behavioral strategies are typically defined as functions from where the player thinks she is in the game tree or game forest (reflected by an information set of possible nodes) to a probability function over available moves, an event model allows for more subtle definitions of strategy where the agent may make her choice of a probability function also depend on other beliefs she has, such as what strategies other players might use. This paper refers to such strategies as *epistemic behavioral strategies*.

The approach of representing strategies in the dynamic component of the semantics is significantly different from previous dynamic epistemic approaches, such as the one in [5], where a strategy is determined from the epistemic structure of the (static) Kripke model. Here, formulas are interpreted on probabilistic "pointed Bayesian Kripke models" which, when using a variation of the PLCC semantics that does not fix a specific event model, need not commit to any strategy for any player. It is essentially the event model that contains the information about the strategies an agent plays. Although PLCC fixes a single finite event model, the event model may reflect finitely many alternative strategies for each player. Fixing the finite event model constrains the reasoning of the logic to only finitely many profiles of strategies, which in a probabilistic setting, may be a significant restriction. It furthermore typically binds points of a Bayesian Kripke model to a certain strategy represented by the fixed event model. To overcome these limitations, this paper considers a variation of PLCC that does not fix an event model (a common approach among dynamic epistemic logics) and that includes new operators for comparing the utility of event models and their induced strategies as well as the utility of an event model with infinitely many others. With this variation of PLCC that does not fix an event model, the points of a Bayesian Kripke model are truly independent of a particular strategy. In this way, the logic becomes strategically dynamic (determined on the fly), whereas other dynamic epistemic logic approaches are strategically static.

An essential technical difference between the dynamics used here and the one used in other dynamic epistemic logic approaches to reasoning about games is simply the involvement of "valuation change" (or fact change), and this small technical extension allows for a very significant change in interpretation. With valuation change, the atomic propositions assigned to a point in the Bayesian Kripke model may commit to a particular node of the game tree without disrupting the possibility of reasoning about the game through time. Past approaches have handled the passage of time by restricting the uncertainty regarding the possible outcomes of the game or the set of strategy profiles that can be played; but in those settings, the points in the model reflected a particular outcome or strategy profile, and hence made the strategies static. The use of dynamic strategies, however, allows us to better model strategies as actions and to describe various consequences of such strategies; it further makes it easier to describe precisely and explicitly what stage of a game we are reasoning about. This versatility is helpful for reasoning about extensive-form games.

This paper is organized as follows: Section 8.2 define the extensive-form games we reason about in this paper. Section 8.3 introduces a variant of the probabilistic logic of communication and change that is slightly weaker than the one defined in [1]. It is also shown how event models with certain constraints can naturally represent some game and strategy structure. Section 8.4 defines classes of event models for given games, including event models that capture different strategies on finitely many copies of a single game. Section 8.5 shows how we can, using our weaker variety of the probabilistic logic of communication and change, reason about payoffs and express notions of Nash equilibrium and rationality relative to a fixed set of strategy profiles. Another variation of the probabilistic logic of communication and change is defined that allows for comparison between any strategies for the game as well as allows us to express rationality that is not relative to a given set of strategy profiles.

8.2 Game Structures

In this section, we define imperfect and incomplete information games, represented as game forests or game trees, and discuss strategies as well as additional structures for reasoning about alternative strategies.

8.2.1 Incomplete and Imperfect Information Games

We adapt the definition of a finite extensive-form game (such as in [13]) to one more relevant to our paper. One significant difference is that we enforce *epistemic synchronicity*, the condition that any two nodes of a game tree or game forest that an agent is uncertain between must represent the same point in time (the same number of actions have been played up until that point). Another difference, which is essentially a difference in formalism and not substance, is that we represent nodes of a game tree as sets of actions rather than sequences of actions as they are done in [13]. We impose constraints in order to ensure that the sets are arranged in a tree-like or forest-like fashion. There is no loss of generality in representing nodes as sets, assuming that we can adjust the names of the sets of actions; for example, we could replace the set of actions Ev with the union of $\{n\} \times \mathsf{Ev}$ for each n representing the depth of the game at which the actions can be played (this is effectively how actions were named in the Urn Example presented in [1–3]). Representing nodes as (unordered) sets is precisely the approach we use in the semantics of the logic we will describe, which is why it is convenient to define the game this way.

We employ the following notation concerning the subsets of any set S. For $x \in S$ and $A, B, X \subseteq S$:

- We write $A \xrightarrow{x} B$ if $A \cup \{x\} = B$. We write $A \not\twoheadrightarrow xB$ otherwise.
- We write $A \xrightarrow{X} B$ if $A \xrightarrow{x} B$ for some $x \in X$. We write $A \not\twoheadrightarrow XB$ otherwise
- We write $A \rightarrow B$ if $A \xrightarrow{S} B$, that is, $B \setminus A$ is a singleton. We write $A \not\twoheadrightarrow B$ otherwise.

We assume a finite set of *agent players* Ag and a finite set of actions or *events* Ev. We highlight these two components, as they are particularly relevant to the language of the probabilistic logic of communication and change.

Definition 1 A *preference-based* $(\mathsf{Ag}, \mathsf{Ev})$ *forest game* is a tuple

$$\mathcal{F} = (\mathcal{X}, H, \iota, f_\nu, \sim, \succeq),$$

1. $\mathcal{X}$ is a set whose subsets index (epistemically possible) of game trees
2. $H \subseteq \mathcal{P}(\mathcal{X} \cup \mathsf{Ev})$ is a set of *histories* (also called *nodes* or *states*), such that every history in H has at most one predecessor, that is, both of the following hold:

   a. if $h \in H$ and $h \nsubseteq X$, then there is an $h' \in H$ such that $h' \xrightarrow{\mathsf{Ev}} h$. ($h$ has a predecessor)

   b. if $h, h', h'' \in H$ and both $h \xrightarrow{\mathsf{Ev}} h''$ and $h' \xrightarrow{\mathsf{Ev}} h''$, then $h = h'$ (Any predecessor of h is unique.)

   We define the notation

   - $E(h) = \{e \in \mathsf{Ev} \mid h \xrightarrow{e} h' \text{ for some } h' \in H\}$ is the set of actions that can be performed at h.
   - $Z = \{h \in H \mid h \not\twoheadrightarrow \mathsf{Ev}h' \text{ for all } h' \in H\}$ is the set of terminal nodes.
   - $\ell(h)$ is the size of $h \cap \mathsf{Ev}$.

3. $\iota : H \setminus Z \rightarrow \mathsf{Ag} \cup \{\nu\}$ is a player function (where ν is *nature*).

   - For each player $\varpi \in \mathsf{Ag} \cup \{\nu\}$, let $H_\varpi = \iota^{-1}[\{\varpi\}]$ be the set of nodes in which ϖ moves.

4. f_ν maps each $h \in H_\nu$ to a probability mass function over $E(h)$. (f_ν may be thought of as a "strategy" for chance or "nature")
5. $\sim \overset{\text{def}}{=} \{\sim_a \mid a \in \mathsf{Ag}\}$ is a collection of "epistemic" equivalence relations $\sim_a$ $\subseteq (\mathsf{Ev} \times \mathsf{Ev}) \cup (\mathcal{P}(\mathcal{X}) \times \mathcal{P}(\mathcal{X}))$ for each agent player a, such that when $\sim_a^H$ is the smallest relation over H such that

- $h \sim_a^H k$ whenever $h, k \in \mathcal{P}(\mathcal{X})$ and $h \sim_a k$, and
- $h \cup \{e\} \sim_a^H k \cup \{f\}$ whenever $h \sim_a^H k$, $e \in E(h)$, $f \in E(k)$, and $e \sim_a f$.

then for each $a \in \mathsf{Ag}$, the following both hold when $h \sim_a^H k$ and $\iota(h) = a$:

a. $\iota(k) = a$
b. $E(h) = E(k)$.

A game does not specify

6. $\succeq = \{\succeq_a \mid a \in \mathsf{Ag}\}$ is a collection of preference relations $\succeq_a \subseteq Z \times Z$, each being reflexive, transitive, and connected.

A preference-based $(\mathsf{Ag}, \mathsf{Ev})$ forest game for which $\mathcal{X} =$ has a tree-like structure, and corresponds to an imperfect information game. The involvement of multiple possible trees allows us to describe uncertainty players have not just about where in the game he/she is, but what the game structure is. In this regard, we are considering incomplete information games as well as imperfect information games. A game forest could be replaced by an equivalent structure that is just a game tree where nature makes the first move over a given probability distribution, picking the root of any of the games in the original forest. This is slightly different from our setting in that it commits us to a particular probability distribution, where we opt to leave that variable. In the Urn Game, we may view the majority white urn and the majority black urn as two different games, where players are uncertain as to which urn in is (this interpretation is not necessary, as there could be a nature move choosing which urn is in the room, but this is the setting used in [1–3], and hence we will adopt it here).

Example 1 (Urn game) This example is an adaptation of one in [1] to the exact notation used here. Let

- $\mathsf{Ag} = \{1, \ldots, n\}$ be a set of agents, and let
- $\mathsf{Ev} = \{\mathsf{dw}_a, \mathsf{db}_a, \mathsf{w}_a, \mathsf{b}_a \mid a \in \mathsf{Ag}\}$ the set of actions
- $\mathcal{X} = \{\mathsf{mw}, \mathsf{mb}\}$ gives indices for the types of game trees: "majority white urn" and "majority black urn" game trees. The game tree indexed by $\emptyset$ and $\mathcal{X}$ will be empty.
- $H = \mathcal{P}(\mathcal{X}) \cup \bigcup_{a \in \mathsf{Ag}} (H^{\mathsf{drew}_i} \cup H^{\mathsf{wrote}_a})$, where
  - $H^{\mathsf{wrote}_0} = \mathcal{X}$,
  - $H^{\mathsf{drew}_a} = \{h \cup \{d\} \mid h \in H^{\mathsf{wrote}_{a-1}}, d \in \{\mathsf{dw}_a, \mathsf{db}_a\}\}$ (for $a \in \mathsf{Ag}$), and
  - $H^{\mathsf{wrote}_a} = \{h \cup \{w\} \mid h \in H^{\mathsf{drew}_a}, w \in \{\mathsf{w}_a, \mathsf{b}_a\}\}$ (for $a \in \mathsf{Ag}$)
- ι maps each agent player $a \in \mathsf{Ag}$ to H^{drew_a} (the positions where the player has just drawn ball and now must write down a guess), and maps the "chance" player ν to $\bigcup_{a \in \mathsf{Ag}} H^{\mathsf{wrote}_{a-1}}$ (the positions where agent player a is about to draw)
- f_ν maps each $h \in H^{\mathsf{wrote}_{a-1}}$ to

$$\begin{cases} \mu_w & \mathsf{mw} \in h \\ \mu_b & \mathsf{mb} \in h \end{cases}$$

where μ_w given weight 2/3 to dw_a and 1/3 to db_a, and μ_b given weight 2/3 to db_a instead, and 1/3 to dw_a.

- $\sim$ is defined by $h \sim_a k$ iff the following two conditions hold:
  - $\ell(h) = \ell(k)$

    (The same amount of time has ellapsed)
  - $e \in h$ iff $e \in k$ for each $e \in H^{\mathsf{drew}_a} \cup \bigcup_{a \in \mathsf{Ag}} H^{\mathsf{wrote}_a}$

    (h and k agree on all actions that a can observe)
- $\succeq$ is defined by $h \succeq_a k$ iff either of the following hold
  - $\mathsf{mw} \in h$ iff $\mathsf{w}_a \in h$ (a correct guess for a in h), or
  - $\mathsf{mw} \in k$ iff $\mathsf{b}_a \in k$ (an incorrect guess for a in k)

One could replace the preference relation with a utility function. Agent a's utility for node h is 1 if a guessed correctly and 0 otherwise. This utility function induced the relation in the example (where $h \succeq_a k$ if and only if h has at least as high a utility for a as k). But this is just one of many examples of how to reward players for certain behavior, thus turning the Urn Example into an urn game. See [1] for more examples.

Definition 2 A *utility-based* (Ag, Ev) *forest game* is a tuple $(\mathcal{X}, H, \iota, f_\nu, \sim, u)$ which is defined exactly as the preference-based (Ag, Ev) forest games, except that $\succeq$ is replaced by a set u of utility functions $u_a : Z \to \mathbb{R}$ for each agent player $a \in \mathsf{Ag}$.

8.2.2 Strategies and Copies of the Same Game Forest

A *behavioral strategy* for agent player a is a function from each information set, ($\sim_a$-equivalence class) belonging to nodes of a game forest for which a can move, to a distribution on actions available from the nodes in the information set ($\sim_a$ is defined in such a way that the available action from any node of an equivalence class is the same for all nodes in the equivalence class). One can imagine nature as a player whose epistemic equivalence relation is the smallest reflexive relation (yielding certainty at each node). The function f_ν can be thought of as a strategy for nature that is built into the definition of the game. But strategies for agent players are not defined by the game and constitute additional structure.

Reasoning about solution concepts, such as Nash equilibrium, involves comparing strategies. To facilitate this, it may be helpful to reason about different copies of the same game forest, where a strategy for each player is associated with each copy of the game. One way to do this is to introduce another index set Σ for strategies, and to define a *duplicated game forest* as tuple $\mathcal{D} = (\Sigma, \mathcal{F}, \sim^{\Sigma})$, where $\mathcal{F} = (\mathcal{X}, H, \iota, f_\nu, \sim^H, \succeq^{\mathcal{F}})$ is a game forest, and $\sim^{\Sigma}$ is a collection of equivalence relations $\sim_a^\Sigma$ over Σ for each $a \in \mathsf{Ag}$. We can thus extend each component of $\mathcal{F}$ as follows:

- The states space of $\mathcal{D}$ is $D = \{\{\sigma\} \cup h \mid \sigma \in \Sigma, h \in H\}$.
  - We let Z^H be defined according to Definition 1, and we let $Z^D = \{\{\sigma\} \cup h \mid \sigma \in \Sigma, h \in Z^H\}$.
- $\iota : D \setminus Z \to \mathsf{Ag} \cup \{c\}$ by $\iota : \{\sigma\} \cup h \mapsto \iota(h)$.
  - Let $D_{\sigma,\varpi} = \{\{\sigma\} \cup h \mid h \in H_\varpi\}$ for each $\sigma \in \Sigma$ and $\varpi \in \mathsf{Ag} \cup \{\nu\}$.
  - Let $D_\varpi = \bigcup_{\sigma \in \Sigma} D_{\sigma,\varpi}$ for each player $\varpi \in \mathsf{Ag} \cup \{\nu\}$.
  - Let $D_\sigma = \{\{\sigma\} \cup h \mid h \in H\}$ for each $\sigma \in \Sigma$.
- $f_\nu : D_\nu \to (\mathsf{Ev} \to [0, 1])$ is defined by $f_\nu : \{\sigma\} \cup h \mapsto f_\nu(h)$.
- $\sim_a^D$ is defined by $(\{\sigma\} \cup h) \sim_a^D (\{\tau\} \cup k)$ if and only if $\sigma \sim_a^\Sigma \tau$ and $h \sim_a^H k$.
- $\succeq_a^D$ is defined by $(\{\sigma\} \cup h) \succeq_a^D (\{\tau\} \cup k)$ iff $h \succeq_a^\mathcal{F} k$.

An *epistemic behavioral strategy* is similar to a behavioral strategy, where the choices depend not just on the information state of the game forest, but also on the information state of a model external to the game forest. This external model, a Bayesian Kripke frame is the basic structure the probabilistic logic of communication and change describes. We next provide details.

8.3 A Variation of PLCC

Here we present a variation of the probabilistic logic of communication and change, almost as it was done in [1], with respect to a set Ag of agents, a set Ev of informational events (such as information about a move of a game), and a set At of atomic propositions. Two key differences between the definition here and that in [1] is that here we assume $\mathsf{Ev} \subseteq \mathsf{At}$ (with equality if we wished to model just a game tree rather than forest) and our semantics has a less general (but more relevant to the game setting here) way of addressing valuation change. With $\mathsf{Ev} \subseteq \mathsf{At}$, an $e \in \mathsf{Ev}$ represents the information about a move (which possibly not all players see/hear), and the same e as an atomic proposition would represent that e had already occurred.

The language of the *Probabilistic Logic for Communication and Change*, denoted $\mathcal{L}_{\mathrm{PLCC}}(\mathsf{Ag}, \mathsf{Ev}, \mathsf{At})$, is given by the following Backus Naur form:

$$\phi ::= true \mid p \mid \neg\phi \mid \phi_1 \wedge \phi_2 \mid [\pi]\phi \mid [e]\phi \mid t_a \geq \beta$$

$$t_a ::= \alpha \cdot P_a(\phi) \mid t_{a1} + t_{a2}$$

$$\pi ::= a \mid \pi_1; \pi_2 \mid \pi_1 \cup \pi_2 \mid \pi^* \mid \phi?$$

where $p \in \mathsf{At}$ is an atomic proposition, $a \in \mathsf{Ag}$ is an agent, α, β are rational numbers, and $e \in \mathsf{Ev}$ is an informational event.

The semantics are given on Bayesian Kripke models

Definition 3 (*Bayesian Kripke models*) Given sets Ag and At, a *Bayesian Kripke model* is a quadruple $M = (S, \sim, \mu, V)$ where:

- S is a nonempty set of states.
- $\sim$ is a family of equivalence relations $\sim_a$ on S, one for each agent $a \in \mathsf{Ag}$.
- μ is a family of functions $\mu_a : S \to (S \to [0, 1])$, one for each agent $a \in \mathsf{Ag}$, whose values are denoted by $\mu_a^s(s')$ and satisfy the conditions:
  - State determined probability (SDP): if $s \sim_a t$ then $\mu_a^s(s') = \mu_a^t(s')$, for all $s' \in S$;
  - Consistency (CONS): $\mu_a^s(t) = 0$ if $s \nsim_a t$;
  - Caution (CAUT): $s \nsim_a t$ if $\mu_a^s(t) = 0$;
  - Probability (PROB): for every $s \in S$, $\sum_{t \in S} \mu_a^s(t) = 1$.
- $V : \mathsf{At} \to \mathcal{P}(S)$ is a *valuation function*.

Given a Bayesian Kripke model $M = (S, \sim, \mu, V)$, for each $s \in S$, let

$$\mathsf{At}(s) \stackrel{\text{def}}{=} \{p \in \mathsf{At} \mid s \in V(p)\}$$

Definition 4 (*Event Models*) An *event model* over $\mathcal{L}_{\text{PLCC}}$ is the quintuple $\mathcal{E} = (\mathsf{Ev}, \sim, \Phi, \text{pre})$ where:

- Ev is a finite nonempty set of *events*.
- $\sim$ is a set of equivalence relations $\sim_a$ for each agent $a \in \mathsf{Ag}$.
- Φ is a finite set of pairwise unsatisfiable formulas called *preconditions*.
- pre is a family of functions $\text{pre}_a : \Phi \to (\mathsf{Ev} \to [0, 1])$ for each $a \in \mathsf{Ag}$ assigning to each precondition $\phi \in \Phi$ a *subjective occurrence probability* function over Ev (i.e., $\sum_{e \in \mathsf{Ev}} \text{pre}_a(\phi)(e) = 1$), such that $\text{pre}_a(\phi)(e) > 0$ if and only if $\text{pre}_b(\phi)(e) > 0$ for every $a, b \in \mathsf{Ag}$ and $e \in \mathsf{Ev}$.

We define PRE : $\mathsf{Ev} \to \mathcal{P}(\Phi)$, such that PRE : $e \mapsto \{\phi \mid \{\text{pre}_a(\phi)(e) > 0\}$ for any (and hence all) $a \in \mathsf{Ag}$.

Given a Bayesian Kripke model, $M = (S, \sim, \mu, V)$, and a state $s \in S$ define

$$\text{pre}_a(e \mid s) \stackrel{\text{def}}{=} \begin{cases} \text{pre}(\phi)(e) & \phi \in \Phi, M, s \models \phi \\ 0 & \text{there is no such } \phi \end{cases} \tag{8.1}$$

Definition 5 (*Product Update*) The update product of a static Bayesian Kripke model $M = (S, \sim, \mu, V)$ with an event model $\mathcal{E} = (\mathsf{Ev}, \sim, \Phi, \text{pre})$ is the weighted epistemic model $M \otimes \mathcal{E} = (S \otimes \mathsf{Ev}, \sim, \mu, V)$ where:

- $S \otimes \mathsf{Ev} \stackrel{\text{def}}{=} \{(s, e) \mid s \in S, e \in \mathsf{Ev}, (M, s) \models \bigvee \text{PRE}(e)\}$.
- $(s, e) \sim_a (s', e')$ iff $s \sim_a s'$ and $e \sim_a e'$.
- Let $D \stackrel{\text{def}}{=} \sum_{(s',e') \sim_a (w,g)} \left(\mu_a^w(s') \cdot \text{pre}_a(e' \mid s')\right)$, and put:

$$\mu_a^{(w,g)}(s,e) \stackrel{\text{def}}{=} \begin{cases} \frac{\mu_a^w(s)\cdot \text{pre}_a(e|s)}{D} & \text{if } (s,e) \sim_a (w,g) \\ 0 & \text{otherwise} \end{cases}$$

(Note that $D \neq 0$ for $(w, g) \in S \otimes Ev$.)

- $V^{M\otimes\mathcal{E}}(p) \stackrel{\text{def}}{=} \{(s,e) \mid e \in V^M(p) \text{ or } p = e\}$.

In a game setting, we think of the atomic propositions p at a state s as including both atomic facts about the situation (whether the urn actually does have a majority of white or a majority of black balls) as well as a history of the actions already performed. Thus after playing e, we retain all of these facts, and add just one more fact, that e has now been played.

Definition 6 (*Semantics of PLCC*) The semantics for $\mathcal{L}_{\text{PLCC}}$ is given by a relation $\models$ between pointed models (M, s), with $M = (S, \sim, \mu, V)$ and $s \in S$, and formulas ϕ, such that

$$\begin{array}{lll} M,s \models \mathit{true} & \text{iff} & \text{always} \\ M,s \models p & \text{iff} & s \in V(p) \\ M,s \models \neg\phi & \text{iff} & M,s \not\models \phi \\ M,s \models \phi \wedge \psi & \text{iff} & M,s \models \phi \text{ and } M,s \models \psi \\ M,s \models [e]\phi & \text{iff} & M,s \models \bigvee \text{PRE}(e) \text{ then } M \times \mathcal{E}, (s,e) \models \phi, \\ & & \text{where e is an event in the event model } \mathcal{E} \\ M,s \models [\pi]\phi & \text{iff} & \text{for all } t \in S: \text{ if } sR_\pi t \text{ then } M,t \models \phi, \\ M,s \models \sum_{j=1}^n \alpha_j P_a(\phi_j) & \text{iff} & \sum_{j=1}^n \alpha_j \cdot \mu_a^s(\phi_j) \geq \beta \end{array}$$

where $\mu_a^s(\phi_j)$ is an abbreviation for $\sum_{s'\in S, s'\models\phi_j} \mu_a^s(s')$, and R_π is a binary relation given by

$$\begin{array}{lll} sR_a t & \text{iff} & s \sim_a t \\ sR_{\pi_1\cup\pi_2} t & \text{iff} & sR_{\pi_1} \cup R_{\pi_2} t \\ sR_{\pi_1;\pi_2} t & \text{iff} & sR_{\pi_1}; R_{\pi_2} t \text{ (there is } w, \text{ such that} sR_{\pi_1} w \text{ and } wR_{\pi_2} t) \\ sR_{\pi^*} t & \text{iff} & s(R_\pi)^* t \text{ (where} (R_\pi)^* \text{ is the reflexive transitive closure of } R_\pi) \\ sR_{\phi?} t & \text{iff} & s = t \text{ and } s \models \phi \end{array}$$

We write $M \models \varphi$ if $M, s \models \varphi$ for every $s \in S$. We write $\models \varphi$ if $M, s \models \varphi$ for every pointed Bayesian Kripke model M, s.

8.3.1 Event Models for Game Structures

With a few constraints that we define in this section, an event model may describe a game forest structure with epistemic relations for each agent. Who plays at which node, nature's probability function, and the payoff functions are not easily extracted from the event model.

Given a set At of atomic propositions, for any $U \subseteq \mathsf{At}$, let

$$\widetilde{U} \stackrel{\text{def}}{=} \bigwedge_{p \in U} p \wedge \bigwedge_{p \in \mathsf{At} \setminus U} \neg p.$$

If $\mathcal{E}$ were an event model over actions Ev, then for each $e \in \mathsf{Ev}$, we define propositional assignments compatible with e by

$$H_e \stackrel{\text{def}}{=} \{U \subseteq \mathsf{At} \not\models \widetilde{U} \wedge \bigvee \text{PRE}(e) \to \mathit{false}\}.$$

We will identify propositional assignments with nodes of a game tree, and hence the nodes in H_e are those in which e could (in the right situation) be played. Given any $U \subseteq \mathsf{At}$, let

$$E(U) \stackrel{\text{def}}{=} \{e \mid U \in H_e\}.$$

Finally, define the event model induced history space by

$$H \stackrel{\text{def}}{=} \{U, U \cup \{e\} \mid e \in E, U \in H_e\}.$$

We then define $Z \stackrel{\text{def}}{=} \{U \in H \mid E(U) = \emptyset\}$. Recall that $\mathsf{Ev} \subseteq \mathsf{At}$; so let $\mathcal{X} = \mathsf{At} \setminus \mathsf{Ev}$.

An *a-epistemic* formula is a formula of the form $[a]\psi$ for any formula ψ of $\mathcal{L}_{\text{PLCC}}(\mathsf{Ag}, \mathsf{Ev}, \mathsf{At})$. An *a-probability* formula is a formula of the form $t_a \geq \beta$ for some a-probability term t_a of $\mathcal{L}_{\text{PLCC}}(\mathsf{Ag}, \mathsf{Ev}, \mathsf{At})$. Let an *a-formula* be a Boolean combination of a-epistemic and a-probability formulas.

We now define a class of event models that are compatible with forest games.

Definition 7 An event model $\mathcal{E} = (\mathsf{Ev}, \sim, \Phi, \text{pre})$ is a *quasi-game event model* if there exist

- a function $\iota : H \setminus Z \to \mathsf{Ag} \cup \{\nu\}$,
- an equivalence relation $\sim_a^{\mathcal{X}} \subseteq \mathcal{P}(\mathcal{X}) \times \mathcal{P}(\mathcal{X})$ for each agent player $a \in \mathsf{Ag}$,
- a set Ψ_a of pairwise unsatisfiable a-formulas for each agent player $a \in \mathsf{Ag}$,

such that if $\sim_a^H$ is the smallest relation extending $\sim_a^{\mathcal{X}}$ such that $U \sim_a^H U'$ whenever there exist $V, V' \in H$ and $e, e' \in \mathsf{Ev}$, such that

- $U = V \cup \{e\}$ and $U' = V' \cup \{e'\}$, and
- $e \sim_a e'$ and $V \sim_a^H V'$,

and for each $\varpi \in \mathsf{Ag} \cup \{\nu\}$,

$$H_\varpi = \iota^{-1}[\{\varpi\}],$$

and the following properties hold:

1. For each $U \in H$, either $U \subseteq \mathcal{X}$ or there exists exactly one $V \in H$, such that $U \setminus V$ is a singleton.
(This gives H a forest-like structure.)

2. $\Phi \stackrel{\text{def}}{=} \{\widetilde{U} \wedge \psi \mid U \in H_\varpi, \psi \in \Psi_\varpi, \varpi \in \mathsf{Ag} \cup \{\nu\}\}$, where $\Psi_\nu \stackrel{\text{def}}{=} \{true\}$.
   (Each precondition is unique to a node of the game and an epistemic condition for the player who moves at that node.)
3. For each $U \in H_\nu$, and $\phi \in \Phi$ such that $\widetilde{U} \wedge \phi$ is satisfiable, it holds that $\text{pre}_a(\phi) = \text{pre}_b(\phi)$ for every two agents $a, b \in \mathsf{Ag}$.
   (Everyone agrees on the probability distribution over nature's potential moves.)
4. For each $U, V \in H_a$, if $\phi = \widetilde{U} \wedge \chi$ and $\psi = \widetilde{V} \wedge \chi$ for some $\chi \in \Psi_a$ and if $U \sim_a^H V$, then $\text{pre}_a(\phi) = \text{pre}_a(\psi)$.
   (Given epistemic condition Ø, an agent plays the same distribution from any indistinguishable node.)
5. For each $e \in \mathsf{Ev}$, $\models \bigvee \text{PRE}(e) \to \neg e$
   (This guarantees that e can never be repeated.)

The definition of a quasi-game event model involves several components of a forest game: the set $\mathcal{X}$, the forest of histories H, the player function ι, and the epistemic relations $\sim_a^H$. With the appropriate interpretation, we can also determine the chance (nature) distribution assignment f_ν and strategies for each player. As for preferences, any reflexive, transitive, and connected relations over Z for each agent would work, or any utility assignment on Z for each agent would work.

To determine chance moves and agent player strategies from a quasi-game event model, we make the following interpretive assumptions: (1) everyone is correct about the actual probabilities used by nature (thus their subjective probabilities about nature are objective), and (2) any player who moves at a node knows correctly the probabilities of her own moves.

For each $U \in H_\nu$, let ϕ_U be the unique element of Φ_ν. In light of the first interpretive condition, the chance moves are given by $f_\nu : U \mapsto \text{pre}_a(\phi_U)$ for $U \in H_\nu$ and any $a \in \mathsf{Ag}$ (note that the definition of a quasi-game event model ensures that each $\text{pre}_a(\phi_U)$ does not depend on the agent). In other words, everyone accurately knows the actual probabilities of nature.

Let $\Phi_a = \{\widetilde{U} \wedge \psi \mid U \in H_a, \psi \in \Psi_a\}$. In light of the second interpretive condition, a strategy for player a is the restriction of pre_a to Φ_a. We call such a strategy an *epistemic behavioral strategy*, since the strategy depends on some epistemic condition $\psi \in \Psi_a$ as well as the equivalence class of nodes she is about to play from (the dependence is on equivalence classes of nodes because of condition 4 of Definition 7).

8.3.1.1 The Bayesian Kripke Models for a Quasi-game Event Model

The epistemic structure of a quasi-game event model provides for each agent player indistinguishability among certain sets of atomic propositions. One might want to restrict the semantics of those Bayesian Kripke models that are in some sense compatible with this indistinguishability relation over subsets of At. This leads to the following definition.

Definition 8 A Bayesian Kripke model $M = (S, \sim, \mu, V)$ *respects* $\sim_a^H$ if for every states $s, t \in S$ and sets of propositions $U, V \in H$, such that $M, s \models \widetilde{U}$ and $M, t \models \widetilde{V}$, we have that $s \sim_a t$ implies $U \sim_a^H V$.

A model that respects $\sim_a^H$ allows agent player a to distinguish any two states that have histories that a can distinguish. However, agent a may be able to distinguish between some states that have the same history. There may be certain epistemic properties that help a distinguish such pairs of states.

The property of a Bayesian Kripke model respecting $\sim_a^H$ for each a can be expressed by the formula

$$\mathsf{Resp} = \bigwedge_{a \in \mathsf{Ag}} \bigwedge_{U \in H} (\widetilde{U} \to [a] \bigvee_{V \sim_a^H U} \widetilde{U}).$$

Let $\mathfrak{R}(\sim_a^H)$ denote the class of Bayesian Kripke models that respect $\sim_a^H$ for each $a \in \mathsf{Ag}$. It is easy to see that a Bayesian Kripke model $M \in \mathfrak{R}(\sim_a^H)$ if and only if $M \models \mathsf{Resp}$. One can also check that if $\mathcal{E}$ is a quasi-game event model and $M \in \mathfrak{R}(\sim_a^H)$, then $M \otimes \mathcal{E} \in \mathfrak{R}(\sim_a^H)$ as well.

8.4 Event Models for a Given Game Structure

In the previous section, we consider what event models are in some sense compatible with some game. In this section we start with a game and then consider the event models that are compatible with it.

8.4.1 Epistemic Behavioral Strategies

Let $\mathcal{F} = (\mathcal{X}, H, \iota, f_\nu, \sim, \succeq)$ be a preference-based $(\mathsf{Ag}, \mathsf{Ev})$ forest game. We define an event model for this game (the case where $\mathcal{F}$ is utility-based is similar). Let $\mathsf{At} = \mathsf{Ev} \cup \mathcal{X}$. But since an event model involves information about strategies as well as the game, let us first look at strategies in light of a given game.

Recall the notion of a-formulas from Sect. 8.3.1. For each $a \in \mathsf{Ag}$, we call a finite set Ψ_a of pairwise unsatisfiable a-formulas a set of *epistemic base-conditions*. For notational convenience, we also define $\Psi_\nu \stackrel{\text{def}}{=} \{true\}$. We define the set of *epistemic-based preconditions* (from the Ψ_ϖ) as follows:

$$\Phi \stackrel{\text{def}}{=} \{\widetilde{h} \wedge \psi \mid h \in H_\varpi, \psi \in \Psi_\varpi, \varpi \in \mathsf{Ag} \cup \{\nu\}\}. \tag{8.2}$$

Note that Φ is pairwise unsatisfiable, since $\widetilde{h}$ and $\widetilde{k}$ are together unsatisfiable when $h \neq k$, and each Ψ_a consists of pairwise unsatisfiable formulas. We furthermore define the following subsets of Φ for each $\varpi \in \mathsf{Ag} \cup \{\nu\}$ and $h \in H$:

$$\Phi_\varpi \stackrel{\text{def}}{=} \{\widetilde{h} \wedge \psi \mid h \in H_\varpi, \psi \in \Psi_\varpi\}$$
$$\Phi_h \stackrel{\text{def}}{=} \{\widetilde{h} \wedge \psi \mid \varpi = \iota(h), \psi \in \Psi_\varpi\}$$

Definition 9 An *epistemic behavioral strategy profile* on Φ is a function strat : $\Phi \to (\mathsf{Ev} \to [0, 1])$ such that

1. strat(φ) is a probability function ($\sum_{e \in E} \text{strat}(\varphi)(e) = 1$)
2. The support of strat(φ) is contained in $E(h)$, whenever $\varphi \in \Phi_h$ for some $h \in H$
3. strat$(\varphi) = f_\nu(h)$ whenever $\varphi \in \Phi_h$ for $h \in H_\nu$
4. If $h \sim_a k$ for $h \in H_a$, and if $\phi = \widetilde{h} \wedge \chi$ and $\psi = \widetilde{k} \wedge \chi$ for some $\chi \in \Psi_a$, then strat$(\phi) =$ strat(ψ).

What makes strat epistemic is the constraint place upon Φ [that it satisfy (8.2)].

Definition 10 Given a forest game $\mathcal{F}$, a set of epistemic-based preconditions Φ, and a strategy profile strat defined on Φ, we define $\mathfrak{E}(\mathcal{F}, \Phi, \text{strat})$ to be the set of event models $\mathcal{E} = (\mathsf{Ev}, \sim, \Phi, \text{pre})$, where

- Ev and $\sim$ are the components already given in $\mathcal{F}$, and
- for each agent $a \in \mathsf{Ag}$, $\text{pre}_a : \Phi \to (\mathsf{Ev} \to [0, 1])$ is an epistemic behavioral profile (Definition 9) additionally satisfying $\text{pre}_a(\phi) = \text{strat}(\phi)$, whenever $\phi \in \Phi_a$.

Let $\mathfrak{E}(\mathcal{F}, \Phi)$ be the set of epistemic event models for strategy profiles defined with respect to Φ (thus strat may vary). Let $\mathfrak{E}^e(\mathcal{F})$ be the set of epistemic event models for strategy profiles defined just with respect to $\mathcal{F}$ [thus Φ may also vary so long is it satisfies (8.2) for appropriate Ψ_ϖ].

An *ordinary behavioral strategy* is a special case of the epistemic behavioral strategy where $\Psi_\varpi = \{\mathit{true}\}$ for each $\varpi \in \mathsf{Ag} \cup \{\nu\}$. We call such Ψ_ϖ the *ordinary base-conditions*, and the set Φ determined from such Ψ_ϖ using (8.2) is called the *ordinary precondition set*. Note that the ordinary precondition set depends solely on the nodes of the forest.

An epistemic behavioral strategy (Definition 9) defined over the ordinary precondition set is called an *ordinary behavioral strategy*. Let $\mathfrak{E}^o(\mathcal{F})$ be the set of event models over $\mathcal{F}$ with ordinary behavioral strategies.

We now give an example of an epistemic behavioral strategy that upon certain input mimics an ordinary behavioral strategy.

Example 2 We now build an event model for the Urn Game of Example 1. This is done essentially as was done in [1], but with notational differences among other minor adjustments. Let $\mathsf{At} = \mathsf{Ev} \cup \mathcal{X}$, where Ev and $\mathcal{X}$ are defined according to Example 1. We consider the following strategy for each player a: if a considers mw

respectively mb more likely, then a chooses to do w_a respectively b_a with probability 1; and if a considers then mw and mb equally likely, then a writes down what she drew.

Following the setup of Sect. 8.4.1, we have the following epistemic base conditions:

- $\Psi_\nu = \{true\}$
- $\Psi_a = \{\psi_a^w, \psi_a^b\}$ where

$$\psi_a^w = P_a(\mathsf{mw}) > P_a(\mathsf{mb}) \vee (P_a(\mathsf{mw}) = P_a(\mathsf{mb}) \wedge [a]\mathsf{db}_a)$$
$$\psi_a^b = P_a(\mathsf{mw}) < P_a(\mathsf{mb}) \vee (P_a(\mathsf{mw}) = P_a(\mathsf{mb}) \wedge [a]\mathsf{dw}_a)$$

Note that for each $a \in \mathsf{Ag}$, the elements of Ψ_a are pairwise unsatisfiable.

Define the event model $\mathcal{E} = (\mathsf{Ev}, \sim, \Phi, \mathrm{pre})$ by

- $\sim$ is define such that for each a, $\sim_a$ is the smallest equivalence relation for which $\mathsf{dw}_b \sim_a \mathsf{db}_b$ for each agent player $b \neq a$.
- Φ is defined according to (8.2) using the Ψ_ϖ defined in this example.
- pre is defined by $\mathrm{pre}_a = \mathrm{strat}$, where strat maps

$$\widetilde{h} \wedge \psi \mapsto \begin{cases} \delta_{\mathsf{w}_a} & \psi = \psi_a^w \\ \delta_{\mathsf{b}_a} & \psi = \psi_a^b \\ \mu_w & \psi = true, \mathsf{mw} \in h \\ \mu_b & \psi = true, \mathsf{mb} \in h \end{cases}$$

  and where for each event e, δ_e is the Dirac distribution on e (assigning weight 1 to e and 0 to everything else), and μ_w and μ_b are defined according to Example 1.

Note that strat does indeed satisfy the conditions of Definition 9, as strat depends only on the depth of the game tree and purely epistemic features for each agent player node.

The situation at the beginning of the game is represented by a Bayesian Kripke model, and the play of the game can be illustrated by the update product of this model with multiple applications of the event model, each application being a move of the game. There is flexibility for the initial Bayesian Kripke model. Following [1], we consider the initial Bayesian Kripke model to be any that satisfied the formula $[(\cup_{a \in \mathsf{Ag}} a)^*]\chi$ (which reads that it is common knowledge that χ holds), where

$$\chi = (\mathsf{mw} \vee \mathsf{mb}) \wedge \neg(\mathsf{mw} \wedge \mathsf{mb}) \wedge \bigwedge_{a \in \mathsf{Ag}} (P_a(\mathsf{mw}) = P_a(\mathsf{mb})) \wedge \bigwedge_{e \in \mathsf{Ev}} \neg e$$

(which reads that precisely one of mw or mb is true, each agent considers either equally likely, and no action has yet been performed). Given an input model satisfying this, the distribution over actions a player uses given the epistemic behavioral strategy strat is actually determined by the node of the game forest.

Thus, although strat is an epistemic behavioral strategy, the extra epistemic condition in strat could, given what agents know about each other's startegies, be determined from the information set of nodes. Using a duplicated forest game, we can capture uncertainty agents have of different player's strategies.

Example 3 Suppose we have an initial input model with two states: majority white and majority black. Each agent is uncertain about these two states, with all but agent 3 giving both states equal probability. The third agent gives extremely high probability that the urn has a majority of black balls (and everyone is aware of this about player 3). Now even using this same epistemic behavioral strategy, player 3 may play differently at a particular node of the game tree in this example as in the previous example. For instance, even if the first two players draw and write white, the outcome of the first two draws would not be enough to overturn player 3's belief that it is more probable that the urn has a majority of black balls.

8.4.2 Involving Multiple Strategy Profiles

Given a duplicated game forest $\mathcal{D} = (\Sigma, \mathcal{F}, \sim^{\Sigma})$, let $\mathsf{At} = \mathsf{Ev} \cup \mathcal{X} \cup \Sigma$. For each $\sigma \in \Sigma$, let Ψ_a^σ be a set of epistemic base-conditions defined as in Sect. 8.4.1 but where $\Psi_a^\sigma = \Psi_a^\tau$ whenever $\sigma \sim_a^\Sigma \tau$ for each $a \in \mathsf{Ag}$. Thus using (8.2), the collection of Ψ_a^σ for all the $a \in \mathsf{Ag}$ together determine a domain Φ^σ for a strategy profile over the forest game $\mathcal{F}$ (Definition 9). For each $\sigma \in \Sigma$, let us define

$$\Delta^\sigma = \{\sigma \wedge \varphi \mid \varphi \in \Phi^\sigma\}.$$

For each $\phi \in \Phi^\sigma$, define a correspondence $D^\sigma : \Phi^\sigma \to \Delta^\sigma$ by

$$D^\sigma(\phi) = \sigma \wedge \bigwedge_{\tau \in \Sigma, \tau \neq \sigma} \neg\tau \wedge \phi\,.$$

For $\varpi \in \mathsf{Ag} \cup \{\nu\}$ and $h \in H$, let Δ_ϖ^σ and Δ_h^σ be the images under D^σ of Φ_ϖ^σ and Φ_h^σ respectively. Let

$$\Delta \stackrel{\text{def}}{=} \bigcup_{\sigma \in \Sigma} \Delta^\sigma, \qquad \Delta_\varpi = \bigcup_{\sigma \in \Sigma} \Delta_\varpi^\sigma, \qquad \Delta_h \stackrel{\text{def}}{=} \bigcup_{\sigma \in \Sigma} \Delta_h^\sigma.$$

We call Δ a set of *epistemic-based preconditions* for $\mathcal{D}$.

The following is very similar to Definition 9, but with the last condition adjusted to ensure agents know their own strategies.

Definition 11 An *epistemic behavioral strategy profile assignment* on Δ is a function $\text{strat}^D : \Delta \to (\mathsf{Ev} \to [0, 1])$ such that

1. $\mathrm{strat}^D(\varphi)$ is a probability function ($\sum_{e \in E} \mathrm{strat}^D(\varphi)(e) = 1$)
2. The support of $\mathrm{strat}^D(\varphi)$ is contained in $E(h)$, where $\varphi \in \Delta_h$
3. $\mathrm{strat}^D(\varphi) = f_\nu(h)$ whenever $\varphi \in \Delta_\nu$
4. If $\sigma \sim_a^\Sigma \tau$ and $h \sim_a k$ for $\sigma \in \Sigma$ and $h \in \iota(a)$, and if $\phi = \sigma \wedge \widetilde{h} \wedge \chi$ and $\psi = \tau \wedge \widetilde{k} \wedge \chi$ for some $\chi \in \Psi_a^\sigma$, then $\mathrm{strat}^D(\phi) = \mathrm{strat}^D(\psi)$.

Given strat^D and $\sigma \in \Sigma$, let $\mathrm{strat}^\sigma : \Phi^\sigma \to (\mathsf{Ev} \to [0, 1])$ be given by $\mathrm{strat}^\sigma(\phi) \stackrel{\text{def}}{=} \mathrm{strat}^D(D^\sigma(\phi))$. By inheriting the first three constraints of Definition 11 as well as much of the fourth constraint, strat^σ is an epistemic behavioral strategy over Φ^σ in the sense of Definition 9. Given strat^D, let strat^D_ϖ (respectively $\mathrm{strat}^\sigma_\varpi$) be the restriction of strat^D (respectively strat^σ) to the domain Δ_ϖ (respectively Φ^σ_ϖ). We sometimes write σ_ϖ for $\mathrm{strat}^\sigma_\varpi$.

We now define a relation $\approx_B$ to use for selecting alternative strategies for players not in B. Given a strategy profile assignment strat, we also define an equivalence relation $\approx_a^\Sigma$ on Σ, such that $\sigma \approx_a^\Sigma \tau$ iff $\sigma_a = \tau_a$. Given a subset $B \subseteq \mathsf{Ag}$, we let $\approx_B^\Sigma = \bigcap_{a \in B} \approx_a^\Sigma$. Note that by our constraint that every player knows her own strategy $\sim_a^\Sigma \subseteq \approx_a^\Sigma$. We can extend $\approx_a^\Sigma$ to all of D by $s \approx_a t$ iff $s \cap (\mathsf{At} \setminus \Sigma) = t \cap (\mathsf{At} \setminus \Sigma)$ and $(s \cap \Sigma) \approx_a^\Sigma (t \cap \Sigma)$. We extend $\approx_B$ similarly.

We now define event models for duplicated game forests.

Definition 12 Given a duplicated game forest $\mathcal{D}$, a set Φ^D of epistemic-based preconditions for $\mathcal{D}$, and an epistemic behavioral strategy assignment strat^D, let $\mathfrak{E}(\mathcal{D}, \Phi^D, \mathrm{strat}^D)$ be the set of event models $\mathcal{E} = (\mathsf{Ev}, \sim^D, \Phi^D, \mathrm{pre})$, where

- Ev is given by $\mathcal{D}$
- $\sim^D$ is given from $\mathcal{D}$ according to Sect. 8.2.2.
- for each agent $a \in \mathsf{Ag}$, pre_a is an epistemic behavioral strategy profile assignment (Definition 11), such that additionally, for each $a \in \mathsf{Ag}$, $\mathrm{pre}_a(\phi) = \mathrm{strat}^D(\phi)$ whenever $\phi \in \Phi_a^D$.

Let $\mathfrak{E}(\mathcal{D}, \Phi^D)$ be the set of epistemic event models for the duplicated forest $\mathcal{D}$ and set of epistemic-based preconditions Φ^D. Let $\mathfrak{E}^e(\mathcal{D})$ be the set of epistemic event models with respect to $\mathcal{D}$ (where Φ^D ranges over all sets of epistemic-based preconditions). Let $\mathfrak{E}^o(\mathcal{D})$ be the set of all ordinary event models with respect to $\mathcal{D}$ (where Φ^D ranges over sets of ordinary precondition).

The following example shows how different input models yield different relationships among the nodes of the game forest and the probabilities agent have over the possible moves they make.

Example 4 We now consider the situation where there are two possible strategies for each player a: the payoff optimizing strategy $\sigma_a^{\max}$ and the minimizing strategy $\sigma_a^{\min}$. The maximizing strategy is essentially the one discussed in Example 2, and assumes the agent receives positive payoff precisely when guessing correctly. The minimizing strategy is where the player makes the opposite choice as for the maximizing strategy. Let $\Sigma = \{(\tau_1, \dots, \tau_n) \mid \tau_a \in \{\sigma_a^{\max}, \sigma_a^{\min}\}\}$ consist of all resulting strategy profiles. Let $\sigma^{\max} = (\sigma_1^{\max}, \dots, \sigma_n^{\max})$ and $\sigma^{\min} = (\sigma_1^{\min}, \dots, \sigma_n^{\min})$. Then let $\sigma^{\text{smn}} = (\sigma^{\text{min}}_{-3}, \sigma^{\text{max}}_{3})$ be the strategy where everyone plays to minimize payoff except for player 3, who plays to maximize. Here smn abbreviates "some minimize."

Let the Ψ_{ϖ} and Φ be the same as in Example 2. Then let

$$\Psi^D = \{\sigma \wedge \varphi \mid \sigma \in \Sigma, \varphi \in \Phi\}$$

The conjunct σ determines which strategy each player uses, the maximizing strategy (as in Example 2) or the minimizing strategy.

We consider an input model M_ϵ (where $0 < \epsilon < 0.25$) given by

- $S = \{s^{\text{max}}_{\mathsf{mw}}, s^{\text{max}}_{\mathsf{mb}}, s^{\text{smn}}_{\mathsf{mw}}, s^{\text{smn}}_{\mathsf{mb}}\}$.
- For $a \neq 3$, let $\sim_a$ is the smallest equivalence relation such that $s^{\text{max}}_{\mathsf{mw}} \sim_a s^{\text{max}}_{\mathsf{mb}}$ and $s^{\text{smn}}_{\mathsf{mw}} \sim_a s^{\text{smn}}_{\mathsf{mb}}$.
  For $a = 3$, $\sim_a$ is $S \times S$.
- For $a = 3$, μ_a given equal weight to each element of each equivalence class.
  For $a = 3$, and each $x \in \{\mathsf{mw}, \mathsf{mb}\}$,

$$\mu_3 : \begin{cases} s^{\text{max}}_x \mapsto \epsilon \\ s^{\text{smn}}_x \mapsto (0.5 - \epsilon) \end{cases}$$

- V assigns

$$\begin{aligned} \sigma^{\text{max}} &\mapsto \{s^{\text{max}}_{\mathsf{mw}}, s^{\text{max}}_{\mathsf{mb}}\} \\ \sigma^{\text{smn}} &\mapsto \{s^{\text{smn}}_{\mathsf{mw}}, s^{\text{smn}}_{\mathsf{mb}}\} \\ \mathsf{mw} &\mapsto \{s^{\text{max}}_{\mathsf{mw}}, s^{\text{smn}}_{\mathsf{mw}}\} \\ \mathsf{mb} &\mapsto \{s^{\text{max}}_{\mathsf{mb}}, s^{\text{smn}}_{\mathsf{mb}}\} \end{aligned}$$

  and all other atomic propositions to $\emptyset$.

Now if ϵ is very large, then starting from a state in $\{s^{\text{max}}_{\mathsf{mw}}, s^{\text{max}}_{\mathsf{mb}}\}$, the choices made by 3 are the same at each node of H_3 as for Example 2. In particular, if the first two players draw white and write down white, then regardless of what 3 draws, she will write white.

However, if ϵ is very small, player 3 will exhibit different choices from certain nodes of the game tree. For example, if the first two players draw white and write down white, then regardless of what 3 draws, she will write down black (since she weighs highly the assumption that the first two players had drawn black and just wrote down white as that was their strategy).

In the previous example, the minimizing strategies could be thought of as strategies only irrational agents would use. But to express rationality, one would need to be able to compare an existing strategy with alternative strategies in light of a payoff structure.

8.5 Payoffs and Rationality

We can express some properties of preferences by quantifying over the valuations that are better than a certain valuation. Much of the reasoning is done external to the formulas, but when working with a fixed game we can pick out optimal valuations for certain agents among sets of valuations.

Preference relations

Let $\mathcal{D} = (\Sigma, \mathcal{F}, \sim^{\Sigma})$ be a duplicated game forest. Let Z be the set of terminal nodes in $\mathcal{F}$ and $\succeq^{\mathcal{F}}$ the set of preference relations over Z, and let Z^D be the set of terminal nodes in $\mathcal{D}$ and $\succeq^{\mathcal{D}}$ the set of preference relations over Z^D defined according to Sect. 8.2.2. There are many choices for how to extend $\succeq^{\mathcal{F}}$ and $\succeq^{\mathcal{D}}$ to H and D respectively. We opt for a conservative approach (this is a rather arbitrary decision, but reflects the view that agents are cautious about considering one node to be at least as good as another and maximally pessimistic about probabilities). For $h, k \in H$, let $\succeq_a^H$ be the smallest relation such that $h \succeq_a^H k$ whenever either of the following hold:

1. $h \succeq_a^{\mathcal{F}} k$ (hence $h, k \in Z$) or
2. $h \cup \{e\} \succeq_a^H k \cup \{f\}$ for all $e, f \in \mathsf{Ev}$, such that $h \cup \{e\}, k \cup \{f\} \in H$

We define $\succeq_a^D$ is a similar manor. Each of $\succeq_a^F$ and $\succeq_a^D$ can induce similar relations on states of a Bayesian Kripke frame as follows: Given a Bayesian Kripke model $M = (S, \sim, \mu, V)$ with respect to $\mathsf{At} = \mathcal{X} \cup \mathsf{Ev}$, let $s \succeq_a t$ iff $\mathsf{At}(s) \succeq_a^H \mathsf{At}(t)$. The case where $\mathsf{At} = \Sigma \cup \mathcal{X} \cup \mathsf{Ev}$ is similar. For $S \in \{H, D\}$, we define

$$\preceq_a^S \stackrel{\text{def}}{=} (\succeq_a^S)^{-1}$$
$$\succ_a^S \stackrel{\text{def}}{=} \succeq_a^S \setminus \preceq_a^S$$
$$\prec_a^S \stackrel{\text{def}}{=} (\succ_a^S)^{-1}$$
$$\not\prec_a^S \stackrel{\text{def}}{=} (S \times S) \setminus \prec_a^S$$

Example 5 In the Urn Game of Example 1, $\{\mathsf{mb}, \mathsf{dw}_1, \mathsf{b}_1\} \succ_1^H \{\mathsf{mb}, \mathsf{dw}_1, \mathsf{w}_1\}$, since regardless of how the plays evolve, ones extending $\{\mathsf{mb}, \mathsf{dw}_1, \mathsf{b}_1\}$ will be preferred to ones extending $\{\mathsf{mb}, \mathsf{dw}_1, \mathsf{w}_1\}$, and there exist an extension of $\{\mathsf{mb}, \mathsf{dw}_1, \mathsf{w}_1\}$ (in this case each extension) that is not preferred to an extension of $\{\mathsf{mb}, \mathsf{dw}_1, \mathsf{b}_1\}$. However, both $\{\mathsf{mb}, \mathsf{dw}_1, \mathsf{b}_1\} \not\succeq_2^H \{\mathsf{mb}, \mathsf{dw}_1, \mathsf{w}_1\}$ and $\{\mathsf{mb}, \mathsf{dw}_1, \mathsf{b}_1\} \not\preceq_2^H \{\mathsf{mb}, \mathsf{dw}_1, \mathsf{w}_1\}$ hold, as there exists an extension of each that is preferable to player 2 over the other.

Utility functions

Utility functions allow us to be more sensitive to probabilities and expected values. Assuming epistemic behavioral strategies are used, these probabilities might not depend on the nodes of the game alone, but also epistemic conditions of an input model. Rather than extending u_a from terminal nodes to all of H or D, we assume an event model $\mathcal{E}$ for the game and induce a function $u_a^{\mathcal{E}}$ on pointed Bayesian Kripke models.

Let $\mathcal{F} = (\mathcal{X}, H, \iota, f_\nu, \sim, u)$ be a utility-based $(\mathsf{Ag}, \mathsf{Ev})$ forest game. We define

$$u_a^{\mathcal{E}}(M, s) \stackrel{\text{def}}{=} \begin{cases} u_a(h) & \mathsf{At}(s) = h \in Z \\ \sum_{e \in E(\mathsf{At}(s))} \mathrm{pre}_a(e \mid s) \cdot u_a^{\mathcal{E}}(M \otimes \mathcal{E}, (s, e)) & \mathsf{At}(s) \in H \setminus Z \\ 0 & \text{otherwise} \end{cases}$$

Recall that $\mathrm{pre}_a(e \mid s)$ is defined by (8.1). For the case where $\mathcal{E} \in \mathfrak{E}^e(\mathcal{D})$, the definition of $u_a^{\mathcal{E}}$ on pointed Bayesian Kripke models is similar.

If $\mathcal{E} \in \mathfrak{E}^o(\mathcal{F})$, where for each $h \in H$, we write ϕ_h for $\widetilde{h} \wedge true \in \Phi$, then we can define u_a on the set H by

$$u_a^{\mathcal{E}}(h) \stackrel{\text{def}}{=} \begin{cases} u_a(h) & h \in Z \\ \sum_{e \in E(h)} \mathrm{pre}_a(\phi_h) \cdot u_a^{\mathcal{E}}(M \otimes \mathcal{E}, (s, e)) & h \in H \setminus Z \end{cases}$$

For the case where $\mathcal{E} \in \mathfrak{E}^o(\mathcal{D})$, the definition of $u_a^{\mathcal{E}}$ on D is similar.

Example 6 Consider the game of Example 1. Let $\mathcal{E}$ be the event model from Example 2, and let M be a model with two states s_w and s_b for which $[(\cup_{a \in \mathsf{Ag}} a)^*]\chi$ is valid (χ coming from Example 2), and where mb is only true at s_b and mw is only true at s_w. Let $N = M \otimes \mathcal{E}$ and $t = (s_b, \mathsf{dw}_1)$. We wish to determine $u_1^{\mathcal{E}}(N, t)$. Now $E(\mathsf{At}(t)) = \{\mathsf{b}_1, \mathsf{w}_1\}$, so we have a summand for each of the two actions. Given that the number of agents is finite, one can calculate that $u_1^{\mathcal{E}}(N \otimes \mathcal{E}, (t, \mathsf{b}_1)) = 1$ and $u_1^{\mathcal{E}}(N \otimes \mathcal{E}, (t, \mathsf{b}_1)) = 0$ by expanding these expressions into numerous summands involving utility of only pointed models each whose points correspond to nodes in Z. This calculation is intuitive, as any play of the game from (t, b_1) results in a play where player 1 made the correct choice (probability 1 that the utility is 1), and any play of the game from (t, w_1) results in a play where player 1 made the incorrect choice (probability 1 that the utility is 0). Now because t reflects that player 1 drew a white ball, she will, according to $\mathcal{E}$ play w_1 with probability 1, that is, $\mathrm{pre}(\mathsf{w}_1 \mid \mathsf{At}(t)) = 1$ and $\mathrm{pre}(\mathsf{b}_1 \mid \mathsf{At}(t)) = 0$. Putting these together, we arrive at $u_1^{\mathcal{E}}(N, t) = 0 \times 1 + 1 \times 0 = 0$.

8.5.1 Comparing Preference or Utility of Nodes

Let $\mathcal{E} \in \mathfrak{E}(\mathcal{D})$ for a preference-based $(\mathsf{Ag}, \mathsf{Ev})$ forest game $\mathcal{D}$. Then we define for each $h \in D$

$$(\not\preceq_a^{\mathsf{node}} \widetilde{h}) \stackrel{\text{def}}{=} \bigvee_{\{k \in D : k \not\preceq_a^D h\}} \widetilde{k}$$

$$(\succeq_a^{\mathsf{node}} \widetilde{h}) \stackrel{\text{def}}{=} \bigvee_{\{k \in D : k \succeq_a^D h\}} \widetilde{k}$$

If instead $\mathcal{E} \in \mathfrak{E}(\mathcal{D})$ for a utility-based $(\mathsf{Ag}, \mathsf{Ev})$ forest game $\mathcal{D}$, then we define for each $h \in D$,

$$(\geq_a^{\mathsf{node}} \widetilde{h}) \stackrel{\text{def}}{=} (\not<_a^{\mathsf{node}} \widetilde{h}) \stackrel{\text{def}}{=} \bigvee_{\{k \in D : u_a(k) \geq u_a(h)\}} \widetilde{k}.$$

We will define more abbreviations in terms of $(\geq_a^{\mathsf{node}} \widetilde{h})$ and $(\not<_a^{\mathsf{node}} \widetilde{h})$; we will not specify whether $\mathcal{D}$ is preference-based or utility-based.

Comparing actions via nodes

We can express that a player a is no worse off playing action e than any other action by

$$\not<_a^{\mathsf{act}}(e) \stackrel{\text{def}}{=} \bigvee_{\{h \in D_a | e \notin h, h \cup \{e\} \in D\}} (\widetilde{h} \wedge \bigwedge_{\{h \cup \{b\} \in D | b \notin h\}} (\not<_a^{\mathsf{node}} \widetilde{h \cup \{b\}}),$$

and that a is at least as well off playing e than any other action by

$$\geq_a^{\mathsf{act}}(e) \stackrel{\text{def}}{=} \bigvee_{\{h \in D_a | e \notin h, h \cup \{e\} \in D\}} (\widetilde{h} \wedge \bigwedge_{\{h \cup \{b\} \in D | b \notin h\}} (\geq_c^{\mathsf{node}} \widetilde{h \cup \{b\}}),$$

But such comparison of actions does not fully account for the epistemics of the game, nor does it allow us to compare randomizations over the immediate actions.

Comparing strategies via nodes

We can express that the current strategy is at least as good for player a than strategy σ_a by

$$(\geq_a^{\mathsf{strat}} \sigma_a) \stackrel{\text{def}}{=} \bigvee_{h \in D} (\widetilde{h} \wedge \bigwedge_{\{h' | h' \approx_{\mathsf{Ag} \setminus \{a\}} h, (h' \cap \Sigma) \approx_a^{\Sigma} \sigma\}} (\geq_a^{\mathsf{node}} \widetilde{h'})).$$

We can express that the current strategy is a best response for player a over the others strategies available for a to choose (given $\mathcal{E}$) by

$$\mathsf{BestResponse}_a \stackrel{\text{def}}{=} \bigvee_{h \in D} (\widetilde{h} \wedge \bigwedge_{k \approx_{\mathsf{Ag} \setminus \{a\}} h} (\geq_a^{\mathsf{node}} \widetilde{k})).$$

In the preference-based games, the requirement that $(\geq_a^{\mathsf{node}} \widetilde{k}))$ is rather strong, though $(\not<_a^{\mathsf{node}} \widetilde{k}))$ may be too weak. With the utility-based games, this may be more reasonable. We can express that the current node is in Nash equilibrium by

$$\mathsf{Nash} \stackrel{\text{def}}{=} \bigwedge_{a \in \mathsf{Ag}} \mathsf{BestResponse}_a.$$

There are a number of different (nonequivalent) possibilities for how to define rationality (with utility values and expectation, it is a bit more straightforward, as risk is converted into expectation). Here is one way. Player a is rational (with respect to the possibilities given by $\mathcal{E}$) if she believes she is playing a best response, that is, if the following holds:

$$\mathsf{Rat}_a \stackrel{\text{def}}{=} [a]\mathsf{BestResponse}_a$$

All these notions of best response and rationality are limited by the fact that we have finite event models and finitary formulas. These are relative to a fixed set of possible strategies (a concept also explored in [6]). But an advantage of these is that, given the right interpretation (with respect to a fixed game), we can express all these notions using the probabilistic logic of communication and change (our setting here is a mild simplification of the one in [1]).

8.5.2 Comparing Event Models

To allow us to compare a strategy with infinitely many alternatives, it may help to add components to the language. It is common among dynamic epistemic logics not to fix a particular event model, and it is this perspective that this section outlines. Toward this goal we define some notions. Let us fix a utility-based $(\mathsf{At}, \mathsf{Ev})$ forest game $\mathcal{F}$. For each $a \in \mathsf{Ag}$, let us define a relation $\approx_a$ over $\mathfrak{E}^o(\mathcal{F})$,[1] such that $\mathcal{E} \approx_a \mathcal{E}_a$ if and only if agent a's strategy is the same in both event models. We similarly define $\approx_B \stackrel{\text{def}}{=} \bigcap_{a \in B} \approx_a$ for each $B \subseteq \mathsf{Ag}$.

We add to the language the following clauses: $\sum_{k=1}^n c_k u_a(\mathcal{E}_k) \geq r$ for $\mathcal{E}_k \in \mathfrak{E}^e(\mathcal{F})$ and $\mathsf{Best}_a(\mathcal{E})$ for $\mathcal{E} \in \mathfrak{E}^o(\mathcal{F})$, where the semantic clauses of these is given by

- $M, s \models \sum_{k=1}^n c_k u_a(\mathcal{E}_k) \geq r$ iff $\sum_{k=1}^n u_a^{\mathcal{E}_k}(M, s) \geq r$
- $M, s \models \mathsf{Best}_a(\mathcal{E})$ iff for all $\mathcal{E}'$ where $\mathcal{E} \approx_{\mathsf{Ag} \setminus \{a\}} \mathcal{E}'$, $M, s \models u_a(\mathcal{E}) - u_a(\mathcal{E}') \geq 0$.

Then we may define $\mathcal{E}$ being rational for a to play by

$$\mathsf{Rat}_a(\mathcal{E}) \stackrel{\text{def}}{=} [a]\mathsf{Best}_a(\mathcal{E}).$$

This notion of rationality is not relative to a restricted set of strategies.

8.6 Conclusion

This paper shows how the probabilistic logic of communication and change can be a foundation for developing logics for extensive-form games with imperfect or incomplete information. Such games include urn games and other games that help

[1] We restrict to $\mathfrak{E}^o(\mathcal{F})$ for simplicity. Without restricting to ordinary behavioral strategies, we might still want to impose further restrictions on $\approx_a$ concerning whether agents b other than a must retain the same probabilistic beliefs about other agents' moves. Such a restriction would be significant in how their epistemic behavioral strategies are played.

us reason about informational cascades and other social phenomena. This differs from other dynamic epistemic logic approaches in that the PLCC allows for a well-developed probabilistic Bayesian reasoning, and that PLCC enables fact changes in the updates, which allows pointed Bayesian Kripke models to reflect in its valuation the precise stage of an extensive-form game.

This paper shows that a weaker version of PLCC (and hence PLCC too) can, for a fixed game, express many concepts important for reasoning about games, such as comparisons of the preferences agents have between nodes of the game trees or between strategies. But with expressing notions of best response, Nash equilibrium, and rationality, PLCC is limited to a finite set of available strategies, given by a fixed event model. We also propose extensions of the logic for reasoning directly about the utility of strategies (via comparing event models). Such a logic, as is common among many dynamic epistemic logics, does not fix an event model, thus making the strategies fully dynamic. Another extension proposed in this paper allows us to quantify over infinitely many strategies, and thus allows for more realistic expressions of game theory concepts that rely on such quantification, such as best response, Nash equilibrium, and rationality.

Future work will hopefully develop, with an axiomatic system, the proposed extension of PLCC that includes operators for comparing the utility of event models as well as for expressing that an event model is in some sense optimal for a certain agent. PLCC and its extensions may help us reason about what epistemic conditions may guarantee certain moves or strategies by players. The fact that the game structure is fixed may seem like a limitation for reasoning about general games, and further extensions of the logic may allow us to reason about arbitrary games.

Acknowledgments I would like to thank the reviewer for the valuable comments.

References

1. Achimescu, A.: Games and Logic for Informational Cascades. Master's of Logic Thesis, ILLC (2014). http://www.illc.uva.nl/Research/Publications/Reports/MoL-2014-04.text.pdf
2. Achimescu, A., Baltag, A., Sack, J.: The probabilistic logic of communication and change. In: Presented at and Published in the Informal Proceedings of the Eleventh Conference on Logic and the Foundations of Game and Decision Theory (LOFT'11) (2014)
3. Achimescu, A., Baltag, A., Sack, J.: The Probabilistic Logic of Communication and Change, Manuscript (2015)
4. Alur, R., Henzinger, T., Kupferman, O.: Alternating-time temporal logic. J. ACM **49**(5), 672–713 (2002)
5. Baltag, A., Smets, S., Zvesper, A.: Keep 'hoping' for rationality: a solution to the backward induction paradox. Synthese **169**, 705–737 (2009)
6. van Benthem, J.: Rational dynamics and epistemic logic in games. Int Game Theory Rev **9**(1), 13–45 (2007). (Erratum reprint, **9**(2), 377–409)
7. van Benthem, J.: Logic in Games. MIT Press, Cambridge (2014)
8. van Benthem, J., van Eijck, J., Kooi, B.: Logics of communication and change. Inf. Comput. **204**(11), 1620–1662 (2006)
9. Chatterjee, K., Henzinger, T., Piterman, N.: Strategy Logic. Inf. Comput. **208**, 677–693 (2010)

10. Chen, T., Lu, J.: Probabilistic alternating-time temporal logic and model checking algorithm. In: Proceedings of the Fourth International Conference on Fuzzy Systems and Knowledge Discovery, 2007. FSKD 2007. pp. 35–39 (2007)
11. Fagin, R., Halpern, J.Y.: Reasoning about knowledge and probability. J. ACM **41**(2), 340–367 (1994)
12. van der Hoek, W., Wooldridge, M.: Cooperation, knowledge, and time: alternating-time temporal epistemic logic and its applications. Studia Logica **75**, 125–157 (2003)
13. Osborne, M.J., Rubinstein, A.: A Course in Game Theory. The MIT Press, Cambridge (1994)
14. Pauly, M., Parikh, R.: Game logic—an overview. Studia Logica **75**, 165–182 (2003)
15. Sack, J., van der Hoek, W.: A modal logic for mixed strategies. Studia Logica **102**, 339–360 (2014)

Chapter 9
Measurement-Theoretic Foundations of Observational-Predicate Logic

Satoru Suzuki

Abstract Vagueness is a ubiquitous feature that we know from many expressions in natural languages. It can invite a serious problem: the Sorites Paradox. The Phenomenal Sorites Paradox is a version of the Sorites Paradox, where observational predicates occur. According to Raffman [15], we can classify perceptual indiscriminability as follows: (1) s-Indiscriminability: perceptual indiscriminability in the statistical sense and (2) d-Indiscriminability: perceptual indiscriminability in the non-statistical (dispositional) sense. The Tolerance Principle on s-Indiscriminability can be false because the objects which are the same may often be recognised discriminable by an examinee A of limited ability of discrimination and the objects which are different may often be recognised indiscriminable by A. The aim of this paper is to propose a new version of logic for observational predicates—Observational-Predicate Logic (OPL)—that can express formally this solution to the Phenomenal Sorites Paradox on s-Indiscriminability and makes it possible to reason about observational predicates. To accomplish this aim, we provide the language of OPL with a statistical model in terms of measurement theory.

Keywords Bounded rationality · Just noticeable difference (JND) · Measurement theory · Observational predicate · Phenomenal Sorites Paradox · Representation theorem · Semiorder

9.1 Motivation

Vagueness is a ubiquitous feature that we know from many expressions in natural languages. It can invite a serious problem: the *Sorites Paradox*. The following argument is an ancient example of this paradox:

S. Suzuki (✉)
Faculty of Arts and Sciences, Komazawa University, 1-23-1, Komazawa,
Setagaya-ku, Tokyo 154-8525, Japan
e-mail: bxs05253@nifty.com

S.C.-M. Yang et al. (eds.), *Structural Analysis of Non-Classical Logics*,
Logic in Asia: Studia Logica Library, DOI 10.1007/978-3-662-48357-2_9

Example 1 (Sorites Paradox)

- 10,00,000 grains of sand make a heap.
- (Inductive Premise): For any n, if n grains of sand make a heap, then $n - 1$ grains of sand do.

---

1 grain of sand makes a heap.[1]

We *specify* the sort of Sorites Paradox we tackle in this paper. Graff [4, p. 907] defines an observational predicate as follows:

Definition 1 (*Observational Predicate*) A predicate is *observational* if its applicability to an object (given a fixed context of evaluation) depends only on the way that object appears.[2]

There are some examples of this predicate:

Example 2 (Observational Predicate) 'looks-red', 'sounds-loud', 'tastes-sweet', etc.

Observational predicates can generate a *special kind* of Sorites Paradox in the following sense. In a sorites series for a *nonobservational* predicate like 'tall', there *must be some difference in height* between any two adjacent members in the series. On the other hand, it is plausible to think that we can arrange a sorites series for *observational* predicates that does not have this feature because the relevant perceptual indiscriminability relation is *nontransitive*. In a sorites series for an observational predicate like 'looks-red', if the relevant perceptual indiscriminability relation is nontransitive, then there *can* be a series of colour patches grading from red to yellow in which there is *no difference in appearance* between any two adjacent patches in the series. This version of Sorites Paradox generated by observational predicates is called the *Phenomenal Sorites Paradox*. By modifying Graff [4, p. 907], we can show the defining features of the Phenomenal Sorites Paradox as follows:

1. the occurrence of some kind of *tolerance principle on perceptual indiscriminability*,
2. the occurrence of some expressions for perceptual indiscriminability—'looks-the-same-as' or 'smells-the-same-as', etc.—in the antecedent of the tolerance principle,
3. the occurrence of *observational predicates* as the other constituents of the argument and
4. the occurrence of some kind of *premises on indiscriminability*.

According to Raffman [15, p. 159], we can classify perceptual indiscriminability as follows:

---

[1]The Sorites Paradox derives its name from the Greek word '$\sigma\omega\rho\acute{o}\varsigma$' for heap.

[2]Here we do not define an observational predicate in a strict sense. For 'predicate', 'object', 'context' and 'the way that object appears' are not yet defined.

1. *s-Indiscriminability*: perceptual indiscriminability in the *statistical* sense, and
2. *d-Indiscriminability*: perceptual indiscriminability in the *non-statistical* (*dispositional*) sense.

The standard model of economics is based on *global rationality* that requires an *optimising behaviour*. But according to Simon [19], cognitive and information-processing constrains on the capabilities of agents, together with the complexity of their environment, render an *optimising behaviour* an *unattainable ideal*. He dismissed the idea that agents should exhibit global rationality and suggested that they in fact exhibit *bounded rationality* that allows a *satisficing behaviour*. If an agent has only a *limited* ability of discrimination, he may be considered to be only *boundedly rational*. We shall discuss *s-Indiscriminability*. If an agent is *boundedly rational*, one possible explanation for this paradox is that the *nontransitivity* of s-Indiscriminability results from the fact that the *cannot generally* discriminate very close quantities. The psychophysicist Fechner [1] explained this inability by the concept of a *threshold of discrimination*, that is, *just noticeable difference* (*JND*). Given a measure function f that an *examiner* could assign to a boundedly rational *examinee* for an object a, its JND δ is the *lowest intensity increment* such that $f(a) + \delta$ is recognised to be higher than $f(a)$ by the examinee. We can consider the notion of noticing a JND from a *statistical* point of view. The JND is usually the difference that a *boundedly rational agent* makes on 50 % of trials. If a different proportion from 50 % is used, then this should be included in the description—for example, '75 % *JND*'. We define the Tolerance Principle on s-Indiscriminability as follows:

Definition 2 (*Tolerance Principle on s-Indiscriminability*) For any object x, y, if an *examiner* B makes a *statistical* judgment that x *looks the same as* y to an *examinee* A in the respect of the property expressed by a *observational predicate* F, then if $F(x)$, then $F(y)$.

The Phenomenal Sorites Paradox on s-Indiscriminability is as follows:

Example 3 (Phenomenal Sorites Paradox on s-Indiscriminability)

- Patch 1 looks red to an examinee A.
- (*Tolerance Principle*): For any patch x, y, if an examiner B makes a statistical judgment that x looks the same as y to A, then if x looks red to A, then y looks red to A.
- (*Premise on Indiscriminability* 1): B makes a statistical judgment that Patch 1 looks the same as Patch 2 to A.
- (*Premise on Indiscriminability* 2): B makes a statistical judgment that Patch 2 looks the same as Patch 3 to A.

  ⋮

- (*Premise on Indiscriminability* 99): B makes a statistical judgment that Patch 99 looks the same as Patch 100 to A.

Patch 100 looks red to A.

In Example 3, s-Indiscriminability relation is *nontransitive* because of an *examinee's limited ability of discrimination*. The Tolerance Principle on s-Indiscriminability can be *false* because the objects which are the *same* may often be recognised *discriminable* by an examinee A of limited ability and the objects that are *different* may often be recognised *indiscriminable* by A. The Premises on s-Indiscriminability are true because of his limited ability. The argument of this example is *valid*. On the other hand, we shall discuss *d-Indiscriminability*. We define the Tolerance Principle on d-Indiscriminability as follows:

Definition 3 (*Tolerance Principle on d-Indiscriminability*) For any object x, y and any *context* z, if an agent A *would make a judgment* that x *looked the same* as y to A in the respect of the property expressed by an *observational predicate* F in z if he were to compare x with y in z, then if $F(x)$, then $F(y)$.

The Phenomenal Sorites Paradox on d-Indiscriminability becomes as follows:

Example 4 (Phenomenal Sorites Paradox on d-Indiscriminability)

- Patch 1 looks red to an agent A in a context C_1.
- (*Tolerance Principle*): For any patch x, y and any context z, if A would make a judgment that x looked the same as y to A in a context z if A were to compare x with y in z, then if x looks red to A, then y looks red to A.
- (*Premise on Indiscriminability* 1): A would make a judgment that Patch 1 looked same as Patch 2 to A in a context C_1 if A were to compare Patch 1 with Patch 2 in C_1.
- (*Premise on Indiscriminability* 2): A would make a judgment that Patch 2 looked same as Patch 3 to A in C_2 if A were to compare Patch 2 with Patch 3 in C_2.

⋮

- (*Premise on Indiscriminability* 99): A would make a judgment that Patch 99 looked same as Patch 100 to A in C_{99} if A were to compare Patch 99 with Patch 100 in C_{99}.

Patch 100 looks red to A in C_{99}.

In Example 4, we agree with Graff [4] and Raffman [15] in thinking that there can occur different d-Indiscriminability relations *relative to contexts* such as $C_1, C_2, \ldots, C_{99}$.[3] This is because, as Raffman [15] argues, if an agent is boundedly rational, he cannot necessarily *attend to* all patches *simultaneously* because of his limited ability of discrimination even if he can have them all *in view simultaneously*. Because, in a *single observation*, the objects that are judged *discriminable* by an agent A are *trivially discriminable* for A and the objects that are judged *indiscriminable* by A are the *trivially indiscriminable* for A, d-Indiscriminability relations relative to $C_1, C_2, \ldots, C_{99}$ are *trivially transitive, respectively* and the Tolerance Principle on d-Indiscriminability is *trivially true*. The Premises on d-Indiscriminability are true because of his *limited ability* of discrimination. The argument of this example is *invalid* because there can occur *different* d-Indiscriminability relations relative to contexts such as $C_1, C_2, \ldots, C_{99}$. The characteristics of the Phenomenal Sorites Paradoxes can be schematised as follows (Table 9.1):

[3]It must be noted that my main concern in this paper is not about the Phenomenal Sorites Paradox on d-Indiscriminability but the Phenomenal Sorites Paradox on s-Indiscriminability.

Table 9.1 Characteristics of Phenomenal Sorites Paradoxes

	s-Indiscriminability	d-Indiscriminability
Indiscriminability Relations	*Same, Nontransitive*	*Possibly Different, Trivially Transitive Respectively*
Tolerance Principle	*Possibly False*	*Trivially True*
Premises on Indiscriminability	True	True
Argument	*Valid*	*Invalid*

Hyde [8] classified responses to the Sorites Paradox in the following four types:

1. denying that logic applies to soritical expressions,
2. denying some premises,
3. denying the validity of the argument and
4. accepting the paradox as sound.

From the consideration above, we conclude that we make a *response (3)* to the Phenomenal Sorites Paradox on d-Indiscriminability but that we make a *response (2)* to the Phenomenal Sorites Paradox on s-Indiscriminability. The *aim of this paper* is to propose a new version of logic for *observational* predicates—Observational-Predicate Logic (OPL)—that makes it possible to reason about observational predicates without inviting the *Phenomenal Sorites Paradox on s-Indiscriminability*. To accomplish this aim, we provide the language of OPL with a *statistical* model in terms of *measurement theory*. Numerous studies (for example, [4, 9, 15]) have been made on the Phenomenal Sorites Paradox on d-Indiscriminability. But only few attempts have so far been made at the Phenomenal Sorites Paradox on s-Indiscriminability. Indeed Hardin [5] discussed the Phenomenal Sorites Paradox on s-Indiscriminability in terms of JNDs, but he dealt with it neither from a logical point of view nor from a measurement-theoretic one. In [25] we also proposed a version of logic for vague predicates—JND-based Vague Predicate Logic (JVL)—that can avoid the Sorites Paradox. In [25], the *scope* of JVL was not clear. On the other hand, in this paper, in terms of *observational predicates*, we make clear the scope of OPL, that is, the *Phenomenal Sorites Paradox*. In JVL, the difference between weak similarity and strong similarity was improperly introduced; whereas in this paper, the difference between s-Indiscriminability and d-Indiscriminability is introduced in order to deal with the Phenomenal Sorites Paradox. In [25], the *completeness* of JVL was mistakenly proved, while in this paper the *first-order undefinability* of an essential property (i.e., $\sim^*$-Connectedness) of the model of the language of OPL is proved.

The structure of this paper is as follows. In Sect. 9.2, we define the Strong Statistical Transitivity (SST) which is one of the most typical conditions for statistical consistency. In Sect. 9.3, we give a measurement-theoretic analysis of JNDs and semiorders. In Sect. 9.4, we define the language $\mathcal{L}_{\mathsf{OPL}}$ of OPL, define a statistical model $\mathfrak{M}$ of OPL, provide OPL with a satisfaction definition and a truth definition and prove first-order undefinability of $\sim^*$-Connectedness. In Sect. 9.5, we discuss higher order vagueness. In Appendix, we touch upon Goodman's conception of JNDs and that of semiorders.

9.2 Statistical Consistency: Strong Statistical Transitivity (SST)

When $\mathcal{I}$ is a nonempty set of individuals, we define a *forced-choice-pair comparison probability function* as follows:

Definition 4 (*Forced-Choice-Pair-Comparison Probability Function* Pr) $Pr : \mathcal{I} \times \mathcal{I} \to [0, 1]$ is called a *forced-choice-pair comparison probability function* if it satisfies the following condition: For any $x, y \in \mathcal{I}$ such that $x \neq y$,

$$Pr(x, y) + Pr(y, x) = 1.$$

Remark 1 (*Relative Frequency*) $Pr(a, b)$ is interpreted as the *relative frequency* with which an agent will choose a rather than b when *forced* to make a choice from $\{a, b\}$.

The following is one of the most typical conditions for *statistical consistency*.

Definition 5 (*Strong Statistical Transitivity (SST)*) Pr is said to satisfy the *Strong Statistical Transitivity (SST)* if for any $x, y, z \in \mathcal{I}$,

If $Pr(x, y) \geq \frac{1}{2}$ and $Pr(y, z) \geq \frac{1}{2}$, then $Pr(x, z) \geq \max\{Pr(x, y), Pr(y, z)\}$.

The following is an example of SST.

Example 5 (*Phenomenal Sorites Paradox on s-Indiscriminability and SST*) Suppose that an examiner observes the relative frequency with which an examinee responds that Patch i $(1 \leq i \leq 100)$ looks different from Patch j $(1 \leq j \leq 100)$. For example, when the relative frequency with which the examinee responds that Patch 50 looks different from Patch 52 is $\frac{3}{4}$ and that with which he responds that Patch 52 looks different from Patch 54 is $\frac{2}{3}$, it is plausible that the relative frequency with which he responds that Patch 50 looks different from Patch 54 should be at least $\frac{3}{4}$. Then these relative frequencies should satisfy *SST*.

9.3 Measurement-Theoretic Analysis of JNDs and Semiorders

Luce [12] introduced the concept of a *semiorder*[4,5] that can provide a *qualitative* counterpart of a JND that is *quantitative*. Scott and Suppes [18, p. 117] defined a semiorder as follows:

Definition 6 (*Semiorder*) Let $\mathcal{I}$ denote a set of individuals. $\succ$ on $\mathcal{I}$ is called a semiorder if, for any $w, x, y, z \in \mathcal{I}$, the following conditions are satisfied:

[4]Van Rooij [34, 35] also argued the relation between the Sorites Paradox and semiorders from a different point of view that does not focus on a representation theorem.

[5]In [23, 24] we proposed a new version of complete and decidable preference logic based on a semiorder on a Boolean algebra.

1. $x \not\succ x$ (*Irreflexivity*),
2. If $w \succ x$ and $y \succ z$, then $w \succ z$ or $y \succ x$ (*Intervality*),
3. If $w \succ x$ and $x \succ y$, then $w \succ z$ or $z \succ y$ (*Semitransitivity*).

There are two main problems with *measurement theory*[6]:

1. the *representation problem*: justifying the assignment of numbers to objects,
2. the *uniqueness problem*: specifying the transformation up to which this assignment is unique.

A solution to the former can be furnished by a *representation theorem*, which establishes that the specified conditions on a qualitative relational system are (necessary and) sufficient for the assignment of numbers to objects that represents (or preserves) all the relations in the system. A solution to the latter can be furnished by a *uniqueness theorem*, which specifies the transformation up to which this assignment is unique. Scott and Suppes [18] proved a representation theorem for semiorders when $\mathcal{I}$ is *finite*. The Scott–Suppes theorem was first extended to *countable* sets by Manders [14]. Because $\mathcal{I}$ of the model $\mathfrak{M}$ of OPL may be countable, the Manders theorem must be considered. A condition (i.e., $\sim^*$-*Connectedness*) is necessary for $\succ$ to have a *positive threshold* even when $\mathcal{I}$ is *countable*. We define $\sim$ by $\succ$ as follows:

Definition 7 ($\sim$) For any $x, y \in \mathcal{I}$, $x \sim y := x \not\succ y$ and $y \not\succ x$.

$\sim^*$ is defined by $\sim$ and $\succ$ as follows:

Definition 8 ($\sim^*$)

$$\begin{aligned} \text{For any } x, y \in \mathcal{I}, x \sim^* y := x \sim y, \text{ or,} \\ x \succ y \text{ and for any } z \in \mathcal{I}, \text{ not } (x \succ z \text{ and } z \succ y), \text{ or,} \\ y \succ x \text{ and for any } z \in \mathcal{I}, \text{ not } (y \succ z \text{ and } z \succ x). \end{aligned}$$

A $\sim^*$-chain is defined by $\sim^*$ as follows:

Definition 9 ($\sim^*$-*Chain*) Let $a_1, \ldots, a_n \in \mathcal{I}$ be such that for any $k < n$, $a_k \sim^* a_{k+1}$. Then we call $(a_1, \ldots, a_n)$ a $\sim^*$-chain between a_1 and a_n.

$\sim^*$-Connectedness is defined by a $\sim^*$-chain as follows:

Definition 10 ($\sim^*$-*Connectedness*) $\sim^*$ on $\mathcal{I}$ is connected if for any $x, y \in \mathcal{I}$, there is a $\sim^*$-chain between x and y.

The Manders theorem can be stated by means of $\sim^*$-*Connectedness* as follows:

Theorem 1 (Representation for Semiorders, Manders [14]) *Suppose that $\succ$ is a binary relation on a countable set $\mathcal{I}$ and that $\sim^*$ is defined by Definition 8 and that δ*

[6] [17] gives a comprehensive survey of measurement theory. The mathematical foundation of measurement had not been studied before Hölder [7] developed his axiomatisation for the measurement of mass. [10, 13, 20] are seen as milestones in the history of measurement theory.

is a positive number. Then $\succ$ *is a semiorder and* $\sim^*$ *is connected iff there is a function* $f : \mathcal{I} \to \mathbb{R}$ *such that for any* $x, y \in \mathcal{I}$,

$$x \succ y \textit{ iff } f(x) > f(y) + \delta.$$

What Theorem 1 says is not how to *construct* f but the *existence* of it. Even if we interpret f as a measure function that an examiner could assign to an examinee and δ as a JND, this interpretation is still not clear. So we consider the notion of a JND from a *statistical* point of view. We define a binary relation Pr^{λ} on $\mathcal{I}$ as follows:

Definition 11 (*Binary Relation* Pr^{λ} *on* $\mathcal{I}$) Pr^{λ} is a binary relation on $\mathcal{I}$ such that for any $x, y \in \mathcal{I}$, $xPr^{\lambda}y$ if $Pr(x, y) > \lambda$.

As we have mentioned before, the JND is usually the difference that a boundedly rational agent makes on $\frac{1}{2}$ of trials. So we consider the JND in terms of $Pr^{\frac{1}{2}}$. In order to prove the *representation theorem for* $Pr^{\frac{1}{2}}$, we define some concepts as follows:

Definition 12 (*Compatibility*) A *semiorder* $\succ$ and a *weak order* $\succeq$ are said to be *compatible* if the following conditions hold: for any $z, y, z \in \mathcal{I}$,

$$\text{If } x \succ y, \text{ then } x \succeq y,$$

and

$$\text{If } x \succeq y \succeq z \text{ and } x \sim z, \text{ then } x \sim y \text{ and } y \sim z.$$

Remark 2 (Motivation of Compatibility) The notion of compatibility is motivated by thinking $\mathcal{I}$ as $\mathbb{R}$, $\succ$ as the relation $x \succ y$ iff $x > y + 1$, and $\succeq$ as $\geq$.

Definition 13 (*Homogeneousness*) The family of semiorders is called *homogeneous* if there is exactly one weak order which is compatible with each member of the family.

Definition 14 (*Discriminatedness*) Pr is called *discriminated* if for any $x, y \in \mathcal{I}$, if $Pr(x, y) = \frac{1}{2}$, then for any $z \in \mathcal{I}$, $Pr(x, z) = Pr(y, z)$.

Roberts [16] proved the following theorem concerning SST and homogeneous family of semiorders:

Theorem 2 (SST and homogeneous family of semiorders, Roberts [16]) *Suppose that* Pr *is a* discriminated *forced-choice-pair-comparison probability function. Then* Pr *satisfies* SST *iff* $\{Pr^{\lambda} : \lambda \in [\frac{1}{2}, 1)\}$ *is a* homogeneous *family of* semiorders.

We have the following corollary of *Theorems* 1 and 2.

Corollary 1 (Representation for $Pr^{\frac{1}{2}}$) *Suppose that* Pr *is a* discriminated *forced-choice-pair-comparison probability function, that* δ *is positive number, and that the relation obtained by Definition 8 from* $Pr^{\frac{1}{2}}$ *is* connected. *Then if* Pr *satisfies* SST, *then there is a function* $f : \mathcal{I} \to \mathbb{R}$ *such that for any* $x, y \in \mathcal{I}$,

$$x Pr^{\frac{1}{2}} y \text{ iff } f(x) > f(y) + \delta,$$

where $\{Pr^{\frac{1}{2}}\}$ *is a* homogeneous *family of* semiorders *with one element.*

Remark 3 (Interpretation of f and δ). We can interpret a measure function f and a JND δ in terms of Corollary 1 that representationally relates a *statistical* variant $Pr^{\frac{1}{2}}$ of semiorder to f and δ by means of SST.

9.4 Observational Predicate Logic (OPL)

9.4.1 Language of OPL

We define the language $\mathcal{L}_{\mathsf{OPL}}$ of OPL as follows:

Definition 15 (*Language of* OPL)

- Let $\mathcal{V}$ denote a set of individual variables, $\mathcal{C}$ a set of individual constants, $\mathcal{P}$ a set of one-place *observational predicate symbols* and $\gtrless_P$ a *s-Discriminability relation symbol* relative to $P \in \mathcal{P}$.
- The language $\mathcal{L}_{\mathsf{OPL}}$ of OPL is given by the following BNF grammar:

$$t ::= x \mid a,$$
$$\varphi ::= P(t) \mid t_i = t_j \mid t_i \gtrless_P t_j \mid \top \mid \neg\varphi \mid \varphi \wedge \psi \mid \forall x \varphi,$$

  where $x \in \mathcal{V}$, $a \in \mathcal{C}$, $P \in \mathcal{P}$.
- $\perp, \vee, \rightarrow, \leftrightarrow$ and $\exists$ are introduced by the standard definitions.
- $t_i \gtrless_P t_j$ means that an *examiner* B makes a *statistical* judgment that an *examinee* A can *discriminate* t_i in P-ness from t_j.
- A *s-Indiscriminability relation symbol* $t_i \approx_P t_j$ relative to P is defined as $\neg t_i \gtrless_P t_j$.
- A *borderline-case predicate symbol* B_P *relative to* P is defined as follows:

$$B_P(t) := \exists x(t \approx_P x \wedge P(t) \wedge \neg P(x)) \vee \exists x(t \approx_P \wedge \neg P(t) \wedge P(x)).$$

- The set of all well-formed formulae of $\mathcal{L}_{\mathsf{OPL}}$ is denoted by $\Phi_{\mathcal{L}_{\mathsf{OPL}}}$.

9.4.2 Semantics of OPL

On the basis of *SST* and $\sim^*$*-Connectedness*, we define a *statistical* model $\mathfrak{M}$ of $\mathcal{L}_{\mathsf{OPL}}$ as follows:

Definition 16 (*Statistical Model $\mathfrak{M}$ of $\mathcal{L}_{\mathsf{OPL}}$*) $\mathfrak{M}$ is a tuple $(\mathcal{I}, a^{\mathfrak{M}}, b^{\mathfrak{M}}, \ldots, F^{\mathfrak{M}}, G^{\mathfrak{M}}, \ldots, Pr^{\frac{1}{2}}_{F^{\mathfrak{M}}}, Pr^{\frac{1}{2}}_{G^{\mathfrak{M}}}, \ldots)$, where:

1. $\mathcal{I}$ is a nonempty set of individuals, called the universe of $\mathfrak{M}$,
2. $a^{\mathfrak{M}}, b^{\mathfrak{M}}, \ldots \in \mathcal{I}$,
3. $F^{\mathfrak{M}}, G^{\mathfrak{M}}, \ldots \subseteq \mathcal{I}$,
4. $Pr_{F^{\mathfrak{M}}} : \mathcal{I} \times \mathcal{I} \to [0, 1]$ is a *discriminated forced-choice-pair-comparison probability function* relative to $F^{\mathfrak{M}}$ that represents the *relative frequency* which an *examiner* B observes and with which an *examinee* A responds relative to $F^{\mathfrak{M}}$ and satisfies the *SST*, ...,
5. $Pr^{\frac{1}{2}}_{F^{\mathfrak{M}}}$ is a binary relation on $\mathcal{I}$ such that for any $x, y \in \mathcal{I}$, $x Pr^{\frac{1}{2}}_{F^{\mathfrak{M}}} y$ if $Pr_{F^{\mathfrak{M}}}(x, y) > \frac{1}{2}$, ..., and
6. The relation obtained by Definition 8 from $Pr^{\frac{1}{2}}_{F^{\mathfrak{M}}}$ is *connected*,....

Remark 4 (Interpretations of Observational Predicate Symbols by Examinee) $F^{\mathfrak{M}}, G^{\mathfrak{M}}, \ldots \subseteq \mathcal{I}$ are the interpretations of *observational predicate symbols* $F, G, \ldots$ *by an examinee* A respectively.

We define an (extended) assignment function as follows:

Definition 17 (*(Extended) Assignment Function*) Let $\mathcal{V}$ denote a set of individual variables, $\mathcal{C}$ a set of individual constants and $\mathcal{I}$ a set of individuals.

- We call $s : \mathcal{V} \to \mathcal{I}$ an assignment function.
- $\tilde{s} : \mathcal{V} \cup \mathcal{C} \to \mathcal{I}$ is defined as follows:

1. For any $x \in \mathcal{V}$, $\tilde{s}(x) = s(x)$,
2. For any $a \in \mathcal{C}$, $\tilde{s}(a) = a^{\mathfrak{M}}$.

We call $\tilde{s}$ an extended assignment function.

By means of Theorem 2, we provide OPL with the following satisfaction definition relative to $\mathfrak{M}$, define the truth in $\mathfrak{M}$ by means of satisfaction and then define validity as follows:

Definition 18 (*Satisfaction*) What it means for $\mathfrak{M}$ to satisfy $\varphi \in \Phi_{\mathcal{L}_{\mathsf{OPL}}}$ with s, in symbols $\mathfrak{M} \models_{\mathsf{OPL}} \varphi[s]$ is inductively defined as follows:

- $\mathfrak{M} \models_{\mathsf{OPL}} P(t)[s]$ iff $\tilde{s}(t) \in P^{\mathfrak{M}}$,
- $\mathfrak{M} \models_{\mathsf{OPL}} t_1 = t_2[s]$ iff $\tilde{s}(t_1) = \tilde{s}(t_2)$,
- $\mathfrak{M} \models_{\mathsf{OPL}} t_i \gtrless_P t_2[s]$ iff $\tilde{s}(t_1) Pr^{\frac{1}{2}}_{P^{\mathfrak{M}}} \tilde{s}(t_2)$ or $\tilde{s}(t_2) Pr^{\frac{1}{2}}_{P^{\mathfrak{M}}} \tilde{s}(t_1)$, where $\{Pr^{\frac{1}{2}}_{P^{\mathfrak{M}}}\}$ is a homogeneous family of semiorders with one element,
- $\mathfrak{M} \models_{\mathsf{OPL}} \top$,
- $\mathfrak{M} \models_{\mathsf{OPL}} \neg\varphi[s]$ iff $\mathfrak{M} \not\models_{\mathsf{OPL}} \varphi[s]$,
- $\mathfrak{M} \models_{\mathsf{OPL}} \varphi \wedge \psi[s]$ iff $\mathfrak{M} \models_{\mathsf{OPL}} \varphi[s]$ and $\mathfrak{M} \models_{\mathsf{OPL}} \psi[s]$,

- $\mathfrak{M} \models_{\mathsf{OPL}} \forall x\varphi[s]$ iff for any $\mathfrak{d} \in \mathcal{I}$, $\mathfrak{M} \models_{\mathsf{OPL}} \varphi[s(x|\mathfrak{d})]$, where $s(x|\mathfrak{d})$ is the function that is exactly like s except for one thing: for the individual variable x, it assigns the individual $\mathfrak{d}$. This can be expressed as follows:

$$s(x|\mathfrak{d})(y) := \begin{cases} s(y) & \text{if } \; y \neq x \\ \mathfrak{d} & \text{if } \; y = x. \end{cases}$$

If $\mathfrak{M} \models_{\mathsf{OPL}} \varphi[s]$ for all s, we write $\mathfrak{M} \models_{\mathsf{OPL}} \varphi$ and say that φ is true in $\mathfrak{M}$. If φ is true in all models of OPL, we write $\models_{\mathsf{OPL}} \varphi$ and say that φ is valid.

Remark 5 (Bivalence of Satisfaction) This definition of satisfaction is *bivalent*.

The next corollary follows directly from Definitions 15 and 18.

Corollary 2 (Satisfaction Condition of $\approx_P$)

$\mathfrak{M} \models_{\mathsf{OPL}} t_1 \approx_P t_2[s]$ *iff not* $\tilde{s}(t_1) Pr^{\frac{1}{2}}_{P^{\mathfrak{M}}} \tilde{s}(t_2)$ *and not* $\tilde{s}(t_2) Pr^{\frac{1}{2}}_{P^{\mathfrak{M}}} \tilde{s}(t_1)$,

where $\{Pr^{\frac{1}{2}}_{P^{\mathfrak{M}}}\}$ *is a homogeneous family of semiorders with one element.*

The next corollary follows from Corollaries 1 and 2 and Definition 18.

Corollary 3 (Relation between s-(In)discriminability, Semiorder $Pr^{\lambda}_{P^{\mathfrak{M}}}$, Measure Function f and JND δ) *Suppose that δ is a positive number. Then there is a function $f : \mathcal{I} \to \mathbb{R}$ that satisfies the following two conditions:*

1. $\mathfrak{M} \models_{\mathsf{OPL}} t_1 \gtrless_P t_2[s]$
   iff $\tilde{s}(t_1) Pr^{\frac{1}{2}}_{P^{\mathfrak{M}}} \tilde{s}(t_2)$ *or* $\tilde{s}(t_2) Pr^{\frac{1}{2}}_{P^{\mathfrak{M}}} \tilde{s}(t_1)$, *where* $\{Pr^{\frac{1}{2}}_{P^{\mathfrak{M}}}\}$ *is a homogeneous family of semiorders with one element*
   iff $f(\tilde{s}(t_1)) > f(\tilde{s}(t_2)) + \delta$ *or* $f(\tilde{s}(t_2)) > f(\tilde{s}(t_1)) + \delta$,
2. $\mathfrak{M} \models_{\mathsf{OPL}} t_1 \approx_P t_2[s]$
   iff not $\tilde{s}(t_1) Pr^{\frac{1}{2}}_{P^{\mathfrak{M}}} \tilde{s}(t_2)$ *and not* $\tilde{s}(t_2) Pr^{\frac{1}{2}}_{P^{\mathfrak{M}}} \tilde{s}(t_1)$, *where* $\{Pr^{\frac{1}{2}}_{P^{\mathfrak{M}}}\}$ *is a homogeneous family of semiorders with one element*
   iff $f(\tilde{s}(t_2)) - \delta \leq f(\tilde{s}(t_1)) \leq f(\tilde{s}(t_2)) + \delta$.

Remark 6 (Relation of Remark 3 *to* OPL*)* This corollary relates Remark 3 of Corollary 1 to the semantics of OPL.

We now return to the Phenomenal Sorites Paradox on s-Indiscriminability. Assume that $\mathfrak{U} := (\mathcal{I}, a^{\mathfrak{U}}_1, \ldots, a^{\mathfrak{U}}_{100}, R^{\mathfrak{U}}, Pr^{\frac{1}{2}}_{R^{\mathfrak{U}}})$ is given, where

- $\mathcal{I} := \{a_1, \ldots, a_{100}\}$,
- a_i denotes the i-th colour patch, for any $i (1 \leq i \leq 100)$ grading from red to yellow,
- R denotes looking red to an examinee A,
- $Pr_{R^{\mathfrak{U}}}$ is a discriminated forced-choice-pair-comparison probability function relative to $R^{\mathfrak{U}}$ that represents the relative frequency which an examiner B observes and with which an examinee A responds relative to $R^{\mathfrak{U}}$ and satisfies SST,

- not $a_1^{\mathfrak{U}} Pr_{R^{\mathfrak{U}}}^{\frac{1}{2}} a_2^{\mathfrak{U}}$ and not $a_2^{\mathfrak{U}} Pr_{R^{\mathfrak{U}}}^{\frac{1}{2}} a_1^{\mathfrak{U}}$, ..., not $a_{99}^{\mathfrak{U}} Pr_{R^{\mathfrak{U}}}^{\frac{1}{2}} a_{100}^{\mathfrak{U}}$ and not $a_{100}^{\mathfrak{U}} Pr_{R^{\mathfrak{U}}}^{\frac{1}{2}} a_{99}^{\mathfrak{U}}$, where $\{Pr_{R^{\mathfrak{U}}}^{\frac{1}{2}}\}$ is a homogeneous family of semiorders with one element,
- $R^{\mathfrak{U}}(a_{50}^{\mathfrak{U}})$ and not $R^{\mathfrak{U}}(a_{51}^{\mathfrak{U}})$, and
- $a_{100}^{\mathfrak{U}} Pr_{R^{\mathfrak{U}}}^{\frac{1}{2}} a_1^{\mathfrak{U}}$, where $\{Pr_{R^{\mathfrak{U}}}^{\frac{1}{2}}\}$ is a homogeneous family of semiorders with one element.

Then we have the following proposition:

Proposition 1 (Non-Tolerance on s-Indiscriminability)

$$\mathfrak{U} \not\models_{\mathsf{OPL}} \forall x \forall y (x \approx_R y \rightarrow (R(x) \rightarrow R(y))).$$

Remark 7 (Avoidance of Phenomenal Sorites Paradox on s-Indiscriminability) This proposition reveals that we can *avoid* the Phenomenal Sorites Paradox on s-Indiscriminability by embodying a *response (2)* of *Motivation*.

The *transitivity* of $\approx_P$ is not valid in OPL:

Proposition 2 (Nontransitivity of $\approx_P$)

$$\not\models_{\mathsf{OPL}} \forall x \forall y \forall z ((x \approx_P y \wedge y \approx_P z) \rightarrow x \approx_P z).$$

Both the *symmetricity* of $\gtrless_P$ and that of $\approx_P$ are valid in OPL:

Proposition 3 (Symmetricity of $\gtrless_P$ and That of $\approx_P$)

- $\models_{\mathsf{OPL}} \forall x \forall y (x \gtrless_P y \rightarrow y \gtrless_P x)$,
- $\models_{\mathsf{OPL}} \forall x \forall y (x \approx_P y \rightarrow y \approx_P x)$.

9.4.3 Metalogic of OPL

We define $\approx_P^*$ that is the *syntactic* counterpart of $\sim^*$ as follows:

Definition 19 ($\approx_P^*$)

$$t_1 \approx_P^* t_2 := t_1 \approx_P t_2 \vee (t_1 \gtrless_P t_2 \wedge \forall x \neg (t_1 \gtrless_P x \wedge x \gtrless_P t_2)).$$

$\sim^*$-Connectedness is necessary, as we have seen, for a semiorder $\succ$ to have a positive threshold even when $\mathcal{I}$ is *countable*. OPL has the following *metalogical* property:

Theorem 3 (First-Order Undefinability of $\sim^*$-Connectedness)
$\sim^*$-Connectedness is not first-order definable.

Proof The following proof is based on [11]. Assume that $\sim^*$-Connectedness is definable in terms of $\approx_P^*$ in $\mathcal{L}_{\mathsf{OPL}}$ by φ. Let $\mathcal{L}'_{\mathsf{OPL}}$ expand $\mathcal{L}_{\mathsf{OPL}}$ with two new individual constants b and c. For any n, let ψ_n be the formula

$$\neg\exists x_1 \exists x_2 \ldots \exists x_n (b \approx^*_P x_1 \wedge x_1 \approx^*_P x_2 \wedge \cdots \wedge x_n \approx^*_P c),$$

saying that there is no $\sim^*$-chain between b and c of length $n+1$. Let $\mathcal{T}$ be the theory

$$\{\psi_n : n > 0\} \cup \{\neg(b = c), \neg(b \approx^*_P c)\} \cup \{\varphi\}.$$

We claim that $\mathcal{T}$ is consistent. By compactness, we have to show that every finite subset $\mathcal{T}' \subseteq \mathcal{T}$ is consistent. Indeed, let m be such that for any $\psi_n \in \mathcal{T}'$, $n < m$. Then a connected graph in which the shortest $\sim^*$-chain between b and c has length $m+1$ is a model of $\mathcal{T}'$. Since $\mathcal{T}$ is consistent, it has a model. Let $\mathfrak{V}$ be a model of $\mathcal{T}$. Then $\mathfrak{V}$ is connected, but there is no $\sim^*$-chain between b and c of length n, for any n. This contradiction shows that $\sim^*$-Connectedness is not first-order definable.

9.5 Discussion: Higher Order Vagueness

9.5.1 *Wright's Argument that Higher Order Vagueness Is per se Paradoxical*

There is little agreement upon *what* higher order vagueness is, whether *there is* higher order vagueness and whether it is a serious *problem*. Wright [36, pp. 129–132] argues that 'higher order vagueness is per se paradoxical' ([36, p. 139]) as follows: What can cause the *first-order* Sorites Paradox is that the vagueness of 'F' implies the truth of the form

$$\neg\exists x(F(x) \wedge \neg F(x')), \tag{9.1}$$

where x' is the immediate successor of x. In order to avoid the first-order Sorites Paradox, Wright introduces an operator Def expressing *definiteness* or *determinacy*. The introduction of Def implies that the vagueness of 'F' does not consist in the truth of (9.1). Instead, what is required is the truth of the form:

$$\neg\exists x(Def(F(x)) \wedge Def(\neg F(x'))). \tag{9.2}$$

But this merely postpones the difficulty. For if the distinction between things which are F and borderline cases of F is *itself* vague, then assent to

$$\neg\exists x(Def(F(x)) \wedge \neg Def(F(x'))) \tag{9.3}$$

would seem to be compelled even if assent to (9.1) is not. If (9.2) rather than (9.1) express the vagueness of 'F', then

$$\neg\exists x(Def(Def(F(x))) \wedge Def(\neg Def(F(x')))) \tag{9.4}$$

rather than (9.3) should express that of $Def(F(x))$. It is very natural to adopt as a rule of inference the following:

$$\text{(DEF)} \quad \frac{\{Def(\varphi_1), \ldots, Def(\varphi_n)\} \vdash \psi}{\{Def(\varphi_1), \ldots, Def(\varphi_n)\} \vdash Def(\psi)} \quad \text{(DEF)}$$

The definitisation of (9.4) by (DEF):

$$Def(\neg\exists x(Def(Def(F(x))) \wedge Def(\neg Def(F(x'))))) \tag{9.5}$$

is as plausible as (9.4). But from (9.5) and so on, by means of (DEF), one can derive

$$\forall x(Def(\neg Def(F(x'))) \rightarrow Def(\neg Def(F(x)))). \tag{9.6}$$

Equation (9.6) can entail that F has no definite instances if it has definite borderline cases of the first order, which is absurd. From (9.2), on the other hand, one can only derive

$$\forall x(Def(\neg F(x')) \rightarrow Def(\neg Def(F(x)))), \tag{9.7}$$

which is innocuous. The trouble is thus distinctively at *higher order*. Heck [6] blocks Wright's derivation by prohibiting the discharge of a premise φ within conditional proof or reductio ad absurdum, when φ occurs as a premise of a line obtained by (DEF). But Heck does not justify this restriction.

9.5.2 Sentential Operator Versus Predicate Symbol

The introduction of the *sentential operator Def* makes it possible to avoid the *first-order* Sorites Paradox. But it has such a harmful consequence as (9.6) in *higher order* vagueness. Since *Def* is a sentential operator, we can apply it *iteratively*. This strong expressive power leads us to derive (9.6). If we adopt this standpoint of Wright in which higher order vagueness is per se paradoxical, what is required will be a logic for vague predicates that is *strong* in expressive power enough to avoid the *first-order* Sorites Paradox and *weak* enough not to have such a harmful consequence as (9.6) in *higher order* vagueness. OPL is such a logic. In OPL, $\neg Def$ corresponds to a *borderline-case predicate symbol* B_P *relative to* P. It was defined in Definition 15 as:

$$\exists x(t \approx_P x \wedge P(t) \wedge \neg P(x)) \vee \exists x(t \approx_P \wedge \neg P(t) \wedge P(x)).$$

Since B_P is a defined *predicate symbol*, we cannot apply it iteratively. So OPL is weak in expressive power enough not to have such a consequence as (9.6).

9.6 Concluding Remarks

In this paper, we have proposed a new version of logic for *observational* predicates—Observational-Predicate Logic (OPL)—that makes it possible to reason about observational predicates without inviting the *Phenomenal Sorites Paradox on s-Indiscriminability*. To accomplish this aim, we have provided the language of OPL with a *statistical* model in terms of *measurement theory*.

This paper is only a part of a larger *measurement-theoretic* study. By means of measurement theory, we constructed or are trying to construct such logics as

1. (dynamic epistemic) preference logic [22, 32],
2. dyadic deontic logic [21],
3. threshold-utility-maximiser's preference logic [23, 24],
4. interadjective-comparison logic [27],
5. gradable-predicate logic [26],
6. logic for better questions and answers [33],
7. doxastic and epistemic logic [31],
8. multidimensional-predicate-comparison logic [29],
9. logic for preference aggregation represented by a Nash collective utility function [30] and
10. modal-qualitative-probability logic [28].

Acknowledgments The author would like to thank an anonymous reviewer of TPLC-2014 for her or his very helpful comments.

Appendix: Goodman on JNDs and Semiorders

Goodman [3] adopts the following four *primitive* predicates:

1. a reflexive, symmetric and *nontransitive* two-place predicate '*overlaps*' **o**,
2. an irreflexive, symmetric and *nontransitive* two-place predicate '*is with*' W,
3. a reflexive, symmetric and *transitive* two-place predicate '*is of equal aggregate size to*' Z and
4. a reflexive, symmetric and *nontransitive* two-place predicate '*match*' M.

Goodman [3, p. 219] defines a three-place predicate '*y is betwixt x and z*' $x/y/z$ by *matching* and other primitive predicates. Goodman [2, p. 469], [3, p. 226] defines '*a is just noticeably different from b*' $JND(a, b)$ by *matching* and *betwixtness* as follows:

Definition 20 (*JND*)

$$JND(a, b) := \neg M(a, b) \wedge \exists x(M(x, a) \wedge M(x, b)) \wedge \forall y(a/y/b \rightarrow (M(y, a) \wedge M(y, b)).$$

Remark 8 (Interpretation of Definition) That a is *just noticeably different* from b means that a does not *match* b, that some element *matches* both a and b, and that every element which is *betwixt* a and b *matches* both a and b.

Goodman [3, p. 227] argues that his definition of JND can satisfy 'the weaker rule (i.e. that *no span between nonmatching elements is enclosed within a span matching elements*)'. Moreover, Goodman [3, p. 213] points out the *anticipation* of *semiorders* as follows:

> This weaker rule was stated, and its use explained, in [2, pp. 434ff]. Publication of it ten years later (i.e. 1951) in the first edition of the present book (i.e. *The Structure of Appearance*) *anticipated* by five years its adoption by R. Duncan Luce as the fundamental principle of his theory of '*semiorders*'. See his article [12] especially axiom $S3$ (i.e. *Semitransitivity*) and $S4$ (i.e. *Intervality*) and the discussion of them on pp. 181–182".

References

1. Fechner, G.T.: Elemente der Psychophysik. Breitkopf und Hartel, Leipzig (1860)
2. Goodman, N.: A Study of Qualities. Ph.D. thesis, Harvard University (1940)
3. Goodman, N.: The Structure of Appearance, 3rd edn. Reidel, Dordrecht (1977)
4. Graff, D.: Phenomenal continua and the sorites. Mind **110**, 905–935 (2001)
5. Hardin, C.L.: Phenomenal colors and sorites. Noûs **22**, 213–234 (1988)
6. Heck Jr., R.G.: A note on the logic of (higher-order) vagueness. Analysis **53**, 201–208 (1993)
7. Hölder, O.: Die Axiome der Quantität und die Lehre vom Mass. Berichte über die Verhandlungen der Königlich Sächsischen Gesellschaft der Wissenschaften zu Leipzig. Mathematisch-Physikalische Klasse **53**, 1–64 (1901)
8. Hyde, D.: Sorites paradox. Stanford Encyclopedia of Philosophy (2005)
9. Keefe, R.: Phenomenal sorites paradoxes and looking the same. Dialectica **65**, 327–344 (2011)
10. Krantz, D.H., et al.: Foundations of Measurement, vol. 1. Academic Press, New York (1971)
11. Libkin, L.: Elements of Finite Model Theory. Springer, Heidelberg (2004)
12. Luce, R.D.: Semiorders and a theory of utility discrimination. Econometrica **24**, 178–191 (1956)
13. Luce, R.D., et al.: Foundations of Measurement, vol. 3. Academic Press, San Diego (1990)
14. Manders, K.L.: On JND representations of semiorders. J. Math. Psychol. **24**, 224–248 (1981)
15. Raffman, D.: Is perceptual indiscriminability nontransitive? Philos. Topics **28**, 153–175 (2000)
16. Roberts, F.S.: Homogeneous families of semiorders and the theory of probabilistic consistency. J. Math. Psychol. **8**, 248–263 (1971)
17. Roberts, F.S.: Measurement Theory. Addison-Wesley, Reading (1979)
18. Scott, D., Suppes, P.: Foundational aspects of theories of measurement. J. Symb. Logic **3**, 113–128 (1958)
19. Simon, H.A.: Models of Bounded Rationality. The MIT Press, Cambridge (1982)
20. Suppes, P., et al.: Foundations of Measurement, vol. 2. Academic Press, San Diego (1989)
21. Suzuki, S.: Measurement-theoretic foundation of preference-based dyadic deontic logic. In: He, X., et al. (eds.) Proceedings of the Second International Workshop on Logic, Rationality, and Interaction (LORI-II). LNCS, vol. 5834, pp. 278–291. Springer, Heidelberg (2009)
22. Suzuki, S.: Prolegomena to dynamic epistemic preference logic. In: Hattori, H., et al. (eds.) New Frontiers in Artificial Intelligence. LNCS, vol. 5447, pp. 177–192. Springer, Heidelberg (2009)
23. Suzuki, S.: Prolegomena to threshold utility maximiser's preference logic. In: Electronic Proceedings of the 9th Conference on Logic and the Foundations of Game and Decision Theory (LOFT 2010) (2010), paper No. 44

24. Suzuki, S.: A measurement-theoretic foundation of threshold utility maximiser's preference logic. J. Appl. Ethics Philos. **3**, 17–25 (2011)
25. Suzuki, S.: Measurement-theoretic foundations of probabilistic model of JND-based vague predicate logic. In: van Ditmarsch, H., et al. (eds.) Proceedings of the Third International Workshop on Logic, Rationality, and Interaction (LORI-III). LNCS, vol. 6953, pp. 272–285. Springer, Heidelberg (2011)
26. Suzuki, S.: Measurement-theoretic foundations of gradable-predicate logic. In: Okumura, M., et al. (eds.) New Frontiers in Artificial Intelligence. LNCS, vol. 7258, pp. 82–95. Springer, Heidelberg (2012)
27. Suzuki, S.: Measurement-theoretic foundations of interadjective-comparison logic. In: Aguilar-Guevara, A., et al. (eds.) Proceedings of Sinn und Bedeutung 16, vol. 2, pp. 571–584. MIT Working Papers in Linguistics, Cambridge (2012)
28. Suzuki, S.: Epistemic modals, qualitative probability, and nonstandard probability. In: Aloni, M., et al. (eds.) Proceedings of the 19th Amsterdam Colloquium (AC 2013), pp. 211–218 (2013)
29. Suzuki, S.: Measurement-theoretic bases of multidimensional-predicate logic (2013)
30. Suzuki, S.: Measurement-theoretic foundations of many-sorted preference aggregation logic for Nash collective utility function (2013)
31. Suzuki, S.: Remarks on decision-theoretic foundations of doxastic and epistemic logic (revised version). Stud. Logic **6**, 1–12 (2013)
32. Suzuki, S.: Measurement-theoretic foundations of dynamic epistemic preference logic. In: Mc-Cready, E., et al. (eds.) Formal Approaches to Semantics and Pragmatics, Studies in Linguistics and Philosophy, vol. 95, pp. 295–324. Springer, Heidelberg (2014)
33. Suzuki, S.: Measurement-theoretic foundations of logic for better questions and answers. In: Zeevat, H., Schmitz, H.C. (eds.) Bayesian Natural Language Semantics and Pragmatics, Language, Cognition, and Mind, vol. 2, pp. 43–69. Springer, Heidelberg (2015)
34. van Rooij, R.: Revealed preference and satisficing behavior. Synthese **179**, 1–12 (2011)
35. van Rooij, R.: Vagueness and linguistics. In: Ronzitti, G. (ed.) Vagueness: A Guide, pp. 123–170. Springer, Heidelberg (2011)
36. Wright, C.: Is higher order vagueness coherent? Analysis **52**, 129–139 (1992)

Chapter 10
Channel Theoretic Reflections on Dynamic Logics of Speech Acts

Tomoyuki Yamada

Abstract We usually succeed in performing illocutionary acts such as commanding, requesting, promising, asserting, conceding, and so on in saying things. There is a systematic relation between what is said and what is achieved in saying it. Yet illocutionary acts may fail to take effect in various ways. You might try to issue a command but fail, for example, because of the lack of suitable authority. The purpose of this paper is to show how the regularities that enable us to perform illocutionary acts and the background conditions that normally support them can be captured in logical terms. For this purpose, we model the relevant kind of regularities in the form of constraints of local logics introduced in channel theory developed by Barwise and Seligman, by building information channels with the language and sets of models of "dynamified" deontic logic $\mathsf{DMDL}^{+}\mathsf{III}$ of acts of commanding and promising developed by Yamada. In doing so, it will be seen that the language of $\mathsf{DMDL}^{+}\mathsf{III}$ needs to be substantially extended in order to talk about the relation between acts of saying things and acts of commanding. We conclude by hinting at how this can be done.

Keywords Illocutionary act · Command · Dynamified deontic logic · Channel theory · Local logic · Background condition · Normal context

10.1 Introduction

In doing things in everyday life, we rely on various regularities that hold normally. For example, by turning the switch of her flashlight on, Judith gets the bulb lit.[1] The relevant regularity may be stated as follows ([1], p. 45):

The switch being on entails the bulb lighting.

[1]A detailed discussion of this example is given by Barwise and Seligman ([1], pp. 4–10, 30, 36–37, 41–45).

T. Yamada (✉)
Hokkaido University, Nishi-7, Kita-10, Kita-ku, Sapporo, Hokkaido 060-0810, Japan
e-mail: yamada@let.hokudai.ac.jp

S.C.-M. Yang et al. (eds.), *Structural Analysis of Non-Classical Logics*,
Logic in Asia: Studia Logica Library, DOI 10.1007/978-3-662-48357-2_10

It will not work, however, if the battery is dead. Thus, we may revise the above statement and get the following:

The switch being on and the battery being live entail the bulb lighting.

What will happen, however, if the bulb is gone? As we know very well, things can go wrong in many different ways.

The same thing can be said about speech acts. We usually succeed in performing illocutionary acts such as commanding, requesting, promising, asserting, conceding, and so on in saying things. There is a systematic relation between what is said and what is achieved in saying it. Yet illocutionary acts may fail to take effect in various ways. You might try to issue a command but fail because of the lack of suitable authority, for example.

The purpose of this paper is to show how the regularities that enable us to perform illocutionary acts and the background conditions that normally support them can be captured in logical terms. For this purpose, we model the relevant kind of regularities in the form of constraints of local logics introduced in channel theory developed by Barwise and Seligman [1], by building information channels with the language and sets of models of "dynamified" deontic logic $\mathsf{DMDL}^{+}\mathsf{III}$ of acts of commanding and promising developed by Yamada [14]. $\mathsf{DMDL}^{+}\mathsf{III}$ is developed by dynamifying a multi-agent variant of deontic logic in a way similar to the way in which PAL (Public Announcement Logic) dynamifies epistemic logic.[2] The procedure we follow in building information channels with the language and models of $\mathsf{DMDL}^{+}\mathsf{III}$ can be applied, mutatis mutandis, to any other dynamified logics that are developed in a similar style, and so may be of some interest even to those who are not particularly interested in speech acts.

The remainder of the paper is structured as follows. In Sect. 10.2, we review how the effects which acts of commanding and promising involve by virtue of their being the very kinds of acts per se can be captured in $\mathsf{DMDL}^{+}\mathsf{III}$.[3] In Sect. 10.3, we review how simple acts of using a flashlight can be modeled by building information channels in channel theory. Then in Sect. 10.4, we build information channels with the language and the models of $\mathsf{DMDL}^{+}\mathsf{III}$ and show how the validities of $\mathsf{DMDL}^{+}\mathsf{III}$ can be restated as the constraints of a local logic that characterizes the core of the channel. For the sake of simplicity, we will concentrate on acts of commanding, and compare them with simple acts of using a flashlight. In the course of this comparison, it will be shown that we need a substantial extension of the language of $\mathsf{DMDL}^{+}\mathsf{III}$ in order to talk about the relation between acts of saying things and acts of commanding. In Sect. 10.5, we make a few observations on what is achieved in $\mathsf{DMDL}^{+}\mathsf{III}$ and what will be needed in order to capture the relevant kind of regularities and the background

[2]PAL is developed by Plaza [4], Gerbrandy and Groeneveld [2], and Kooi and van Benthem [3] among others.

[3]Since actions of each type α can bring about not only the effects that are definitive or essential to their being acts of type α but also various further consequences including very remote ones, it is not safe to talk about "the effects" simpliciter. In this paper, however, we will only talk about their definitive effects, and usually refer to them as "the effects" for the sake of simplicity.

conditions that support them in the suggested extension of DMDL^{+}III from the point of view of channel theory.

10.2 Logical Dynamics of Speech Acts

Inspired by the development of systems of DEL (Dynamic Epistemic Logic), of which PAL was the earliest, a series of dynamified logics that deal with various specific speech acts have been developed by Yamada [12–17].[4] The general methodology can be summarized in the form of a recipe as follows:

1. Carefully identify the aspects affected by the speech acts you want to study.
2. Find a modal logic that characterizes these aspects, and use it as the base logic.
3. Add dynamic modalities that represent types of those speech acts.
4. Expand truth definition by adding clauses that interpret the speech acts under study as what updates the very aspects.
5. Find (if possible) a complete set of recursion axioms for the resulting dynamic logic, and derive its completeness from that of the base logic.[5]

DMDL^{+}III (Dynamified Multi-agent Deontic Logic plus alethic modalities) is one of the logics developed in this way, and MDL^{+}III is its static base logic. The choice of deontic logic as the base logic reflects the view that acts of commanding and promising change the deontic status of the possible courses of action. The language of MDL^{+}III is defined as follows ([14], p. 98):

Definition 1 Take a countably infinite set **Aprop** of proposition letters and a finite set I of agents, with p ranging over **Aprop** and i, j, k over I. The language $\mathcal{L}_{\mathsf{MDL^{+}III}}$ of MDL^{+}III is given by the following syntax:

$$\varphi ::= \top \mid p \mid \neg\varphi \mid (\varphi \wedge \psi) \mid \Box\varphi \mid O_{(i,\, j,\, k)}\varphi\ .$$

The formula of the form $O_{(i,\, j,\, k)}\varphi$ means that it is obligatory for i with respect to j by the name of k, where i is the agent who owes the obligation (sometimes called "obligor"), j is the agent to whom the obligation is owed (sometimes called "obligee"), and k is the agent who creates the obligation. We will illustrate how these indices are used to differentiate obligations created by acts of commanding from those created by acts of promising later on.

---

[4] A detailed textbook exposition of the development of PAL and other systems of DEL can be found in van Ditmarsch et al. [9].

[5] Recursion axioms are also known as "reduction axioms" in the literature. Here we follow van Benthem's advice to refer to them as "recursion axioms".

The language of $\mathsf{DMDL^{+}III}$ is defined by adding dynamic modalities as follows ([14], p. 100):

Definition 2 Take the same countably infinite set **Aprop** of proposition letters and the same finite set I of agents, with p ranging over **Aprop** and i, j, k over I. The language $\mathcal{L}_{\mathsf{DMDL^{+}III}}$ of $\mathsf{DMDL^{+}III}$ is given by the following syntax:

$$\begin{aligned}\varphi &::= \top \mid p \mid \neg\varphi \mid (\varphi \wedge \psi) \mid \Box\varphi \mid O_{(i,\, j,\, k)}\varphi \mid [\pi]\varphi \\ \pi &::= \mathsf{Com}_{(i,\, j)}\varphi \mid \mathsf{Prom}_{(i,\, j)}\varphi\ .\end{aligned}$$

The expressions of the form $\mathsf{Com}_{(i,\, j)}\varphi$ and those of the form $\mathsf{Prom}_{(i,\, j)}\varphi$ are terms that stand for types of speech acts, and the expressions of the form $[\mathsf{Com}_{(i,\, j)}\varphi]$ and those of the form $[\mathsf{Prom}_{(i,\, j)}\varphi]$ are dynamic modalities. The formula of the form $[\mathsf{Com}_{(i,j)}\varphi]\psi$ means that ψ holds after i commands j to see to it that φ, and the formula of the form $[\mathsf{Prom}_{(i,j)}\varphi]\psi$ means that ψ holds after i promises j that i will see to it that φ.[6]

Truth definitions for $\mathsf{MDL^{+}III}$ and $\mathsf{DMDL^{+}III}$ are given with reference to $\mathcal{L}_{\mathsf{MDL^{+}III}}$-models ([14], pp. 98–99, 101).[7]

Definition 3 By an $\mathcal{L}_{\mathsf{MDL^{+}III}}$-model, we mean a tuple

$$\mathcal{M} = \langle W^{\mathcal{M}},\ A^{\mathcal{M}},\ \{D^{\mathcal{M}}_{(i,\, j,\, k)} \mid i, j, k \in I\},\ V^{\mathcal{M}}\rangle$$

where

1. $W^{\mathcal{M}}$ is a nonempty set (heuristically, of "possible worlds"),
2. $A^{\mathcal{M}} \subseteq W^{\mathcal{M}} \times W^{\mathcal{M}}$,
3. $D^{\mathcal{M}}_{(i,\, j,\, k)} \subseteq A^{\mathcal{M}}$ for each $i, j, k \in I$,
4. $V^{\mathcal{M}}$ is a function that assigns a subset $V^{\mathcal{M}}(p)$ of $W^{\mathcal{M}}$ to each proposition letter $p \in$ **Aprop**.

$A^{\mathcal{M}}$ here is the alethic accessibility relation to be used in interpreting $\Box$, and $D^{\mathcal{M}}_{(i,\, j,\, k)}$ is the deontic accessibility relation to be used in interpreting $O_{(i,\, j,\, k)}$. When no confusion is likely, we will omit the superscript.

For the sake of simplicity, no frame conditions are imposed on the alethic accessibility relation. Each deontic accessibility relation, on the other hand, is required to be a subset of the alethic accessibility relation. Together with the truth definition, this means that only possible things are permitted. Note that deontic accessibility relations are not assumed to be serial. This allows for the possibility of conflicts of

---

[6] The formulas of the form $\langle\mathsf{Com}_{(i,j)}\varphi\rangle\psi$ and those of the form $\langle\mathsf{Prom}_{(i,j)}\varphi\rangle\psi$ are introduced as the abbreviations for $\neg[\mathsf{Com}_{(i,j)}\varphi]\neg\psi$ and $\neg[\mathsf{Prom}_{(i,j)}\varphi]\neg\psi$, respectively, but according to the semantics given below, they are equivalent to $[\mathsf{Com}_{(i,j)}\varphi]\psi$ and $[\mathsf{Prom}_{(i,j)}\varphi]\psi$, respectively.

[7] In what follows, the definition and the notation are slightly simplified, but there is no substantial difference.

obligations, but indexing on the deontic accessibility relations minimizes the possibility of deontic explosion.[8]

Truth definition for $\mathsf{MDL}^{+}\mathsf{III}$ is completely standard. The clause for deontic modality, for example, reads as follows:

$$\mathcal{M}, w \models_{\mathsf{MDL}^{+}\mathsf{III}} O_{(i,\,j,\,k)}\varphi \text{ iff for any } v \text{ such that } \langle w, v\rangle \in D^{\mathcal{M}}_{(i,\,j,\,k)}, \mathcal{M}, v \models_{\mathsf{MDL}^{+}\mathsf{III}} \varphi\,.$$

Truth definition for $\mathsf{DMDL}^{+}\mathsf{III}$ is given by adding clauses for dynamic modalities to the set of clauses in the truth definitions for $\mathsf{MDL}^{+}\mathsf{III}$ reproduced mutatis mutandis. The clauses for dynamic modalities read as follows:

$$\mathcal{M}, w \models_{\mathsf{DMDL}^{+}\mathsf{III}} [\mathsf{Com}_{(i,\,j)}\varphi]\psi \text{ iff } \mathcal{M}_{\mathsf{Com}_{(i,\,j)}\varphi}, w \models_{\mathsf{DMDL}^{+}\mathsf{III}} \psi\,,$$
$$\mathcal{M}, w \models_{\mathsf{DMDL}^{+}\mathsf{III}} [\mathsf{Prom}_{(i,\,j)}\varphi]\psi \text{ iff } \mathcal{M}_{\mathsf{Prom}_{(i,\,j)}\varphi}, w \models_{\mathsf{DMDL}^{+}\mathsf{III}} \psi\,,$$

where $\mathcal{M}_{\mathsf{Com}_{(i,\,j)}\varphi}$ is the $\mathcal{L}_{\mathsf{MDL}^{+}\mathsf{III}}$-model obtained from $\mathcal{M}$ by replacing $D_{(j,\,i,\,i)}$ with its subset $\{(x, y) \in D_{(j,\,i,\,i)} \mid \mathcal{M}, y \models_{\mathsf{DMDL}^{+}\mathsf{III}} \varphi\}$ while keeping the other things unchanged, and $\mathcal{M}_{\mathsf{Prom}_{(i,\,j)}\varphi}$ is the $\mathcal{L}_{\mathsf{MDL}^{+}\mathsf{III}}$-model obtained from $\mathcal{M}$ by replacing $D_{(i,\,j,\,i)}$ with its subset $\{(x, y) \in D_{(i,\,j,\,i)} \mid \mathcal{M}, y \models_{\mathsf{DMDL}^{+}\mathsf{III}} \varphi\}$ while keeping the other things unchanged.

Thus defined, $D^{\mathcal{M}_{\mathsf{Com}_{(i,\,j)}\varphi}}_{(j,\,i,\,i)} \subseteq D^{\mathcal{M}}_{(j,\,i,\,i)}$ but $D^{\mathcal{M}_{\mathsf{Com}_{(i,\,j)}\varphi}}_{(k,\,l,\,m)} = D^{\mathcal{M}}_{(k,\,l,\,m)}$ if $D^{\mathcal{M}}_{(k,\,l,\,m)} \neq D^{\mathcal{M}}_{(j,\,i,\,i)}$, and $D^{\mathcal{M}_{\mathsf{Prom}_{(i,\,j)}\varphi}}_{(i,\,j,\,i)} \subseteq D^{\mathcal{M}}_{(i,\,j,\,i)}$ but $D^{\mathcal{M}_{\mathsf{Prom}_{(i,\,j)}\varphi}}_{(k,\,l,\,m)} = D^{\mathcal{M}}_{(k,\,l,\,m)}$ if $D^{\mathcal{M}}_{(k,\,l,\,m)} \neq D^{\mathcal{M}}_{(i,\,j,\,i)}$. This guarantees that updated models satisfy Clause 3 of Definition 3; they remain to be $\mathcal{L}_{\mathsf{MDL}^{+}\mathsf{III}}$-models. Since the updated deontic accessibility relations are subsets of the original deontic accessibility relations, they are subsets of the alethic accessibility relation as well. This will hold even if we impose some additional frame conditions on the alethic accessibility relation in Definition 3. $\mathsf{MDL}^{+}\mathsf{III}$ and $\mathsf{DMDL}^{+}\mathsf{III}$ are completely axiomatized in [14].

Based on the above truth definition, the following two principles are seen to hold ([14], p. 102):

Proposition 1 (The CUGO Principle) *If φ is a formula of* $\mathsf{MDL}^{+}\mathsf{III}$ *and is free of modal operators of the form $O_{(j,\,i,\,i)}$, the following formula is valid:*

$$[\mathsf{Com}_{(i,\,j)}\varphi]O_{(j,\,i,\,i)}\varphi\,.$$

Proposition 2 (The PUGO Principle) *If φ is a formula of* $\mathsf{MDL}^{+}\mathsf{III}$ *and is free of modal operators of the form $O_{(i,\,j,\,i)}$, the following formula is valid:*

$$[\mathsf{Prom}_{(i,\,j)}\varphi]O_{(i,\,j,\,i)}\varphi\,.$$

[8]For more on deontic explosion, see [15], pp. 308–311.

These principles partially characterize the effects of acts of commanding and promising, respectively: [c]ommands and [p]romises [u]sually [g]enerate [o]bligations.

Note the difference in the order of indices on the deontic operators occurring in the formulas mentioned in the two principles. In the case of obligations generated by an act of commanding, the creator of the obligation is the agent who issues the command and the commandee is the agent who owes the obligations. By contrast, in the case of the obligations generated by an act of promising, the creator and the agent who owes the obligations are both the agent who makes the promise, and the promisee is the agent to whom the obligations are owed (the obligee). The sameness of the agent who creates the obligation and the agent who owes the obligation in the case of an act of promising indicates that the agent who promises commits herself to the action she promises to do.[9]

Yamada ([14], p. 96) gives an example of a professor who receives a letter from his political guru in which she (the guru) commands him to join an important political demonstration in Tokyo next year. Unfortunately, the day on which the demonstration is scheduled is the very same day on which the conference his former student is organizing is to be held in São Paulo. He has already promised her (his former student) that he will give an invited talk in that conference. Although the time in São Paulo is 12 h behind the time in Tokyo, no available means of transportation are fast enough to enable him to attend both events. It is possible for him to join the demonstration in Tokyo, but if he chooses to do so, he will not be able to keep his promise. It is also possible for him to attend the conference in São Paulo, but if he chooses to do so, he will not be able to obey his guru's command. Let p be the proposition that he will attend the conference in São Paulo, say, on July 7, 2016, and q be the proposition that he will join the demonstration in Tokyo on July 7, 2016. Let, in addition, a, b, c be the professor, his former student, and his guru, respectively. Then by CUGO Principle and PUGO Principle the following holds in the situation before he made his promise:

$$[\mathsf{Prom}_{(a,b)}p][\mathsf{Com}_{(c,a)}q](O_{(a,b,a)}p \wedge O_{(a,c,c)}q) .$$

Let $(\mathcal{M}, w)$ be that situation. Then we have

$$(\mathcal{M}_{\mathsf{Prom}_{(a,b)}p})_{\mathsf{Com}_{(c,a)}q}, w \models_{\mathsf{DMDL+III}} O_{(a,b,a)}p \wedge O_{(a,c,c)}q .$$

Moreover, we also have

$$(\mathcal{M}_{\mathsf{Prom}_{(a,b)}p})_{\mathsf{Com}_{(c,a)}q}, w \models_{\mathsf{DMDL+III}} \Diamond p \wedge \Diamond q \wedge \neg\Diamond(p \wedge q) .$$

Thus $((\mathcal{M}_{\mathsf{Prom}_{(a,b)}p})_{\mathsf{Com}_{(c,a)}q}, w)$ is exactly the situation in which the professor finds himself when he receives the letter from his guru.

[9]Whether the index for an obligee plays any substantial role in the case of acts of commanding may be disputable, but even if it is just an idle wheel, it is harmless.

10.3 Actions in Channel Theory

In this section, we review how simple acts of using a flashlight can be modeled in channel theory. We first reproduce definitions of the notions we need from Part I of Barwise and Seligman [1].[10]

The most basic building blocks of channel theory are classifications and infomorphisms. A classification is a system defined as follows ([1], pp. 28, 69):

Definition 4 A *classification* $\mathbf{A} = \langle \text{tok}(A), \text{typ}(A), \models_A \rangle$ consists of

1. a set, tok($\mathbf{A}$), of objects to be classified, called *tokens* of $\mathbf{A}$,
2. a set, typ($\mathbf{A}$), of objects used to classify the tokens, called the *types* of $\mathbf{A}$, and
3. a binary relation, $\models_{\mathbf{A}}$, between tok($\mathbf{A}$) and typ($\mathbf{A}$).

If $a \models_{\mathbf{A}} \alpha$, then a is said to be *of type* α *in* $\mathbf{A}$. A classification can be represented by the diagram of the following form:

$$\begin{array}{c} \text{typ}(\mathbf{A}) \\ \Big| \models_{\mathbf{A}} \\ \text{tok}(\mathbf{A}) \end{array}$$

A simple form of regularity can be captured in terms of the relation that holds between sets of types of a classification. By a *sequent* we just mean a pair $\langle \Gamma, \Delta \rangle$ of sets of types. Then we can define the notion of constraints ([1], p. 29).

Definition 5 Let $\mathbf{A}$ be a classification and let $\langle \Gamma, \Delta \rangle$ be a sequent of $\mathbf{A}$. A token a of $\mathbf{A}$ satisfies $\langle \Gamma, \Delta \rangle$ provided that if a is of type α for every $\alpha \in \Gamma$ then a is of type α for some $\alpha \in \Delta$. We say that Γ *entails* Δ in $\mathbf{A}$, written $\Gamma \vdash_{\mathbf{A}} \Delta$, if every token a of $\mathbf{A}$ satisfies $\langle \Gamma, \Delta \rangle$. If $\Gamma \vdash_{\mathbf{A}} \Delta$ then the pair $\langle \Gamma, \Delta \rangle$ is called a *constraint* supported by the classification $\mathbf{A}$.

Now an infomorphism captures an interesting relation between classifications ([1], p. 32).

Definition 6 If $\mathbf{A} = \langle \text{tok}(\mathbf{A}), \text{typ}(\mathbf{A}), \models_{\mathbf{A}} \rangle$ and $\mathbf{C} = \langle \text{tok}(\mathbf{C}), \text{typ}(\mathbf{C}), \models_{\mathbf{C}} \rangle$ are classifications, then an *infomorphism* from $\mathbf{A}$ to $\mathbf{C}$ is a pair $f = \langle f^{\wedge}, f^{\vee} \rangle$ of functions

[10] Although the rigorous development of channel theory is given in Part II of the book, the simpler and more intuitive exposition in Part I is enough for our purposes here. We sometimes use the notation of Part II, however, even in presenting the definitions from Part I when it is more convenient to do so.

$$\begin{array}{ccc}
\mathrm{typ}(\mathbf{A}) & \xrightarrow{f^{\wedge}} & \mathrm{typ}(\mathbf{C}) \\
\Big| \models_{\mathbf{A}} & & \Big| \models_{\mathbf{C}} \\
\mathrm{tok}(\mathbf{A}) & \xleftarrow[f^{\vee}]{} & \mathrm{tok}(\mathbf{C})
\end{array}$$

satisfying the biconditional

$$f^{\vee}(c) \models_{\mathbf{A}} \alpha \quad \textit{iff} \quad c \models_{\mathbf{C}} f^{\wedge}(\alpha)$$

for all tokens c of **C** and all types α of **A**.

This biconditional is called *the fundamental property of infomorphisms*.

The infomorphism f from **A** to **C** is sometimes written as $f : \mathbf{A} \rightleftarrows \mathbf{C}$ or even represented by a single arrow from **A** to **C**. Note that the direction of the infomorphism f is the same as the direction of the function $f^{\wedge}$ on types.

Given an infomorphism, we can reason about how things are in one classification in terms of how things are in another classification. Let arbitrary classifications **A**, **B** and an infomorphism $f : \mathbf{A} \rightleftarrows \mathbf{B}$ are given. We write Γ^f for the set of translations of types in Γ when Γ is a set of types of **A**. If Γ is a set of types of **B**, we write Γ^{-f} for the set of types whose translations are in Γ. Then we can consider the following two inference rules ([1], p. 38):

$$f\text{-Intro} : \frac{\Gamma^{-f} \vdash_{\mathbf{A}} \Delta^{-f}}{\Gamma \vdash_{\mathbf{B}} \Delta}$$

$$f\text{-Elim} : \frac{\Gamma^{f} \vdash_{\mathbf{B}} \Delta^{f}}{\Gamma \vdash_{\mathbf{A}} \Delta}$$

The rule f-Intro preserves validity in the sense that if Γ^{-f} entails Δ^{-f} in **A**, Γ entails Δ in **B**, since, by the fundamental property of infomorphism, if $b \in \mathrm{tok}(\mathbf{B})$ were a counterexample to $\langle\Gamma, \Delta\rangle$ in **B**, $f^{\vee}(b)$ would be a counterexample to $\langle\Gamma^{-f}, \Delta^{-f}\rangle$ in **A**. By contrast, f-Elim does not preserve validity. Since there may be a token $a \in \mathrm{tok}(\mathbf{A})$ for which there is no token $b \in \mathrm{tok}(\mathbf{B})$ such that $f^{\vee}(b) = a$, it can be a counterexample to $\langle\Gamma, \Delta\rangle$ in **A** even if there is no counterexample to $\langle\Gamma^{f}, \Delta^{f}\rangle$ in **B**.

From this we can also see that f-Intro does not preserve nonvalidity in the sense that even if Γ^{-f} does not entail Δ^{-f} in **A**, Γ may entail Δ in **B**. If the only counterexamples to $\langle\Gamma^{-f}, \Delta^{-f}\rangle$ in **A** are those tokens a for which there are no tokens b in **B** such that $f^{\vee}(b) = a$, Γ may entail Δ in **B**. By contrast, f-Elim preserves nonvalidity. By the fundamental property of infomorphism again, if $b \in \mathrm{tok}(\mathbf{B})$ is a counterexample to $\langle\Gamma^{f}, \Delta^{f}\rangle$ in **B**, $f^{\vee}(b)$ is a counterexample to $\langle\Gamma, \Delta\rangle$ in **A** ([1], pp. 38–39).

Now let us turn to information channels ([1], pp. 34–35).

Definition 7 An *information channel* consists of an indexed family $\mathcal{C} = \{f_i : \mathbf{A}_i \rightleftarrows \mathbf{C}\}$ of infomorphisms with a common codomain **C** called the *core* of the channel.

We can model the relation between various parts of a flashlight and the flashlight as a whole by building an information channel. Let **Flashlight**, **Bulb**, and **Switch** be classifications that classify instances of flashlights f_t, bulbs b_t, and switches s_t at various times t. Then we can define infomorphisms $f_{\mathbf{Bulb}}$ from **Bulb** to **Flashlight**, and $f_{\mathbf{Switch}}$ from **Switch** to **Flashlight**. The pair of these two infomorphisms forms an information channel depicted by the following diagram:

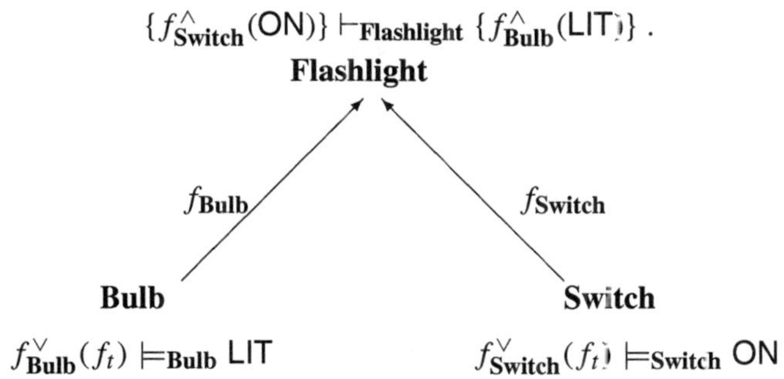

Given a particular flashlight f_t at a particular time t, $f^{\vee}_{\mathbf{Bulb}}(f_t)$ is the bulb of f_t at time t, and the formula $f^{\vee}_{\mathbf{Bulb}}(f_t) \models_{\mathbf{Bulb}} \mathsf{LIT}$ means that $f^{\vee}_{\mathbf{Bulb}}(f_t)$ is lit. By the fundamental property of infomorphisms, it entails $f_t \models_{\mathbf{Flashlight}} f^{\wedge}_{\mathbf{Bulb}}(\mathsf{LIT})$. This means that f_t has the property of having its bulb lit. Moreover, $f^{\vee}_{\mathbf{Switch}}(f_t)$ is the switch of f_t at time t, and the formula $f^{\vee}_{\mathbf{Switch}}(f_t) \models_{\mathbf{Switch}} \mathsf{ON}$ means that $f^{\vee}_{\mathbf{Switch}}(f_t)$ is on. By the fundamental property of infomorphisms again, it entails $f_t \models_{\mathbf{Flashlight}} f^{\wedge}_{\mathbf{Switch}}(\mathsf{ON})$. It means that f_t has the property of having its switch turned on.

Suppose, for the sake of simplicity, every token of **Flashlight** is in good working order. Then we have

$$\{f^{\wedge}_{\mathbf{Switch}}(\mathsf{ON})\} \vdash_{\mathbf{Flashlight}} \{f^{\wedge}_{\mathbf{Bulb}}(\mathsf{LIT})\} .$$

This captures the regularity we discussed at the beginning of this paper. We can think of this as a constraint in a local logic defined as follows ([1], p. 40):

Definition 8 A *local logic* $\mathfrak{L} = \langle \mathbf{A}, \vdash_{\mathfrak{L}}, N_{\mathfrak{L}} \rangle$ consists of a classification **A**, a set $\vdash_{\mathfrak{L}}$ of sequents (satisfying certain structural rules) involving the types of **A**, called the *constraints* of $\mathfrak{L}$, and a subset $N_{\mathfrak{L}}$ of the set of all the tokens of **A**, called the *normal tokens* of $\mathfrak{L}$, which satisfy all the constraints of $\vdash_{\mathfrak{L}}$.

A local logic $\mathfrak{L}$ is *sound* if every token is normal; it is *complete* if every sequent that holds of all normal tokens is in the consequence relation $\vdash_{\mathfrak{L}}$.[11]

In the above example, **Flashlight** is assumed to have only normal tokens, but we can expand **Flashlight** by adding more tokens. Let **Flashlight**, **Bulb**, and **Switch** be abbreviated as **F**, **B**, and **S**. Let $\mathbf{F}'$ be the expanded classification, and suppose the tokens of the bulbs and the switches of added tokens of flashlights are all in tok(**B**), and tok(**S**), respectively. Then we can define more infomorphisms such that the following diagram commutes ([1], pp. 43–44):

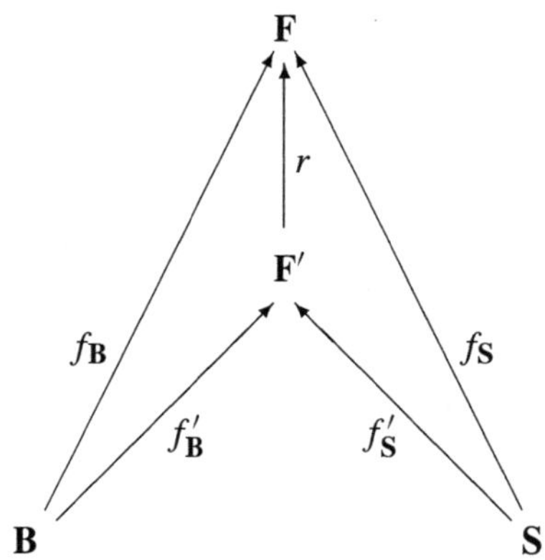

Note that we have an infomorphism r from $\mathbf{F}'$ to $\mathbf{F}$ such that the diagram commutes. When we have such an infomorphism, $\mathbf{F}'$ is said to be a refinement of **F**.

Since the rule r-ELIM is not sound, even if we have $\{f_{\mathbf{S}}^{\wedge}(\mathsf{ON})\} \vdash_{\mathbf{F}} \{f_{\mathbf{B}}^{\wedge}(\mathsf{LIT})\}$, it may be the case that we do not have $\{f_{\mathbf{S}}'^{\wedge}(\mathsf{ON})\} \vdash_{\mathbf{F}'} \{f_{\mathbf{B}}'^{\wedge}(\mathsf{LIT})\}$. This happens if tok($\mathbf{F}'$) includes a non-normal token with a dead battery, for example. Since all tokens of **F** are normal, we can think of **F** as an idealization of $\mathbf{F}'$.

We now look at how actions can be modeled in channel theory. Generally speaking, actions can be considered as connections that connect initial states and final states of actions, and so they can be modeled by constructing an information channel $\mathcal{C}_{\text{Act}} = \{f_{\text{init}} : \mathbf{C}_{\text{init}} \rightleftarrows \mathbf{C}_{\text{Act}},\ f_{\text{fin}} : \mathbf{C}_{\text{fin}} \rightleftarrows \mathbf{C}_{\text{Act}}\}$ such that $\mathbf{C}_{\text{Act}}$ classifies action tokens, and $\mathbf{C}_{\text{init}}$ and $\mathbf{C}_{\text{fin}}$ classify initial states and final states, respectively.

[11] In Part II of Barwise and Seligman [1], the structural rules mentioned in Definition 8 are discussed as the conditions for a theory to be regular ([1], p. 119), and the notion of local logic is defined in terms of the notion of a regular theory ([1], p. 150).

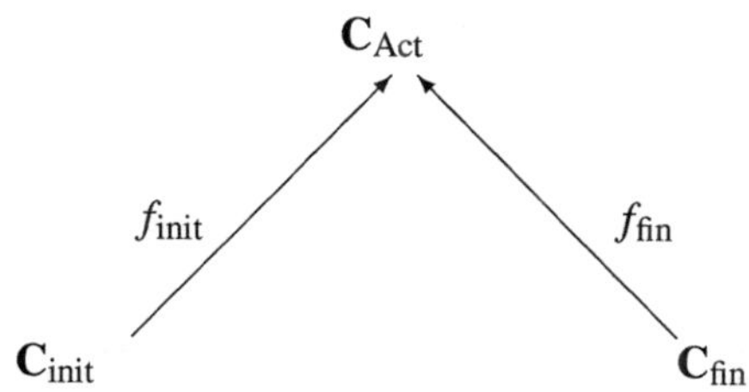

Then, the local logic on $\mathbf{C}_{\mathrm{Act}}$ can be defined. We do this for acts of commanding in the next section.

10.4 Acts of Commanding in Channel Theory

In this section, we construct information channels with models and the language of $\mathsf{DMDL}^{+}\mathsf{III}$ in order to model acts of commanding in channel theory. For the sake of simplicity, we ignore alethic modalities and acts of promising. We will work not with the whole class of $\mathsf{MDL}^{+}\mathsf{III}$-models but with its subset that includes only an arbitrary chosen $\mathsf{MDL}^{+}\mathsf{III}$-model $\mathcal{M}$ and any $\mathsf{MDL}^{+}\mathsf{III}$-models that can be obtained by updating $\mathcal{M}$ finite times.

Definition 9 Given a language $\mathfrak{L}_{\mathsf{DMDL}^{+}\mathsf{III}}$ of $\mathsf{DMDL}^{+}\mathsf{III}$, an arbitrary model $\mathcal{M}$ of the static base logic $\mathsf{MDL}^{+}\mathsf{III}$, and the truth in relation $\models_{\mathsf{DMDL}^{+}\mathsf{III}}$, deontic state classification $\mathbf{D}^{\mathcal{M}} = \langle \mathrm{tok}(\mathbf{D}^{\mathcal{M}}), \mathrm{typ}(\mathbf{D}^{\mathcal{M}}), \models_{\mathbf{D}^{\mathcal{M}}} \rangle$ based on $\mathcal{M}$ is defined as follows:

1. Let σ be a possibly empty finite sequence $\pi_0, \pi_1, \ldots, \pi_n$ of types of acts of commanding from the language $\mathfrak{L}_{\mathsf{DMDL}^{+}\mathsf{III}}$, $\mathcal{M}_\sigma$ be the model $(\cdots((\mathcal{M}_{\pi_0})_{\pi_1})\cdots)_{\pi_n}$, and w be a world of $\mathcal{M}$. $\mathrm{tok}(\mathbf{D}^{\mathcal{M}})$ is the set of model world pair of the form $\langle \mathcal{M}_\sigma, w \rangle$.
2. $\mathrm{typ}(\mathbf{D}^{\mathcal{M}})$ is the set of formulas of $\mathfrak{L}_{\mathsf{DMDL}^{+}\mathsf{III}}$.
3. $\langle \mathcal{M}_\sigma, w \rangle \models_{\mathbf{D}^{\mathcal{M}}} \varphi$ iff $\mathcal{M}_\sigma, w \models_{\mathsf{DMDL}^{+}\mathsf{III}} \varphi$.

Note that $\mathcal{M}_\sigma$ is an $\mathfrak{L}_{\mathsf{DMDL}^{+}\mathsf{III}}$-model obtained from $\mathcal{M}$ by sequentially updating $\mathcal{M}$ with acts of commanding of type π_i in σ in the order in σ. $\mathcal{M}_\sigma = \mathcal{M}$ if σ is empty.

This classification can be used both as the initial state classification $\mathbf{D}^{\mathcal{M}}_{\mathrm{init}}$ and as the final state classification $\mathbf{D}^{\mathcal{M}}_{\mathrm{fin}}$. Then we can define an information channel that models acts of commanding depicted by the following diagram:

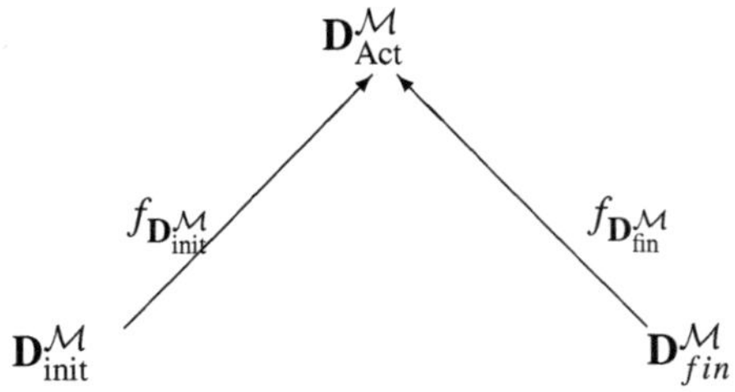

Definition 10 $\mathcal{D}^{\mathcal{M}} = \{f_{\mathbf{D}^{\mathcal{M}}_{\text{init}}} : \mathbf{D}^{\mathcal{M}}_{\text{init}} \rightleftarrows \mathbf{D}^{\mathcal{M}}_{\text{Act}},\ f_{\mathbf{D}^{\mathcal{M}}_{\text{fin}}} : \mathbf{D}^{\mathcal{M}}_{\text{fin}} \rightleftarrows \mathbf{D}^{\mathcal{M}}_{\text{Act}}\}$ with a core $\mathbf{D}^{\mathcal{M}}_{\text{Act}}$ is defined by the following conditions:

1. $\text{tok}(\mathbf{D}^{\mathcal{M}}_{\text{Act}})$ is a set of particular utterances in some natural language, say English, that possibly count as acts of commanding.
2. Let $f^{\vee}_{\mathbf{D}^{\mathcal{M}}_{\text{init}}}$ and $f^{\vee}_{\mathbf{D}^{\mathcal{M}}_{\text{fin}}}$ be functions that map each token utterance $u \in \text{tok}(\mathbf{D}^{\mathcal{M}}_{\text{Act}})$ to its initial state $f^{\vee}_{\mathbf{D}^{\mathcal{M}}_{\text{init}}}(u) \in tok(\mathbf{D}^{\mathcal{M}}_{\text{init}})$ and its final state $f^{\vee}_{\mathbf{D}^{\mathcal{M}}_{\text{fin}}}(u) \in tok(\mathbf{D}^{\mathcal{M}}_{\text{fin}})$, respectively.
3. $\text{typ}(\mathbf{D}^{\mathcal{M}}_{\text{Act}})$ of the classification $\mathbf{D}^{\mathcal{M}}_{\text{Act}}$ consists of translations $f^{\wedge}_{\mathbf{D}^{\mathcal{M}}_{\text{init}}}(\varphi) = \langle \varphi, 1\rangle$ and $f^{\wedge}_{\mathbf{D}^{\mathcal{M}}_{\text{fin}}}(\varphi) = \langle \varphi, 2\rangle$ of each formula φ of $\mathfrak{L}_{\mathsf{DMDL^{+}III}}$ given by the two functions $f^{\wedge}_{\mathbf{D}^{\mathcal{M}}_{\text{init}}}$ and $f^{\wedge}_{\mathbf{D}^{\mathcal{M}}_{\text{fin}}}$, respectively, and action types of the language $\mathfrak{L}_{\mathsf{DMDL^{+}III}}$.
4. The classification relation $\models_{\mathbf{D}^{\mathcal{M}}_{\text{Act}}}$ is defined by the following three conditions:

   a. $u \models_{\mathbf{D}^{\mathcal{M}}_{\text{Act}}} \langle \varphi, 1\rangle$
   iff for some $\langle \mathcal{M}_{\sigma}, w\rangle \in tok(\mathbf{D}^{\mathcal{M}}_{\text{init}})$, $f^{\vee}_{\mathbf{D}^{\mathcal{M}}_{\text{init}}}(u) = \langle \mathcal{M}_{\sigma}, w\rangle$ and $\langle \mathcal{M}_{\sigma}, w\rangle \models_{\mathbf{D}^{\mathcal{M}}_{\text{init}}} \varphi$,

   b. $u \models_{\mathbf{D}^{\mathcal{M}}_{\text{Act}}} \langle \varphi, 2\rangle$
   iff for some $\langle \mathcal{M}_{\tau}, w\rangle \in tok(\mathbf{D}^{\mathcal{M}}_{\text{fin}})$, $f^{\vee}_{\mathbf{D}^{\mathcal{M}}_{\text{fin}}}(u) = \langle \mathcal{M}_{\tau}, w\rangle$, and $\langle \mathcal{M}_{\tau}, w\rangle \models_{\mathbf{D}^{\mathcal{M}}_{\text{fin}}} \varphi$,

   c. $u \models_{\mathbf{D}^{\mathcal{M}}_{\text{Act}}} \mathsf{Com}_{(i,\,j)}\varphi$
   iff for some $\langle \mathcal{M}_{\sigma}, w\rangle \in tok(\mathbf{D}^{\mathcal{M}}_{\text{init}})$, for some $\langle \mathcal{M}_{\tau}, w\rangle \in tok(\mathbf{D}^{\mathcal{M}}_{\text{fin}})$, $f^{\vee}_{\mathbf{D}^{\mathcal{M}}_{\text{init}}}(u) = \langle \mathcal{M}_{\sigma}, w\rangle$, $f^{\vee}_{\mathbf{D}^{\mathcal{M}}_{\text{fin}}}(u) = \langle \mathcal{M}_{\tau}, w\rangle$, and $\mathcal{M}_{\tau} = (\mathcal{M}_{\sigma})_{\mathsf{Com}_{(i,\,j)}\varphi}$.

Note that the pairs $f_{\mathbf{D}^{\mathcal{M}}_{\text{init}}} = \langle f^{\wedge}_{\mathbf{D}^{\mathcal{M}}_{\text{init}}}, f^{\vee}_{\mathbf{D}^{\mathcal{M}}_{\text{init}}}\rangle$ and $f_{\mathbf{D}^{\mathcal{M}}_{\text{fin}}} = \langle f^{\wedge}_{\mathbf{D}^{\mathcal{M}}_{\text{fin}}}, f^{\vee}_{\mathbf{D}^{\mathcal{M}}_{\text{fin}}}\rangle$ satisfy the fundamental property of infomorphisms. Thus $\mathcal{D}^{\mathcal{M}}$ is an information channel.

Now we can consider the local logic $\mathfrak{L}_{\mathbf{D}^{\mathcal{M}}_{\text{Act}}} = \langle \mathbf{D}^{\mathcal{M}}_{\text{Act}}, \vdash_{\mathfrak{L}_{\mathbf{D}^{\mathcal{M}}_{\text{Act}}}}, N_{\mathfrak{L}_{\mathbf{D}^{\mathcal{M}}_{\text{Act}}}}\rangle$ on $\mathbf{D}^{\mathcal{M}}_{\text{Act}}$. Constraints in $\vdash_{\mathfrak{L}_{\mathbf{D}^{\mathcal{M}}_{\text{Act}}}}$ can be derived from the valid formulas of $\mathsf{DMDL^{+}III}$. For example, as

$$[\mathsf{Com}_{(i,\,j)}\varphi](\psi \wedge \xi) \to [\mathsf{Com}_{(i,\,j)}\varphi]\psi$$

is valid in $\mathsf{DMDL^{+}III}$, the following two analogues hold.

$$\{ f^{\wedge}_{\mathbf{D}^{\mathcal{M}}_{\text{init}}}([\mathsf{Com}_{(i,\,j)}\varphi](\psi \wedge \xi)) \} \vdash_{\mathcal{L}_{\mathbf{D}^{\mathcal{M}}_{\text{Act}}}} \{ f^{\wedge}_{\mathbf{D}^{\mathcal{M}}_{\text{init}}}([\mathsf{Com}_{(i,\,j)}\varphi]\psi) \} ,$$

$$\{ f^{\wedge}_{\mathbf{D}^{\mathcal{M}}_{\text{fin}}}([\mathsf{Com}_{(i,\,j)}\varphi](\psi \wedge \xi)) \} \vdash_{\mathcal{L}_{\mathbf{D}^{\mathcal{M}}_{\text{Act}}}} \{ f^{\wedge}_{\mathbf{D}^{\mathcal{M}}_{\text{fin}}}([\mathsf{Com}_{(i,\,j)}\varphi]\psi) \} .$$

Generally speaking, if φ is valid in $\mathsf{DMDL}^{+}\mathsf{III}$, the following analogues hold.

$$\emptyset \vdash_{\mathcal{L}_{\mathbf{D}^{\mathcal{M}}_{\text{Act}}}} \{ f^{\wedge}_{\mathbf{D}^{\mathcal{M}}_{\text{init}}}(\varphi) \} ,$$

$$\emptyset \vdash_{\mathcal{L}_{\mathbf{D}^{\mathcal{M}}_{\text{Act}}}} \{ f^{\wedge}_{\mathbf{D}^{\mathcal{M}}_{\text{fin}}}(\varphi) \} .$$

And more interestingly, the following two hold.

$$\{ f^{\wedge}_{\mathbf{D}^{\mathcal{M}}_{\text{init}}}([\mathsf{Com}_{(i,\,j)}\varphi]\psi), \mathsf{Com}_{(i,\,j)}\varphi \} \vdash_{\mathcal{L}_{\mathbf{D}^{\mathcal{M}}_{\text{Act}}}} \{ f^{\wedge}_{\mathbf{D}^{\mathcal{M}}_{\text{fin}}}(\psi) \} ,$$

$$\{ \mathsf{Com}_{(i,\,j)}\varphi, f^{\wedge}_{\mathbf{D}^{\mathcal{M}}_{\text{fin}}}(\psi) \} \vdash_{\mathcal{L}_{\mathbf{D}^{\mathcal{M}}_{\text{Act}}}} \{ f^{\wedge}_{\mathbf{D}^{\mathcal{M}}_{\text{init}}}([\mathsf{Com}_{(i,\,j)}\varphi]\psi) \} .$$

The former means that if $[\mathsf{Com}_{(i,\,j)}\varphi]\psi$ holds in the initial situation, and an act of commanding of type $\mathsf{Com}_{(i,\,j)}\varphi$ is performed, ψ holds in the final situation. The latter means that if an act of commanding of type $\mathsf{Com}_{(i,\,j)}\varphi$ is performed and ψ holds in the final situation, $[\mathsf{Com}_{(i,\,j)}\varphi]\psi$ holds in the initial situation. Together, they state the intuition behind the clause for the command modality in the truth definition.

As regards CUGO Principle, there may be tokens of type $\mathsf{Com}_{(i,\,j)}\varphi$ but not of type $f^{\wedge}_{\mathbf{D}^{\mathcal{M}}_{\text{fin}}}(O_{(j,\,i,\,i)}\varphi)$ if $O_{(j,\,i,\,i)}$ occurs in φ. The problem of characterizing the set of formulas φ such that

$$[\mathsf{Com}_{(i,\,j)}\varphi]O_{(j,\,i,\,i)}\varphi$$

is valid is still open. It is possible, however, to construct a sound local logic that includes an analogue of CUGO Principle as its constraint. Let us say the content φ of a command of form $\mathsf{Com}_{(i,\,j)}\varphi$ is non-deontic when no deontic operators occur in φ. Then imagine a context where people only try to issue commands with non-deontic contents. Let $(\mathbf{D}^{\mathcal{M}}_{\text{Act}})^{-}$ be a classification that models such a context. Then we can safely suppose that $\text{typ}((\mathbf{D}^{\mathcal{M}}_{\text{Act}})^{-}) = \text{typ}(\mathbf{D}^{\mathcal{M}}_{\text{Act}})$, $\text{tok}((\mathbf{D}^{\mathcal{M}}_{\text{Act}})^{-}) \subseteq \text{tok}(\mathbf{D}^{\mathcal{M}}_{\text{Act}})$, and the classification relation $\models_{(\mathbf{D}^{\mathcal{M}}_{\text{Act}})^{-}}$ is the restriction of $\models_{\mathbf{D}^{\mathcal{M}}_{\text{Act}}}$ to $\text{tok}((\mathbf{D}^{\mathcal{M}}_{\text{Act}})^{-}) \times \text{typ}((\mathbf{D}^{\mathcal{M}}_{\text{Act}})^{-})$. Since the operator $O_{(j,\,i,\,i)}$ does not occur in φ if φ is nondeontic, we have

$$\{ \mathsf{Com}_{(i,\,j)}\varphi \} \vdash_{\mathcal{L}_{(\mathbf{D}^{\mathcal{M}}_{\text{Act}})^{-}}} \{ f^{\wedge}_{\mathbf{D}^{\mathcal{M}}_{\text{fin}}}(O_{(j,\,i,\,i)}\varphi) \} .$$

Now, note that commands with nondeontic contents are quite ordinary. $(\mathbf{D}^{\mathcal{M}}_{\text{Act}})^{-}$, however, may include a token that fails to count as an act of commanding. Even if $O_{(j,\,i,\,i)}$ does not occur in φ, an attempted command of the form $\mathsf{Com}_{(i,\,j)}\varphi$ may fail if i lacks the suitable authority. Consider the following slightly odd scenario:

A private: Clean the room!
A sergeant: You don't have the authority to give me a command.

This scenario is odd because a private normally would not say such a thing to a sergeant.[12] By contrast, the following scenario looks normal.

A sergeant: Clean the room!
A private: Yes, sir.

Since $\mathsf{DMDL}^{+}\mathsf{III}$ is sound and complete with respect to $\mathcal{L}_{\mathsf{DMDL}^{+}\mathsf{III}}$-models, if we include only sequents that are derived from the validities of $\mathsf{DMDL}^{+}\mathsf{III}$ in $\vdash_{\mathcal{L}_{\mathbf{D}^{\mathcal{M}}_{\mathrm{Act}}}}$, we will have no non-normal tokens. Yet the regularities we rely on in performing illocutionary acts seem to have exceptions. In order to capture the regularities involved here, the language and the model of $\mathsf{DMDL}^{+}\mathsf{III}$ have to be extended substantially.

It seems instructive here to look more closely at the failures in using a flashlight in order to find out what kind of things our failures are. Consider the two information channels $\mathfrak{F}$ with the core $\mathbf{F}_{\mathrm{Act}}$ and $\mathfrak{F}'$ with the core $\mathbf{F}'_{\mathrm{Act}}$ depicted by the following diagram:

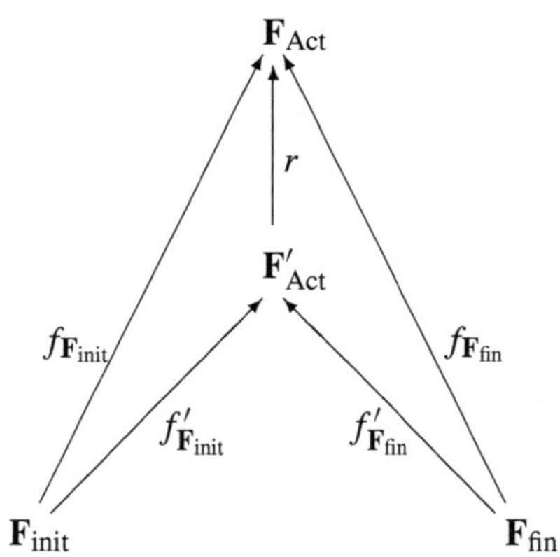

$\mathbf{F}_{\mathrm{init}}$ and $\mathbf{F}_{\mathrm{fin}}$ here are copies of the enriched flashlight classification $\mathbf{F}'$ in Sect. 10.3. $\mathbf{F}_{\mathrm{Act}}$ models a normal context in which all the flashlight tokens involved are in good working order, and $\mathbf{F}'_{\mathrm{Act}}$ models a larger context that includes flashlight tokens with dead batteries.

[12]If the private is the sergeant's father, however, he may say things like this to the sergeant. See the discussion of authority and organizations in Sect. 10.5.

Let TSO and GBL be the type of acts of turning the switch on and the type of acts of getting the bulb lit. Then the following sequent holds in $\mathbf{F}_{\mathrm{Act}}$ but fails in $\mathbf{F}'_{\mathrm{Act}}$.

$$\langle\{\mathsf{TSO}\}, \{\mathsf{GBL}\}\rangle\ .$$

Note that a counterexample to this sequent is a token of type TSO but not of type GBL. If an agent attempted, but failed, to get the bulb lit by turning the switch of the flashlight on, her act of turning the switch on can be said to be a failed attempt of getting the bulb lit, but it is neither an act of getting the bulb lit nor is it a non-normal token of an act of getting the bulb lit. It is a non-normal token of the local logic on $\mathbf{F}'_{\mathrm{Act}}$ if the above sequent is in $\vdash_{\mathfrak{L}_{\mathbf{F}'_{\mathrm{Act}}}}$.

Note also that we can distinguish preconditions, postconditions, and background conditions of normal cases as follows:

preconditions: the switch being off and the bulb being unlit.
postconditions: the switch being on and the bulb being lit.
background conditions: the battery being live, the bulb not being gone, ...

In the initial situation of each action token of type TSO in $\mathrm{F}_{\mathrm{Act}}$, these background conditions are satisfied, but they are not satisfied in some cases in $\mathbf{F}'_{\mathrm{Act}}$. They are the conditions to be satisfied if tokens of type TSO are to be of type GBL as well.

Now let us go back to the failed attempt of acts of commanding. It is not a non-normal token of an act of commanding, either. But then what kind of act is it a token of? The above scenarios suggest that it is a token of an act of saying “Clean the room” seriously. Let p be the proposition that a particular room r is clean, and $\mathsf{Say}_{(i,j)}\mathsf{CTR}$ be the type of acts of i’s saying “Clean the room” to j seriously and while saying this, referring to r with a definite description “the room”. Then the following sequent can be said to be a rough first approximation of the relevant regularity that holds normally[13]:

$$\langle\{\mathsf{Say}_{(i,j)}\mathsf{CTR}\}, \{\mathsf{Com}_{(i,j)}\,p\}\rangle\ .$$

In order to talk about such constraints in a logic that extends $\mathsf{DMDL}^{+}\mathsf{III}$, we need a language much richer than $\mathcal{L}_{\mathsf{DMDL}^{+}\mathsf{III}}$, as is indicated by the fact that we have already informally added $\mathsf{Say}_{(i,j)}\mathsf{CTR}$ to the set of types of the core of the channel $\mathcal{D}^{\mathcal{M}}$. If we are to talk about sequents of this kind in a systematic way, we have to be able to deal with the relation between expressions and their interpretations for some fragment of a natural language. In doing so, we will have to be able to deal with subsentential expressions, and this will require us to use quantified modal logic as

[13] Saying “Clean the room” seriously can be a way of performing various kinds of illocutionary acts other than commanding. We here only note that such multiplicity of performable illocutionary acts can be nicely captured in channel theory since the set Δ of the sequent $\langle\Gamma, \Delta\rangle$ is treated disjunctively (see Definition 5), and leave the issues that this multiplicity raises aside for further study.

the static base.[14] We will not try to develop such an extended system in this paper, however. Instead, we will make a few observations on $\mathsf{DMDL}^{+}\mathsf{III}$ and its possible extensions from the point of view of channel theory in the next section.[15]

10.5 Channel Theoretic Reflections on DMDL^{+}III and Its Possible Extensions

Note that the private's utterance in the first scenario is a counterexample to the sequent

$$\langle\{\mathsf{Say}_{(i,\, j)}\mathsf{CTR}\}, \{\mathsf{Com}_{(i,\, j)}p\}\rangle\ ,$$

but the sergeant's utterance in the second scenario is not. Since people normally do not try to issue commands for which they lack suitable authority, we can rely on constraints like this in normal circumstances. Thus we can think of a local logic that only deals with normal cases. Then the above sequent can be a constraint of such a local logic.

Note also that the agent i's having suitable authority for issuing a command of the form $\mathsf{Com}_{(i,\, j)}p$ is a condition that has to be satisfied in order for an act of type $\mathsf{Say}_{(i,\, j)}\mathsf{CTR}$ to be of type $\mathsf{Com}_{(i,\, j)}p$ as well. It is not a condition that has to be satisfied in order for an act of commanding of type $\mathsf{Com}_{(i,\, j)}p$ to have the effect of making it obligatory for j to see to it that p. This shows why $\mathsf{DMDL}^{+}\mathsf{III}$ is sound although it does not deal with the conditions on the authority of utterers. It characterizes the effects of acts of commanding, and utterances are acts of commanding only if the utterers have suitable authority. The private's failed attempt of commanding is not a counterexample to the validities of $\mathsf{DMDL}^{+}\mathsf{III}$.

This means that if we only wish to characterize how acts of commanding change situations, we do not have to take background conditions for acts of commanding into account. If we wish to talk about the relation between acts of saying things and acts of commanding performed in saying these things, however, we have to be able to take them into account, and thus we need to have a way for talking about the conditions on authority. This requires us to add some more structure to the models. One way of doing this is the following. We model each organization by a function org_k indexed by a finite indexing set K that assigns a (possibly empty) subset of the

---

[14] Note that we require $\mathsf{Say}_{(i,\, j)}\mathsf{CTR}$ to represent an intuitively very complex action type. We do so partly because we do not have a way of dealing with subsentential expressions such as "the room" in propositional modal logic, and partly because we do not have a way of combining two action types α and β to form a complex action type such as $\alpha \cap \beta$ of IPDL in $\mathcal{L}_{\mathsf{DMDL}^{+}\mathsf{III}}$ either. In order to treat complex action types in a systematic way, we will have to allow some such constructions. For IPDL, see Sect. 4.4 of Troquard and Balbiani [7].

[15] Yamada [11] presents a rough outline of an account that states the relation between the types of utterances, the types of contexts, the types of illocutionary acts performed, and the types of background conditions in the form of conditional constraints in situation theory. It seems possible to restate it in channel theory.

set of action types to each pair $\langle i, j\rangle \in I \times I$ for each world w. The set $org_k(i, j, w)$ is the set of acts that org_k authorize i to do to j in w. Then, we define

$$\mathcal{M}, w \models \mathsf{Auth}_{(i,j,k)}\varphi \text{ iff } \mathsf{Com}_{(i,j)}\varphi \in org_k(i, j, w)\ .$$

The formula of the form $\mathsf{Auth}_{(i,j,k)}\varphi$ means that k authorizes i to command j to see to it that φ.[16] For the sake of discussion, we will informally (and partially) imagine an extended language $\mathcal{L}_{\mathsf{EDMDL^{+}III}}$ to be obtained from $\mathcal{L}_{\mathsf{DMDL^{+}III}}$ by adding formulas of the form $\mathsf{Auth}_{(i,j,k)}\varphi$ and a set of action types that stand for acts of saying things such as $\mathsf{Say}_{(i,j)}\mathsf{CTR}$. As regards the models, let us add the functions org_k for all $k \in K$ to $\mathsf{DMDL^{+}III}$-models.

For comparison, we also imagine (again, informally and partially) two extended classifications $\mathbf{E}^{\mathcal{N}}_{\mathrm{init}}$ and $\mathbf{E}^{\mathcal{N}}_{\mathrm{fin}}$ constructed from the $\mathcal{L}_{\mathsf{EDMDL^{+}III}}$-model $\mathcal{N}$ that extends $\mathcal{L}_{\mathsf{DMDL^{+}III}}$-model $\mathcal{M}$ in the same way as $\mathbf{D}^{\mathcal{M}}_{\mathrm{init}}$ and $\mathbf{D}^{\mathcal{M}}_{\mathrm{fin}}$ are constructed from $\mathcal{M}$.

In addition to them, let $\mathbf{E}^{\mathcal{N}}_{\mathrm{Act}}$ and $(\mathbf{E}^{\mathcal{N}}_{\mathrm{Act}})'$ be classifications whose tokens are connections that connect tokens from $\mathbf{E}^{\mathcal{N}}_{\mathrm{init}}$ with tokens from $\mathbf{E}^{\mathcal{N}}_{\mathrm{fin}}$ and whose set of types includes the action types of $\mathcal{L}_{\mathsf{EDMDL^{+}III}}$ and translations of types from $\mathbf{E}^{\mathcal{N}}_{\mathrm{init}}$ and $\mathbf{E}^{\mathcal{N}}_{\mathrm{fin}}$ with suitably extended classification relations. Suppose $\mathbf{E}^{\mathcal{N}}_{\mathrm{Act}}$ models a normal context, which includes the sergeant's utterance in the second scenario and other similar ones, while $(\mathbf{E}^{\mathcal{N}}_{\mathrm{Act}})'$ models a wider context where the private's utterance in the first scenario and other similar failures due to the lack of suitable authority are included.

Then we can consider two channels such that the following diagram commutes:

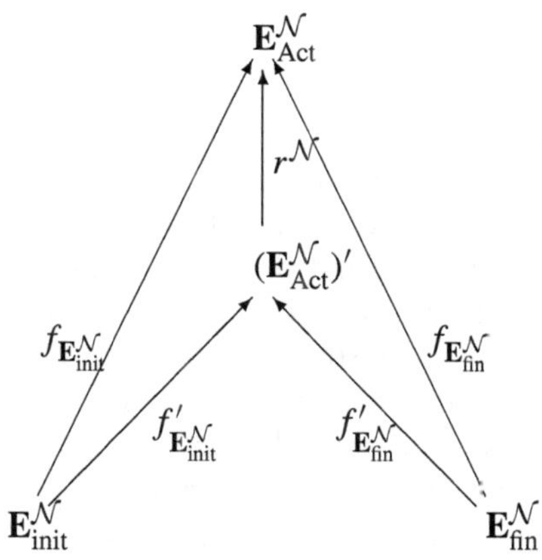

[16] Since people usually belong to a few or more organizations, there may be cases in which a person i is authorized to give a set of commands to another person j by an organization k_1 while j is authorized by another organization k_2 to give i another (possibly conflicting) set of commands. For example, there may be a case in which you are a coach of a local football team, and your boss is a player in the team.

Note that the private's utterance in the first scenario is not included in $\mathrm{tok}(\mathbf{E}^{\mathcal{N}}_{\mathrm{Act}})$, whereas the sergeant's utterance in the second scenario is included both in $\mathrm{tok}(\mathbf{E}^{\mathcal{N}}_{\mathrm{Act}})$ and in $\mathrm{tok}((\mathbf{E}^{\mathcal{N}}_{\mathrm{Act}})')$.

Now consider two sound and complete local logics $\mathfrak{L}_{\mathbf{E}^{\mathcal{N}}_{\mathrm{Act}}}$ and $\mathfrak{L}_{(\mathbf{E}^{\mathcal{N}}_{\mathrm{Act}})'}$ on $\mathbf{E}^{\mathcal{N}}_{\mathrm{Act}}$ and $(\mathbf{E}^{\mathcal{N}}_{\mathrm{Act}})'$, respectively. We have

$$\{\mathsf{Say}_{(i,\,j)}\mathsf{CTR}\} \vdash_{\mathfrak{L}_{\mathbf{E}^{\mathcal{N}}_{\mathrm{Act}}}} \{\mathsf{Com}_{(i,\,j)}p\}\,, \tag{10.1}$$

$$\{\mathsf{Say}_{(i,\,j)}\mathsf{CTR}\} \nvdash_{\mathfrak{L}_{(\mathbf{E}^{\mathcal{N}}_{\mathrm{Act}})'}} \{\mathsf{Com}_{(i,\,j)}p\}\,, \tag{10.2}$$

$$\{f^{\wedge}_{\mathbf{E}^{\mathcal{N}}_{\mathrm{init}}}(\mathsf{Auth}_{(i,j,k)}p), \mathsf{Say}_{(i,\,j)}\mathsf{CTR}\} \vdash_{\mathfrak{L}_{\mathbf{E}^{\mathcal{N}}_{\mathrm{Act}}}} \{\mathsf{Com}_{(i,\,j)}p\}\,, \tag{10.3}$$

$$\{f^{\wedge}_{\mathbf{E}^{\mathcal{N}}_{\mathrm{init}}}(\mathsf{Auth}_{(i,j,k)}p), \mathsf{Say}_{(i,\,j)}\mathsf{CTR}\} \vdash_{\mathfrak{L}_{(\mathbf{E}^{\mathcal{N}}_{\mathrm{Act}})'}} \{\mathsf{Com}_{(i,\,j)}p\}\,. \tag{10.4}$$

Let us examine whether it is possible to say what these statements say in $\mathcal{L}_{\mathsf{EDMDL+III}}$.

Consider (10.1) first. It seems clear that no formula in $\mathcal{L}_{\mathsf{EDMDL+III}}$ could say exactly what (10.1) says. (10.1) says that $\{\mathsf{Say}_{(i,\,j)}\mathsf{CTR}\}$ entails $\{\mathsf{Com}_{(i,\,j)}p\}$ in $\mathbf{E}^{\mathcal{N}}_{\mathrm{Act}}$, but it does not make sense to try to refer to the classification $\mathbf{E}^{\mathcal{N}}_{\mathrm{Act}}$ in $\mathcal{L}_{\mathsf{EDMDL+III}}$.

Let us put this point aside for the moment, however. Even if it does not make sense to say that $\{\mathsf{Say}_{(i,\,j)}\mathsf{CTR}\}$ entails $\{\mathsf{Com}_{(i,\,j)}p\}$ in $\mathbf{E}^{\mathcal{N}}_{\mathrm{Act}}$ in $\mathcal{L}_{\mathsf{EDMDL+III}}$, is it not possible to say simply that $\{\mathsf{Say}_{(i,\,j)}\mathsf{CTR}\}$ entails $\{\mathsf{Com}_{(i,\,j)}p\}$ in $\mathcal{L}_{\mathsf{EDMDL+III}}$?

Now, since the entailment relation here is understood as a relation between sets of action types, we might wish to extend $\mathcal{L}_{\mathsf{EDMDL+III}}$ by introducing formulas of the form

$$\Gamma \Rightarrow \Delta\,,$$

and let it say that Γ entails Δ. In order to do so, however, we have to extend the truth definition by adding a clause for formulas of this form. Here we have to face another difficulty. In channel theory, we can define the entailment relation by saying that Γ entails Δ in a given classification iff every token of that classification that is of type α for every $\alpha \in \Gamma$ is of type β for some $\beta \in \Delta$, but in $\mathcal{L}_{\mathsf{EDMDL+III}}$, we have no way of talking about tokens. Is there a formula of $\mathcal{L}_{\mathsf{EDMDL+III}}$ that can virtually capture the relation between $\{\mathsf{Say}_{(i,\,j)}\mathsf{CTR}\}$ and $\{\mathsf{Com}_{(i,\,j)}p\}$?

What we should note here is the following. If $\{\mathsf{Say}_{(i,\,j)}\mathsf{CTR}\}$ entails $\{\mathsf{Com}_{(i,\,j)}p\}$ in $\mathbf{E}^{\mathcal{N}}_{\mathrm{Act}}$, we can say that every token of type $\mathsf{Say}_{(i,\,j)}\mathsf{CTR}$ is normally of type $\mathsf{Com}_{(i,\,j)}p$. This implies that after an act of type $\mathsf{Say}_{(i,\,j)}\mathsf{CTR}$ is performed, all the formulas that characterize the effects of an act of type $\mathsf{Com}_{(i,\,j)}p$ normally hold.

Now, this consideration might seem to suggest the following:

$$[\mathsf{Com}_{(i,\,j)}p]\varphi \rightarrow [\mathsf{Say}_{(i,\,j)}\mathsf{CTR}]\varphi\,.$$

Unfortunately, however, this is not correct. We need to note that truth of $[\mathsf{Com}_{(i,\,j)}p]\varphi$ at w in $\mathcal{M}$ does not guarantee that φ characterizes the effects of acts of type $\mathsf{Com}_{(i,\,j)}p$. Take an $\mathsf{MDL}^{+}\mathsf{III}$-model $\mathcal{M}$ with four worlds $w, v, u, t \in W^{\mathcal{M}}$ such that $D^{\mathcal{M}}_{(j,\,i,\,i)} = \{\langle w, v\rangle, \langle w, u\rangle, \langle w, t\rangle\}$, $V^{\mathcal{M}}(p) = \{v, u\}$, and $V^{\mathcal{M}}(q) = \{u, t\}$. Then it is not very hard to see that we have

$$\mathcal{M}, w \models_{\mathsf{MDL}^{+}\mathsf{III}} [\mathsf{Com}_{(i,\,j)}p]\neg O_{(j,\,i,\,i)}q \wedge \neg[\mathsf{Com}_{(i,\,j)}(p \wedge q)]\neg O_{(j,\,i,\,i)}q\ ,$$

but intuitively $\mathsf{Com}_{(i,\,j)}(p \wedge q)$ entails $\mathsf{Com}_{(i,\,j)}p$. The formula $\neg O_{(j,\,i,\,i)}q$ happens to be true at w in $\mathcal{M}_{\mathsf{Com}_{(i,\,j)}p}$, but is not made so by i's act of commanding j to see to it that p. It holds at w in $\mathcal{M}$ and survives the update by $\mathsf{Com}_{(i,\,j)}p$.

This shows that we should count a formula among the formulas that characterize the effects of an act of type $\mathsf{Com}_{(i,\,j)}p$ only if its truth in the situation brought about by that act is essential for the very act to be of type $\mathsf{Com}_{(i,\,j)}p$. Here, CUGO Principle suggests the formula $O_{(j,\,i,\,i)}p$. If an act of type $\mathsf{Say}_{(i,\,j)}\mathsf{CTR}$ performed in a normal situation is also of type $\mathsf{Com}_{(i,\,j)}p$, we surely have

$$[\mathsf{Say}_{(i,\,j)}\mathsf{CTR}]O_{(j,\,i,\,i)}p$$

there. Is there a formula or a set of formulas of $\mathcal{L}_{\mathsf{EDMDL}^{+}\mathsf{III}}$ that could say that the situation is normal in such a way that $[\mathsf{Say}_{(i,\,j)}\mathsf{CTR}]O_{(j,\,i,\,i)}p$ holds in it?

Now (10.4) suggests the formula $\mathsf{Auth}_{(i,j,\,k)}p$. Thus if the following formula is valid, it can be said to be a way of saying something close to what (10.1) says in $\mathcal{L}_{\mathsf{EDMDL}^{+}\mathsf{III}}$.

$$\mathsf{Auth}_{(i,j,\,k)}p \rightarrow [\mathsf{Say}_{(i,\,j)}\mathsf{CTR}]O_{(j,\,i,\,i)}p\ . \tag{10.5}$$

Unfortunately, however, (10.5) is not valid. Even if the agent i has the suitable authority for commanding j to see to it that p, her act of type $\mathsf{Say}_{(i,\,j)}\mathsf{CTR}$ might fail to be of type $\mathsf{Com}_{(i,\,j)}p$. For example, j might suddenly become faint and fail to hear what is said. There are various ways things can go wrong.

This does not mean that we should abandon dynamified modal logics of speech acts, however. First, as we have seen, if our goal is to characterize how acts of commanding change situations, we only have to take utterances that count as commands into account. Failed attempts of issuing commands do not affect the validity of the formulas provable in $\mathsf{DMDL}^{+}\mathsf{III}$.[17] Second, we may try to incorporate ideas from modal logics that deal with laws that hold only normally or ceteris paribus.[18] And

[17]This does not mean that we do not have to extend $\mathsf{DMDL}^{+}\mathsf{III}$. If we wish to differentiate what Rescher calls "do-it-always commands" from "do-it-now commands" ([5], pp. 21–22), for example, we need quantification. This, however, is another issue.

[18]For normality, see Veltman [10], and for the normality reading of ceteris paribus conditions, see van Benthem et al. [8].

finally, we may try to extend $\mathsf{EDMDL}^{+}\mathsf{III}$ further so as to take more background conditions into account.

Whether it is possible to have a complete list of background conditions seems disputable, however. Although the kind of regularities relevant in the case of acts of commanding are mostly noncausal ones, the regularities that relate to the securing of uptake (the addressee's understanding of the force and content) include causal laws that can fail in various ways. Searle offers a set of conditions that are meant to be necessary and jointly sufficient for an act of promising, but it includes "[n]ormal input and output conditions" that are meant to "cover the large and indefinite range of conditions under which any kind of serious and literal linguistic communication is possible" ([6], p. 57). To say that they obtain is just to say that the context is normal with respect to "the conditions for intelligent speaking" and "the conditions for understanding" (ibid.).

Now, one of the virtues of channel theory is that it enables us to model the regularities that only hold normally even if we are not able to enumerate all the conditions jointly sufficient for the case being normal. Moreover, it enables us to model our everyday reasoning across contexts as well. The sergeant's utterance in the first scenario moves us from $\mathcal{L}_{\mathbf{E}^{\mathcal{N}}_{\mathrm{Act}}}$ to $\mathcal{L}_{(\mathbf{E}^{\mathcal{N}}_{\mathrm{Act}})'}$ by raising the issue of authority. A theorist of speech acts may also proceed in the same way from relatively simple regularities to less simple ones by raising issues of yet to be studied background conditions step by step. In order to do this in the dynamified logic of speech acts, we need to assume "everything else being normal" at each step. Thus one way of saying something close to what (10.1) says is to further extend $\mathcal{L}_{\mathsf{EDMDL}^{+}\mathsf{III}}$ by introducing modal operator "Normally" and say

$$\text{Normally}\,[\mathsf{Say}_{(i,\,j)}\mathsf{CTR}]\,O_{(j,\,i,\,i)}\,p\ .$$

What this says is not exactly what (10.1) says can be seen from the fact that something close to both what (10.3) and (10.4) say is expressed by a formula of the following form:

$$\text{Normally}\,(\mathsf{Auth}_{(i,j,k)}\,p \rightarrow [\mathsf{Say}_{(i,\,j)}\mathsf{CTR}]\,O_{(j,\,i,\,i)}\,p)\ .$$

Formulas of this form cannot differentiate what (10.3) says from what (10.4) says. Since it does not make sense to talk about classifications in the object language of $\mathcal{L}_{\mathsf{EDMDL}^{+}\mathsf{III}}$ nor in its suggested extension, this is unavoidable. It does not seem harmful, however, and we can say that the suggested "step by step" treatment seems to be a reasonable way of dealing with background conditions for extending $\mathsf{EDMDL}^{+}\mathsf{III}$ in order to capture the kind of regularities supported by them.

Acknowledgments This work is supported by the Grant-in-Aid for Scientific Research on Innovative Areas: Prediction and Decision Making (23120002, MEXT Japan). Various parts of earlier versions of this paper were presented at the 2014 Taiwan Philosophical Logic Colloquium (October 24–25, 2014, National Taiwan University, Taipei, Taiwan), the 2014 Autumn Research Meeting of the Japan Association for Philosophy of Science (November 1, 2014, Komaba Campus, the University of Tokyo, Tokyo, Japan), Hokkaido-Bucharest Joint Philosophy Workshop (November

3, 2014, Hokkaido University, Sapporo, Japan), and Workshop on Correlated Information Change (November 24–26, 2014, University of Amsterdam, Amsterdam, The Netherlands). I am grateful to the participants of these meetings for their helpful comments and critical discussions. I would also like to thank Chin-mu Yang, Makoto Kikuchi, Shunzo Majima, and Sonja Smet for inviting me to these meetings.

References

1. Barwise, J., Seligman, J.: Information Flow: The Logic of Distributed Systems. Cambridge University Press, Cambridge (1997)
2. Gerbrandy, J., Groeneveld, W.: Reasoning about information change. J. Logic Lang. Inform. **6**, 147–169 (1997)
3. Kooi, B.P., van Benthem, J.: Reduction axioms for epistemic actions. In: Schmidt, R., Pratt-Hartmann, I., Reynolds, M., Wansing, H. (eds.) Preliminary Proceedings of AiML-2004: Advances in Modal Logic. Technical Report Series, vol. UMCS-04-9-1, pp. 197–211. Department of Computer Science, University of Manchester (2004)
4. Plaza, J.: Logics of public communications. In: Emrich, M., Pfeifer, M., Hadzikadic, M., Ras, Z. (eds.) Proceedings of the 4th International Symposium on Methodologies for Intelligent Systems, pp. 201–216 (1989). Reprinted in Synthese **158**, 165–179 (2007)
5. Rescher, N.: The Logic of Commands. Routledge & Kegan Paul Ltd. (1966)
6. Searle, J.R.: Speech Acts: An Essay in the Philosophy of Language. Cambridge University Press, Cambridge (1969)
7. Troquard, N., Balbiani, P.: Propositional dynamic logic. In: Zalta, E.N. (ed.) The Stanford Encyclopedia of Philosophy. Spring 2015 Edition (2015). http://plato.stanford.edu/archives/spr2015/entries/logic-dynamic/
8. van Benthem, J., Girard, P., Roy, O.: Everything else being equal: a modal logic approach to ceteris paribus preferences. J. Philos. Logic **38**(1), 83–125 (2009)
9. van Ditmarsch, H., van der Hoek, W., Kooi, B.: Dynamic Epistemic Logic. Synthese Library, vol. 337. Springer, Dordrecht (2007)
10. Veltman, F.: Defaults in update semantics. J. Philos. Logic **25**, 221–261 (1996)
11. Yamada, T.: An ascription-based theory of illocutionary acts. In: Vanderveken, D., Kubo, S. (eds.) Essays in Speech Act Theory. Pragmatics & Beyond, New Series, vol. 77, pp. 151–174. John Benjamins, Amsterdam (2002)
12. Yamada, T.: Acts of commanding and changing obligations. In: Inoue, K., Sato, K., Toni, F. (eds.) Computational Logic in Multi-Agent Systems, 7th International Workshop, CLIMA VII, Hakodate, Japan, May 2006, Revised Selected and Invited Papers. Lecture Notes in Artificial Intelligence, vol. 4371, pp. 1–19. Springer, Berlin (2007)
13. Yamada, T.: Logical dynamics of commands and obligations. In: Washio, T., Satoh, K., Takeda, H., Inokuchi, A. (eds.) New Frontiers in Artificial Intelligence, JSAI 2006 Conference and Workshops, Tokyo, Japan, June 2006, Revised Selected Papers. Lecture Notes in Artificial Intelligence, vol. 4384, pp. 133–146. Springer, Berlin (2007)
14. Yamada, T.: Acts of promising in dynamified deontic logic. In: Sato, K., Inokuchi, A., Nagao, K., Kawamura, T. (eds.) New Frontiers in Artificial Intelligence, JSAI 2007 Conference and Workshops, Miyazaki, Japan, June 18–22, 2007, Revised Selected Papers. Lecture Notes in Artificial Intelligence, vol. 4914, pp. 95–108. Springer, Berlin (2008)
15. Yamada, T.: Logical dynamics of some speech acts that affect obligations and preferences. Synthese **165**, 295–315 (2008)
16. Yamada, T.: Acts of requesting in dynamic logic of knowledge and obligation. Eur. J. Anal. Philos. **7**(2), 59–82 (2011)
17. Yamada, T.: Dynamic logic of propositional commitments. In: Trobok, M., Miščvić, N., Žarnić, B. (eds.) Between Logic and Reality: Modeling Inference, Action, and Understanding, pp. 183–200. Springer, Berlin (2012)

References

Chapter 11
Constructive Embedding from Extensions of Logics of Strict Implication into Modal Logics

Sakiko Yamasaki and Katsuhiko Sano

Abstract Dyckhoff and Negri (Arch Math Logic 51:71–92 (2012), [8]) give a constructive proof of Gödel–Mckinsey–Tarski embedding from intermediate logics to modal logics via labelled sequent calculi. Then, they regard a monotonicity of atomic propositions in intuitionistic logic as an initial sequent, i.e., an axiom. However, we regard the monotonicity as an additional inference rule and employ a modified translation sending an atomic variable P to $P\&\Box P$ to generalize their result to an embedding from extensions of Corsi's **F** of logic of strict implication to normal extensions of modal logics **K**. In this process, we provide a **G3**-style labelled sequent calculi for extensions of **F** and show that our calculi admit the cut rule and enjoy soundness and completeness for Kripke semantics.

Keywords Labelled sequent calculus · Modal logic · Intermediate logic · Gödel–Mckinsey–Tarski embedding · Cut elimination · Completeness · Kripke semantics · Strict implication

11.1 Introduction

The Gödel–Mckinsey–Tarski translation sends a formula of intuitionistic logic to a formula of modal logic **S4** by the following mapping:

$$\begin{aligned} P^{\Box} &:= \Box P \\ \bot^{\Box} &:= \bot \\ (A\&B)^{\Box} &:= A^{\Box}\&B^{\Box} \\ (A \vee B)^{\Box} &:= A^{\Box} \vee B^{\Box} \\ (A \supset B)^{\Box} &:= \Box(A^{\Box} \supset B^{\Box}). \end{aligned}$$

S. Yamasaki (✉)
Graduate School of Humanities, Tokyo Metropolitan University, Tokyo, Japan
e-mail: megumegu.world8008@gmail.com

K. Sano
School of Information Science, Japan Advanced Institute of Science and Technology, Nomi, Japan
e-mail: v-sano@jaist.ac.jp

S.C.-M. Yang et al. (eds.), *Structural Analysis of Non-Classical Logics*,
Logic in Asia: Studia Logica Library, DOI 10.1007/978-3-662-48357-2_11

By this translation, it holds that

A is a theorem of intuitionistic logic if, and only if, $A^{\Box}$ is a theorem of **S4**.

The left-to-right direction was first shown by Gödel [9]. In addition, he conjectured that the opposite direction (faithfulness) also holds. The proof of faithfulness was first established by Mckinsey and Tarski by an algebraic method [19]. However, the algebraic proof given by Mckinsey and Tarski is not constructive in the sense that their proof does not provide an effective procedure for rewriting a derivation of $A^{\Box}$ in **S4** into a corresponding derivation of A in intuitionistic logic.

There have been several approaches to give a constructive proof of the direction of faithfulness. Troelstra and Schwichitenberg [31] employed the idea of sequent calculi without structural rules (called **G3**-style calculi) to show the faithfulness by a proof-theoretic method. Mints [20] outlined a constructive proof via **G3**-style sequent calculus, though he employed a different translation that prefixes $\Box$ to all subformulas of a formula of intuitionistic logic. Dyckhoff and Negri [8] established a constructive embedding uniformly from intermediate logics, namely logics between intuitionistic logic and classical logic, into modal logics between **S4** and **S5**. A key idea of them is to employ *labelled* sequent calculi which internalize the notion of Kripke semantics into the syntax. For example, the expressions $x{:}A$ (read "A holds at x") and $x\mathsf{R}y$ (read "we can access from x to y") form a sequent.

We may take a weaker logic, called *subintuitionistic logic*, than intuitionistic logic and ask what kind of subintuitionistic logic we can embed to modal logic **K4** by the same translation above. Visser's basic propositional logic is an answer to the question. However, as far as the authors know, there is no constructive proof of this embedding result. We may also change only the atomic clause of the translation to the clause sending P to P and ask what kind of subintuitionistic logic we can embed to modal logic **K**. Then, Corsi's logic **F** of strict implication [5] becomes an answer.

One of the motivations of this paper is to provide a uniform constructive embedding from extensions of Corsi's logic **F** of strict implication to modal logics by generalizing Dyckhoff and Negri's labelled sequent calculi. However, it seems not straightforward to generalize Dyckhoff and Negri's result, because there are at least two difficulties. First, their proof of the direction of faithfulness of the translation seemingly depends on the assumption of reflexivity of an accessibility relation in Kripke semantics for intuitionistic logic. This becomes an obstacle to generalize Dyckhoff and Negri's result to Visser's basic propositional logic. Second, they expressed the monotonicity of atomic variables in Kripke semantics of intuitionistic logic in terms of an initial sequent (an axiom of the form $x\mathrm{R}y, x{:}P, \Gamma \Rightarrow \Delta, y{:}P$) and derive the identity sequent $x{:}P, \Gamma \Rightarrow \Delta, x{:}P$ of atomic variables by the axiom of monotonicity and the rule of reflexivity. This second point becomes an obstacle to generalize their result to, say, Corsi's **F** of strict implication.[1]

For the first difficulty, we change the original translation into the one sending P to $P\&\Box P$ and remove the dependency on the reflexivity from Dyckhoff and Negri's

[1] In the last moment of revising this paper, we were informed that Sara Negri [23] also proposed a different translation of ours to obtain a similar result for subintuitionistic logic *without* the requirement of monotonicity. However, her result did not cover Visser's basic propositional logic.

argument for the faithfulness. We note that this revised translation was already proposed in [32] by Visser and he used this translation to embed his basic propositional logic also to modal logic **K4**. For the second difficulty, we simply take the identity sequent $x{:}P, \Gamma \Rightarrow \Delta, x{:}P$ as an initial sequent and regard the property of monotonicity as an additional inference rule rather than an axiom. By these modifications, we can establish constructive embedding uniformly from extensions of Corsi's logic **F** of strict implications to modal logics. Although we modify the translation, we note that our result implies the result by Dyckhoff and Negri, because $\Box P$ and $P \& \Box P$ becomes equivalent in (normal) modal logics containing **T**. To sum up, our revised translation sending P to $P \& \Box P$ can be regarded as a "unification" of the original Gödel–Mckinsey–Tarski translation sending P to $\Box P$ and Corsi's translation sending P to P so that we can prove the uniform constructive embedding results from logics of strict implications to intermediate logics.[2]

The following is the outline of this paper. Section 11.2 first reviews the syntax for Corsi's logics of strict implication and its Kripke semantics, and then introduces the notion of geometric implication for describing several frame properties. In Sect. 11.3, we introduce the notion of labelled formalism to define a labelled sequent calculus for the logic of strict implication and extend it to rules corresponding to a set of geometric implications. Section 11.4 demonstrates that our labelled sequent calculus with rules for geometric implications captures several existing intermediate logics, subintuitionistic logics including Visser's Basic Propositional Logic, and extensions of Corsi's logic **F** of strict implication. After establishing the admissibility of cut in our sequent calculi in a uniform manner in Sect. 11.5, Section 11.6 establishes our constructive embedding results from logics of strict implication into modal logics via our labelled calculi. In Sect. 11.7, we uniformly prove the soundness and completeness of our labelled sequent calculi for logics of strict implication with respect to Kripke semantics.

11.2 Kripke Semantics for Extensions of F

The syntax $\mathcal{L}$ of Corsi's logic **F** of strict implication is the same as intuitionistic logic. That is, $\mathcal{L}$ consists of a countably infinite set Atom of atomic variables (denoted by P, Q, etc.), $\bot$ as well as the logical connectives $\&$, $\vee$, $\supset$. The set $\mathsf{Form}_{\mathcal{L}}$ of all *$\mathcal{L}$-formulas* is inductively defined as follows:

$$\mathsf{Form}_{\mathcal{L}} \ni A ::= P \mid \bot \mid A \& A \mid A \vee A \mid A \supset A.$$

where $P \in \mathsf{Atom}$. We denote $\mathcal{L}$-formulas by A, B, C, etc.

Let us move to Kripke semantics for $\mathcal{L}$. We say that $\mathfrak{F} = (S, R)$ is a *frame* if S is a nonempty set and $R \subseteq S \times S$. $\mathfrak{M} = (S, R, V)$ is a *model* if (S, R) is a frame and V is

[2] The revised translation sending P to $P \& \Box P$ was recently also employed by the second author and Ma [28] for providing a topological semantics for Visser's basic propositional logic.

a mapping $\mathsf{Atom} \to \mathcal{P}(S)$, called a *valuation*. We say that a valuation V is *monotone* if sRs' and $s \in V(P)$ jointly imply $s' \in V(P)$ for all $s, s' \in S$ and $P \in \mathsf{Atom}$. $\mathfrak{M} = (S, R, V)$ is said to be *monotone* if a valuation V is monotone. Given a model $\mathfrak{M} = (S, R, V)$, a state $s \in S$ and a formula A, the *satisfaction relation* $\mathfrak{M}, s \models A$ is defined by:

$$
\begin{array}{ll}
\mathfrak{M}, s \models P & \text{iff } s \in V(P), \\
\mathfrak{M}, s \models \bot & \quad\text{Never}, \\
\mathfrak{M}, s \models A\&B & \text{iff } \mathfrak{M}, s \models A \text{ and } \mathfrak{M}, s \models B, \\
\mathfrak{M}, s \models A \vee B & \text{iff } \mathfrak{M}, s \models A \text{ or } \mathfrak{M}, s \models B, \\
\mathfrak{M}, s \models A \supset B & \text{iff for all } s' \in S \text{ with } sRs' : \mathfrak{M}, s' \models A \text{ implies } \mathfrak{M}, s' \models B.
\end{array}
$$

We denote classes of models by $\mathbb{M}$, $\mathbb{N}$, etc. Given a model $\mathfrak{M} = (S, R, V)$, we say that A is *valid in* $\mathfrak{M}$ if $\mathfrak{M}, s \models A$ for all states $s \in S$. Given a class $\mathbb{M}$ of models, A is *valid in* $\mathbb{M}$ if A is valid in $\mathfrak{M}$ for all models $\mathfrak{M} \in \mathbb{M}$.

In order to talk about a property of frames, we can also use the first-order syntax whose signature is $\{\mathsf{R}\}$. With the help of this, let us introduce the syntactic notion of *geometric implication* and the semantic notion of *geometric frame*.

Definition 1 (*Geometric Implication*) A *geometric implication* is a first-order sentence of the following form:

$$\forall \boldsymbol{x}\, (S_1 \& \cdots \& S_m \supset \exists \boldsymbol{y} \bigvee_{1 \leqslant j \leqslant n} (T_{j1} \& \cdots \& T_{jn_j})),$$

where $\boldsymbol{x}$ and $\boldsymbol{y}$ are finite tuples of pairwise distinct variables of the first-order syntax and we assume that no variable occurs in both $\boldsymbol{x}$ and $\boldsymbol{y}$, $S_1, \ldots, S_m$ and $T_{j1}, \ldots, T_{jn_j}$ are atomic predicates of the form $x\mathsf{R}y$ and we use R from $\mathfrak{F} = (S, R)$ to interpret our binary predicate R.

In what follows in this paper, we always assume for simplicity that the length of $\boldsymbol{y}$ is *one* as in [8, 21]. Table 11.1 provides examples of geometric implications, which allow us to capture several classes of models. When we have no disjunct in the antecedent of a geometric implication, the form becomes $\forall \boldsymbol{x}\, (S_1 \& \cdots \& S_m \supset \bot)$.

11.3 Labelled Sequent Calculus for F

11.3.1 Labelled Formalism

Now we introduce the labelled formalism for our sequent calculus. Let Var be a countably infinite set of *labels* (denoted by x, y, z, etc.). Given a label $x \in \mathsf{Var}$ and an $\mathcal{L}$-formula, we say that an expression $x{:}A$ is a *labelled formula*. It corresponds to the satisfaction relation "$\mathfrak{M}, x \models A$" in Kripke semantics. A *relational atom* is an expression $x\mathsf{R}y$, where x and y are labels, where $x\mathsf{R}y$ means that "there is an edge from x to y" or "y is accessible from x" in Kripke semantics. We say that a *labelled expression* (denoted by φ, ψ, etc.) is an expression of the form $x{:}A$ or an expression of the form $x\mathsf{R}y$. We say that φ is a *labelled atomic formula* if φ is a labelled formula $x{:}A$ and A is atomic. Given finite multisets Γ and Δ of labelled expressions, we say that $\Gamma \Rightarrow \Delta$ is a *sequent* if the succedent Δ does not contain any relational atoms. Table 11.2 presents a **G3**-style labelled sequent calculus **G3F** for Corsi's logic **F**.[3] The logical rules of Table 11.2 for each connective reflect the satisfaction relation defined in the previous section. For example, let us take the satisfaction relation for the implication, i.e.,

$$\mathfrak{M}, s \models A \supset B \text{ iff for all } s' \in S \text{ with } sRs' : \mathfrak{M}, s' \models A \text{ implies } \mathfrak{M}, s' \models B.$$

The left-to-right direction of this clause is translated into the left rule $(L{\supset})$ and the right-to-left direction is into the right rule $(R{\supset})$.

Moreover, we may equip **G3F** with additional inference rules. In this paper, we are concerned with the following two kinds of rules: the rule of monotonicity of atomic variables and the rules for geometric implications of Definition 1.

First, to capture monotone valuations, we introduce the following rule:

$$\frac{x\mathsf{R}y, x{:}P, y{:}P, \Gamma \Rightarrow \Delta}{x\mathsf{R}y, x{:}P, \Gamma \Rightarrow \Delta}\,(Mon).$$

We note that Dyckhoff and Negri [8] regarded this property of valuations as an axiom $x\mathsf{R}y, x{:}P, \Gamma \Rightarrow \Delta, y{:}P$.

Second, recall from Definition 1 that the following geometric implication σ:

Table 11.1 Examples of geometric implications

Name	Frame property
Reflexivity	$\forall x(x\mathsf{R}x)$
Transitivity	$\forall x, y, z(x\mathsf{R}y \& y\mathsf{R}z \supset x\mathsf{R}z)$
Symmetry	$\forall x, y(x\mathsf{R}y \supset y\mathsf{R}x)$
Connectedness	$\forall x, y, z((x\mathsf{R}y \& x\mathsf{R}z) \supset (y\mathsf{R}z \vee z\mathsf{R}y))$
Seriality	$\forall x \exists y(x\mathsf{R}y)$
Directedness	$\forall x, y, z((x\mathsf{R}y \& x\mathsf{R}z) \supset \exists w(y\mathsf{R}w \& z\mathsf{R}w))$
Euclidean	$\forall x, y, z(x\mathsf{R}y \& x\mathsf{R}z \supset y\mathsf{R}z)$
Emptiness	$\forall x, y(x\mathsf{R}y \supset \bot)$

[3] **G3**-style sequent calculus, which was first developed by Kleene in [15], is the sequent calculus that does not contain any structural rule: rules of weakening, contraction and exchange, while it has an axiom with a context: $A, \Gamma \Rightarrow \Delta, A$. In [7], Dragalin showed that rules of weakening and contraction are height-preserving admissible. A general introduction to **G3**-style sequent calculus can be found in [24, 31].

Table 11.2 Labelled sequent calculus **G3F**

(Axioms)	$\dfrac{}{x{:}P, \Gamma \Rightarrow \Delta, x{:}P}\ (Id) \quad \dfrac{}{x{:}\bot, \Gamma \Rightarrow \Delta}\ (L\bot)$
(Logical rules)	$\dfrac{x{:}A, x{:}B, \Gamma \Rightarrow \Delta}{x{:}A\&B, \Gamma \Rightarrow \Delta}\ (L\&) \quad \dfrac{\Gamma \Rightarrow \Delta, x{:}A \quad \Gamma \Rightarrow \Delta, x{:}B}{\Gamma \Rightarrow \Delta, x{:}A\&B}\ (R\&)$ $\dfrac{x{:}A, \Gamma \Rightarrow \Delta \quad x{:}B, \Gamma \Rightarrow \Delta}{x{:}A \vee B, \Gamma \Rightarrow \Delta}\ (L\vee) \quad \dfrac{\Gamma \Rightarrow \Delta, x{:}A, x{:}B}{\Gamma \Rightarrow \Delta, x{:}A \vee B}\ (R\vee)$ $\dfrac{x\mathsf{R}y, x{:}A{\supset}B, \Gamma \Rightarrow \Delta, y{:}A \quad x\mathsf{R}y, x{:}A{\supset}B, y{:}B, \Gamma \Rightarrow \Delta}{x\mathsf{R}y, x{:}A{\supset}B, \Gamma \Rightarrow \Delta}\ (L{\supset})$ $\dfrac{x\mathsf{R}y, y{:}A, \Gamma \Rightarrow \Delta, y{:}B}{\Gamma \Rightarrow \Delta, x{:}A{\supset}B}\ (R{\supset})^{a}$

[a] y is fresh in the conclusion

$$\sigma := \forall \boldsymbol{x}\ (S_1\& \cdots \&S_m \supset \exists \boldsymbol{y} \bigvee_{1 \leqslant j \leqslant n} (T_{j1}\& \cdots \&T_{jn_j})),$$

where we note that we always assume for simplicity that the length of $\boldsymbol{y}$ is one as in [8, 21]. Then, any geometric implication σ can be transformed to an inference rule called *Geometric Rule Scheme* (GRS):

$$\dfrac{\overline{T_1}[z_1/y_1], \overline{S}, \Gamma \Rightarrow \Delta \quad \cdots \quad \overline{T_n}[z_n/y_n], \overline{S}, \Gamma \Rightarrow \Delta}{\overline{S}, \Gamma \Rightarrow \Delta}\ (GRS),$$

where $[z_i/y_i]$ is a substitution of z_i to y_i, $z_1, \ldots, z_n$ are fresh in the conclusion, $\overline{S}$ denotes the multisets of atomic formulas $S_1, \ldots, S_m$ of the form $x\mathsf{R}y$, and $\overline{T_j}$ denotes the multisets of atomic formulas $T_{j1}, \ldots, T_{jk_j}$ of the form $x\mathsf{R}y$. When a geometric rule scheme is of the from $\forall \boldsymbol{x}\ (S_1\& \cdots \&S_m \supset \bot)$, the corresponding rule takes the following form:

$$\dfrac{}{\overline{S}, \Gamma \Rightarrow \Delta}\ (GRS)$$

and the rule is called a *zero-premise* geometric rule scheme. Table 11.3 provides geometric rule schemes for frame properties of Table 11.1. Note that (Emp) in Table 11.3 is a zero-premise geometric rule scheme, i.e., an inference rule with no premise.

Definition 2 We denote by **G3F*** an extension of **G3F** by a *finite* set $*$ of geometric rule schemes. We use **G3Fm*** to mean the extension of **G3F*** by the rule (Mon) of monotonicity of atomic variables. By **G3F(m)***, we mean **G3F*** or **G3Fm***.

In what follows, when we want to refer any inference rule r (possibly not in **G3F(m)***), we often employ the following notation for the rule:

$$\dfrac{\Gamma_1 \Rightarrow \Delta_1 \quad \cdots \quad \Gamma_n \Rightarrow \Delta_n}{\Gamma \Rightarrow \Delta}\ r.$$

Table 11.3 Examples of geometric rule schemes

Frame property	Geometric rule scheme
Reflexivity	$\dfrac{x\mathsf{R}x, \Gamma \Rightarrow \Delta}{\Gamma \Rightarrow \Delta}(Ref)$
Transitivity	$\dfrac{x\mathsf{R}y, y\mathsf{R}z, x\mathsf{R}z, \Gamma \Rightarrow \Delta}{x\mathsf{R}y, y\mathsf{R}z, \Gamma \Rightarrow \Delta}(Tran)$
Symmetry	$\dfrac{x\mathsf{R}y, y\mathsf{R}x, \Gamma \Rightarrow \Delta}{x\mathsf{R}y, \Gamma \Rightarrow \Delta}(Sym)$
Connectedness	$\dfrac{x\mathsf{R}y, x\mathsf{R}z, y\mathsf{R}z, \Gamma \Rightarrow \Delta \quad x\mathsf{R}y, x\mathsf{R}z, z\mathsf{R}y, \Gamma \Rightarrow \Delta}{x\mathsf{R}y, x\mathsf{R}z, \Gamma \Rightarrow \Delta}(Con)$
Seriality	$\dfrac{x\mathsf{R}y, \Gamma \Rightarrow \Delta}{\Gamma \Rightarrow \Delta}(Ser)$ y is fresh
Directedness	$\dfrac{x\mathsf{R}y, x\mathsf{R}z, y\mathsf{R}w, z\mathsf{R}w, \Gamma \Rightarrow \Delta}{x\mathsf{R}y, x\mathsf{R}z, \Gamma \Rightarrow \Delta}(Dir)$ w is fresh
Euclidean	$\dfrac{x\mathsf{R}y, x\mathsf{R}z, y\mathsf{R}z, \Gamma \Rightarrow \Delta}{x\mathsf{R}y, x\mathsf{R}z, \Gamma \Rightarrow \Delta}(Euc)$
Emptiness	$\dfrac{}{x\mathsf{R}y, \Gamma \Rightarrow \Delta}(Emp)$

Definition 3 (*Context and Principal Formula*) The Γ and Δ in an inference rule of **G3F(m)*** are called the *context*. In the conclusion of each rule of **G3F(m)***, the formula(s) not in the context is called the *principal formula(s)*.

Definition 4 (*Derivation*) A *derivation* $\mathcal{D}$ in **G3F(m)*** is inductively defined as a tree generated by the axioms and the rules of **G3F(m)***. We say that the *end sequent* of $\mathcal{D}$ is the sequent in the root node of $\mathcal{D}$. The *height* of a derivation is the maximum length of branches in the derivation from the end sequent to an axiom. A sequent $\Gamma \Rightarrow \Delta$ is *derivable* in **G3F(m)*** (notation: **G3F(m)**$^* \vdash \Gamma \Rightarrow \Delta$) if it has a derivation $\mathcal{D}$ in **G3F(m)*** whose end sequent is $\Gamma \Rightarrow \Delta$. We write **G3F(m)**$^* \vdash_n \Gamma \Rightarrow \Delta$ to mean that $\Gamma \Rightarrow \Delta$ has a derivation whose height is at most n.

If it is clear from the context, we often omit "**G3F(m)***" from the expression of "**G3F(m)**$^* \vdash \Gamma \Rightarrow \Delta$."

Proposition 1 *For any formula A, $x{:}A, \Gamma \Rightarrow \Delta, x{:}A$ is derivable in* **G3F(m)***.

11.4 Extensions of Logic of Strict Implications

11.4.1 Intermediate Logics

Intermediate logics are logics between intuitionistic logic and classical logic. In our setting, intuitionistic logic can be captured by the extension **G3Fm** with (Ref) and $(Tran)$ of Table 11.3. Let us write this extension as **G3Int**. Dyckhoff and Negri [8] presented intuitionistic logic as a sequent calculus denoted by **G3I** but there are several differences between their formulation and our formulation. Let us comment on one important difference. Instead of (Id) of **G3Fm***, **G3I** has an axiom for monotonicity of atomic variables:

$$x\mathsf{R}y, x{:}P, \Gamma \Rightarrow \Delta, y{:}P$$

where the axiom (Id) of **G3Int** is derivable from the rule (Ref) and this monotonicity axiom. In contrast, **G3Int** explicitly includes (Id) as an axiom and treat monotonicity of atomic variables as the rule (Mon). Of course, two formulations, **G3Int** and **G3I**, are equipollent, because:

- $x\mathsf{R}y, x{:}P, \Gamma \Rightarrow \Delta, y{:}P$ is derivable in our **G3Int**,
- (Id) is derivable and (Mon) is admissible in Dyckhoff and Negri's **G3I**.

Then, as Dyckhoff and Negri did in [8], we can also cover several intermediate logics with the help of geometric rule schemes. Here we list some examples from [8].

1. Jankov logic **KC**: *Jankov logic* or the logic of *weak excluded middle* is characterized by the axiom $\neg P \vee \neg\neg P$ (cf. [4]). We obtain the corresponding labelled sequent calculus **G3Jan** by adding the rule (Dir) of Table 11.3 to **G3Int**.
2. Gödel-Dummett logic **LC**: *Gödel-Dummett logic* **LC** is axiomatized by $(P \supset Q) \vee (Q \supset P)$ (cf. [4]). We obtain the corresponding labelled sequent calculus **G3GD** by adding the rule (Con) of Table 11.3 to **G3Int**.
3. Classical Logic **CL**: When we extend intuitionistic logic with $\neg\neg P \supset P$ or $P \vee \neg P$, we obtain classical logic. When we add to **G3Int** (Sym) or (Euc) (these are equivalent with each other, when we assume reflexivity of R), we obtain the labelled sequent calculus for classical logic.

Compared to Dyckhoff and Negri's **G3I**, we stress that our formulation **G3F**(**m**)* is more modular so that we can also cover *subintuitionistic logic* such as Visser's basic propositional logic [32] and Corsi's logics of strict implication [5], as we will see below.

11.4.2 Extensions of Basic Propositional Logic

Basic propositional logic (**BPL**) is first introduced by Visser in [32]. **BPL** is a proper sublogic of intuitionistic logic, whose Kripke semantics is given by dropping the property of reflexivity from Kripke semantics of intuitionistic logic. For example, neither $p\&(p\supset q)\supset q$ nor $(p\supset(p\supset q))\supset(p\supset q)$ belongs to **BPL** as theorems, while they are easily seen to be theorems of intuitionistic logic. The first proof system of **BPL** is given by Visser [32] in the style of natural deduction. There are also Gentzen-style sequent calculi [2, 12, 14, 27] and Hilbert-style axiomatizations [13, 29, 30]. We can provide a labelled sequent calculus **G3B** of **BPL** by extending **G3Fm** with $(Tran)$. We also demonstrate two extensions of **BPL** as follows:

1. Extension **DNT** by seriality: As far as the authors know, the extension **DNT** of **BPL** by $\neg\neg\top$ ($\top$ is defined as $\bot\supset\bot$) was first studied by Ishigaki and Kashima [11], where they provide a sequent calculus for this extension and showed that the calculus is complete with respect to finite transitive and serial Kripke models with monotone valuations and the calculus also enjoys cut-elimination theorem. Recently, Ma and the second author [18] showed that A is a theorem of **DNT** iff A is a theorem of **CL**, for all constant formulas A, i.e., formulas without any atomic variables. Since intuitionsitic logic and classical logic also have the same set of theorems for the constant formulas [4, p. 35], their result implies that **DNT** and intuitionistic logic have the same constant theorems. We note that we cannot establish the same result for **BPL**, since **BPL** does not have the following property: $A \leftrightarrow \top$ or $A \leftrightarrow \bot$ is a theorem of **BPL** for any constant formula A, where $A \leftrightarrow B$ is defined as $(A\supset B)\&(B\supset A)$. $\top \supset \bot$ becomes a counterexample of this property [18, Theorem 5.1]. Finally, a labelled sequent calculus for **DNT** can be obtained by adding the rule (Ser) of Table 11.3 to **G3B**.
2. Extension **Log(•)** by emptiness: Ma and the second author [18] recently provided a sound and complete natural deduction calculus of the extension of **BPL** by the condition of Emptiness in Table 11.1 and showed that the set **Log(•)** of all theorems of the extension satisfies the following properties. First, any implicational formula $A \supset B$ belongs to **Log(•)**, while the implication free fragment of **Log(•)** is empty. Second, **Log(•)** is *not* closed under taking modus ponens because $\top \supset \bot, \top \in$ **Log(•)** but $\bot \notin$ **Log(•)**. A labelled sequent calculus for **Log(•)** can be obtained by adding the rule (Emp) of Table 11.3 to **G3B**.

11.4.3 Logics of Strict Implication

The notion of *strict implication* is proposed by Lewis [16] to overcome the paradoxes of *material implication*. The several systems of logics of strict implication are first presented in [17] (see [10] for more details of the systems). From modern viewpoints, strict implication is regarded as a boxed implication in the syntax of modal logic, i.e., $A\supset B := \Box(A \rightarrow B)$, where $\rightarrow$ stands for material implication. Later, a family of logics of strict implication is studied by Corsi [5] under the name of weak logic with strict implication, where she also provided Hilbert-style axiomatization for the family of logics of strict implication. Then, Ishigaki and Kashima [11] study non-labelled Gentzen-style sequent calculi for Corsi's logics of strict implication. Hilbert-style axiomatizations of logics of strict implication are presented also in [6, 26]. Moreover, natural deduction systems for logics of strict implication are proposed in [3].

Logics of strict implication are sometimes also called subintuitionistic logics, which are characterized by classes of Kripke models. Kripke semantics for logics of strict implication keep the same satisfaction relation as Kripke semantics for intuitionistic logic but it does not always satisfy the property of monotonicity. Logics of strict implication can be captured by combinations of frame properties. We demonstrate that several extensions in the previous studies are captured by our labelled sequent calculi.

1. Extension **FD** [5] by seriality: **FD** is obtained by adding to **F** the axiom $\neg\neg\top$ and it is characterized by the class of Kripke models satisfying seriality. This logic is also studied by Došen under the name of $\mathbf{D}_\sigma$ [6] and Ishigaki and Kashima under the name of $\mathbf{GKD}^I$ [11]. We can obtain the corresponding labelled sequent calculus **G3FD** by adding (Ser) of Table 11.3 to **G3F**.
2. Extension **FC** [5] by connectedness: Corsi [5] defines **FC** as the extension of **F** with the axiom $((C\&(A\supset B))\supset D) \vee ((A\&(C\supset D))\supset B)$. The labelled sequent calculus **G3FC** for **FC** is obtained by adding (Con) of Table 11.3 to **G3F**.
3. Extension **FT** [5] by transitivity: Corsi [5] defines **FT** as the extension of **F** with $(A\supset B)\supset(C\supset(A\supset B))$, and Ishigaki and Kashima [11] provide a non-labelled sequent calculus $\mathbf{GK4}^I$ of this logic. Restall [26] also presented this logic under the name of b. The labelled sequent calculus **G3FT** is obtained by adding $(Tran)$ of Table 11.3 to **G3F**.
4. Extension **FR** [5] by reflexivity: Corsi [5] defines **FR** as the extension of **F** with $A\&(A\supset B)\supset B$, and Ishigaki and Kashima [11] provide a non-labelled sequent calculus $\mathbf{GKT}^I$ of this logic. When we add (Ref) of Table 11.3 to **G3F**, we obtain the corresponding labelled sequent calculus **G3FR**.
5. Extension by reflexivity and transitivity **FRT** [5]: Corsi defines **FRT** [5] as the extension of **FT** with $A\&(A\supset B)\supset B$. This logic is studied also by Restall [26] under the name of bw. Ishigaki and Kashima [11] provide a non-labelled sequent calculus $\mathbf{GS4}^I$ of this logic. The corresponding labelled sequent calculus **G3FRT** is obtained by adding both (Ref) and $(Tran)$ of Table 11.3 to **G3F**.
6. Extension **FS** [5] by symmetry: Corsi [5] defines **FS** as the extension of **F** with $A\supset(B\vee\neg(A\supset B))$ and Ishigaki and Kashima [11] provide a non-labelled sequent calculus $\mathbf{GKB}^I$ of this logic. We can obtain the corresponding labelled sequent calculus **G3FS** by adding (Ser) of Table 11.3 to **G3F**. While the admissibility of cut in $\mathbf{GKB}^I$ is not shown in [11], **G3FS** admits the cut rule as shown in the next section.
7. Extension $\mathbf{GK5}^I$ [11] by Euclidean: $\mathbf{GK5}^I$ is the non-labelled sequent calculus of the logic of Kripke models whose accessibility relation is Euclidean. The corresponding labelled sequent calculus **G3FE** to this logic is obtained by adding the rule (Euc) of Table 11.3 to **G3F**. While the admissiblity of cut in $\mathbf{GK5}^I$ is not shown in [11], **G3FE** admits the cut rule as shown in the next section.

11.5 Admissibility of Cut

In this section, we establish admissibility of the cut rule in **G3F**(**m**)*, following the standard argument of **G3**-style sequent calculus such as [8, 21, 22].

Definition 5 The cut rule is

$$\frac{\Gamma \Rightarrow \Delta, x{:}A \quad x{:}A, \Pi \Rightarrow \Sigma}{\Gamma, \Pi \Rightarrow \Delta, \Sigma}\ (Cut),$$

where we say that $x{:}A$ is the *cut labelled formula.*

First, we define the notion of substitution for labelled expressions as follows. The substitution $z[y/x]$ of label x to label y in label z is defined as:

$$z[y/x] \equiv \begin{cases} y & \text{if } z \equiv x; \\ z & \text{if } z \not\equiv x. \end{cases}$$

Then, we naturally define the substitution $[y/x]$ in labelled expression φ by: $(z{:}A)[y/x] \equiv z[y/x] : A$ and $(z\mathsf{R}w)[y/x] \equiv z[y/x]\mathsf{R}w[y/x]$.

Lemma 1 *If* $\Gamma \Rightarrow \Delta$ *is derivable in* **G3F(m)***, *then* $\Gamma[y/x] \Rightarrow \Delta[y/x]$ *is also height-preserving derivable, i.e., if* **G3F(m)*** $\vdash_n \Gamma \Rightarrow \Delta$, *then* **G3F(m)*** $\vdash_n \Gamma[y/x] \Rightarrow \Delta[y/x]$.

We call Lemma 1 *height-preserving substitution* (*hp-substitution*).

Definition 6 (*Admissibility*) A rule is said to be *admissible* in **G3F(m)*** if, whenever the premise(s) of the rule is derivable in **G3F(m)***, the conclusion of the rule is also derivable in **G3F(m)***. A rule is said to be *height-preserving admissible* (*hp-admissible*) in **G3F(m)*** if, whenever the premise(s) of the rule is derivable in **G3F(m)*** with height at most n, the conclusion of the rule is also derivable in **G3F(m)*** with height at most n.

Lemma 2 (Weakening) *The rules of weakening are hp-admissible in* **G3F(m)***, *i.e.,*

(i) *If* $\vdash_n \Gamma \Rightarrow \Delta$, *then* $\vdash_n x{:}A, \Gamma \Rightarrow \Delta$.
(ii) *If* $\vdash_n \Gamma \Rightarrow \Delta$, *then* $\vdash_n \Gamma \Rightarrow \Delta, x{:}A$.
(iii) *If* $\vdash_n \Gamma \Rightarrow \Delta$, *then* $\vdash_n x\mathsf{R}y, \Gamma \Rightarrow \Delta$.

We can show each item of this lemma by induction on height n of derivation.

Definition 7 (*Invertibility*) A rule is said to be *height-preserving invertible* (*hp-invertible*) in **G3F(m)*** if, whenever the conclusion of the rule is derivable in **G3F(m)*** with height at most n, the premise(s) of the rule is also derivable in **G3F(m)*** with height at most n.

Lemma 3 (Inversion) *All the rules of* **G3F(m)*** *are hp-invertible.*

Proof We distinguish three cases: (i) left and right rules of & and $\vee$; (ii) $(L\supset)$, (GRS) and (Mon); (iii) $(R\supset)$. For (i), in the case of $(L\&)$, it is enough to show that $\vdash_n x{:}A\&B, \Gamma \Rightarrow \Delta$ implies $\vdash_n x{:}A, x{:}B, \Gamma \Rightarrow \Delta$. If $x{:}A\&B, \Gamma \Rightarrow \Delta$ is an axiom or a zero-premise geometric rule scheme, then $x{:}A, x{:}B, \Gamma \Rightarrow \Delta$ is also an axiom or a zero-premise geometric rule scheme. If $n > 0$, (1) if $x{:}A\&B$ is the principal formula, then it is obvious. (2) Otherwise, apply induction hypothesis to the premise(s) of the original derivation, and then apply the rule.

For (ii), in the case of $(L\supset)$, it is enough to show that $\vdash_n x\mathsf{R}y, x{:}A\supset B, \Gamma \Rightarrow \Delta$ implies $\vdash_n x\mathsf{R}y, x{:}A\supset B, \Gamma \Rightarrow \Delta, y{:}A$ and $\vdash_n x\mathsf{R}y, x{:}A\supset B, y{:}B, \Gamma \Rightarrow \Delta$. If $n > 0$, consider whether $x{:}A\supset B$ is the principal formula. (1) If $x{:}A\supset B$ is the principal formula, then it is obvious. (2) Otherwise, apply hp-weakening to $\vdash_n x\mathsf{R}y, x{:}A\supset B, \Gamma \Rightarrow \Delta$, then we can obtain $\vdash_n x\mathsf{R}y, x{:}A\supset B, \Gamma \Rightarrow \Delta, y{:}A$ and $\vdash_n x\mathsf{R}y, x{:}A\supset B, y{:}B, \Gamma \Rightarrow \Delta$.

For $(R\supset)$, it is enough to show that $\vdash_n \Gamma \Rightarrow \Delta, x{:}A\supset B$ implies $\vdash_n x\mathsf{R}y, y{:}A, \Gamma \Rightarrow \Delta, y{:}B$. If $n > 0$, (1) if $x{:}A\supset B$ is the principal formula, then similar to the former cases. (2) Otherwise, we divide our argument depending on the last rule r of the derivation. If r is any rule except $(R\supset)$, then apply induction hypothesis to the premise, and then the same rule r. If r is $(R\supset)$ and another implication formula, say $z{:}C\supset D$, is the principal formula, then the last step of the derivation is

$$\dfrac{\vdots \atop z\mathsf{R}w, w{:}C, \Gamma \Rightarrow \Delta', w{:}D, x{:}A\supset B}{\Gamma \Rightarrow \Delta', z{:}C\supset D, x{:}A\supset B}\ (R\supset) .$$

Then, apply induction hypothesis to the premise, and then apply $(R\supset)$ for $z{:}C\supset D$.

$$\dfrac{\vdots \atop z\mathsf{R}w, w{:}C, x\mathsf{R}y, y{:}A, \Gamma \Rightarrow \Delta', y{:}B, w{:}D}{x\mathsf{R}y, y{:}A, \Gamma \Rightarrow \Delta', z{:}C\supset D, y{:}B}\ (R\supset) .$$

□

Lemma 4 (Contraction) *The rules of contraction are hp-admissible in* **G3F(m)***, *i.e.,*

(i) *If* $\vdash_n x{:}A, x{:}A, \Gamma \Rightarrow \Delta$, *then* $\vdash_n x{:}A, \Gamma \Rightarrow \Delta$.
(ii) *If* $\vdash_n \Gamma \Rightarrow \Delta, x{:}A, x{:}A$, *then* $\vdash_n \Gamma \Rightarrow \Delta, x{:}A$.
(iii) *If* $\vdash_n x\mathsf{R}y, x\mathsf{R}y, \Gamma \Rightarrow \Delta$, *then* $\vdash_n x\mathsf{R}y, \Gamma \Rightarrow \Delta$.

Proof By simultaneous induction on height n of derivations. If $n = 0$, then each sequent assumed is an axiom or a zero-premise geometric rule scheme. It is clear that the desired sequents are also an axiom or a zero-premise geometric rule scheme. Let $n > 0$. We focus on item (i) and then we need to use argument by cases. If the contracted formula is not one of the principal formula(s) of the last rule of the derivation, then it is obvious. Otherwise, then we distinguish further cases: (1) $(L\supset)$, (Mon); (2) left rules of & and $\vee$. Note that we take only these four rules as the last rule. In the first case, consider (Mon). The original derivation is

$$\dfrac{\vdots \atop x\mathsf{R}y, x{:}P, x{:}P, y{:}P, \Gamma \Rightarrow \Delta}{x\mathsf{R}y, x{:}P, x{:}P, \Gamma \Rightarrow \Delta}\ (Mon).$$

Apply induction hypothesis for (i) to the premise:

$$\dfrac{\vdots \atop x\mathsf{R}y, x{:}P, y{:}P, \Gamma \Rightarrow \Delta}{x\mathsf{R}y, x{:}P, \Gamma \Rightarrow \Delta}\ (Mon).$$

For the second case, consider $(L\&)$. The last step of the derivation is

$$\dfrac{\vdots \atop x{:}B, x{:}C, x{:}B\&C, \Gamma \Rightarrow \Delta}{x{:}B\&C, x{:}B\&C, \Gamma \Rightarrow \Delta}\ (L\&).$$

We apply hp-invertibility to the premise, so we obtain $x{:}B, x{:}C, x{:}B, x{:}C, \Gamma \Rightarrow \Delta$. And apply induction hypothesis of (i) to the result of the application, and then apply $(L\&)$. □

Recall that the *cut labelled formula* of (Cut)

$$\dfrac{\Gamma \Rightarrow \Delta, x{:}A \quad x{:}A, \Pi \Rightarrow \Sigma}{\Gamma, \Pi \Rightarrow \Delta, \Sigma}\ (Cut),$$

is the formula $x{:}A$, which is eliminated in applying the cut rule.

Definition 8 The *weight* of the cut labelled formula $x{:}A$ is the number of logical connectives in A, and the *cut-height* of (Cut) is the sum of heights of derivations of the two premises of (Cut).

Theorem 1 (Cut Elimination) *The cut rule is admissible in* **G3F(m)***.

Proof By induction on the weight of the cut labelled formula $x{:}A$, with subinduction on the cut-height of (Cut). Our proof is organized as follows. First, we consider the cases ((i) and (ii) below) where at least one of the premises of cut is an axiom or a zero-premise geometric rule scheme and show how cut is eliminated. For the rest, there are three cases: (iii) the cut labelled expression is not principal in the left premise; (iv) the cut labelled expression is principal in the left premise only; (v) the cut labelled formula is principal in both premises of cut.

(i) The left premise of cut is an axiom or a zero-premise geometric rule scheme: We omit the proof of this case.
(ii) The right premise of cut is an axiom or a zero-premise geometric rule scheme: First, suppose that the right premise $x{:}A, \Pi \Rightarrow \Sigma$ is the axiom (Id). That is, we have one of the following cases: the right premise is of the form $x{:}A, y{:}P, \Pi' \Rightarrow$

$\Sigma', y{:}P$ or of the form $x{:}P, \Pi \Rightarrow \Sigma', x{:}P$ where $A \equiv P$ in the latter case. For the former case, we note that $\Gamma, \Pi \Rightarrow \Delta, \Sigma$ is also an axiom (Id). For the latter case, we need to obtain $\Gamma, \Pi \Rightarrow \Delta, \Sigma', x{:}P$, which is derivable from the left premise $\Gamma \Rightarrow \Delta, x{:}P$ by hp-weakening. Second, suppose that the right premise is the axiom $(L\bot)$. If $A \not\equiv \bot$ in the cut labelled expression $x{:}A$, we can find a $w{:}\bot$ in Π and so $\Gamma, \Pi \Rightarrow \Delta, \Sigma$ is also an axiom $(L\bot)$. Otherwise, i.e., if $A \equiv \bot$, we need to check the last rule of the left premise $\Gamma \Rightarrow \Delta, x{:}\bot$. If the last rule is an axiom, this case is reduced to the case (i). Otherwise, this case becomes a special case of (iii). Finally, suppose that the right premise is a zero-premise geometric rule scheme. If the right premise of the cut is a zero-premise geometric rule scheme which is of the form $x{:}A, \overline{S}, \Pi' \Rightarrow \Sigma$, then the conclusion of the cut is also a zero-premise geometric rule scheme.

(iii) The cut labelled expression is not principal in the left premise: We divide our argument into cases, depending on the last applied rule of the left premise of (Cut). That is, there are eight cases including all logical rules, (Mon) and (GRS). Here we just demonstrate the case of $(R\supset)$. Then, we have the following derivation:

$$\dfrac{\dfrac{\begin{array}{c}\vdots\\ y\mathsf{R}z, z{:}B, \Gamma \Rightarrow \Delta', z{:}C, x{:}A\end{array}}{\Gamma \Rightarrow \Delta', y{:}B\supset C, x{:}A}(R\supset)\dagger \quad \begin{array}{c}\vdots\\ x{:}A, \Pi \Rightarrow \Sigma\end{array}}{\Gamma, \Pi \Rightarrow \Delta', y{:}B\supset C, \Sigma}(Cut)$$

where z is fresh in the lower sequent $\Gamma \Rightarrow \Delta', y{:}B\supset C, x{:}A$. We first apply hp-substitution with $[w/z]$ to $y\mathsf{R}z, z{:}B, \Gamma \Rightarrow \Delta', z{:}C, x{:}A$ to avoid the variable clash, where we assume that w is not in the conclusion of (Cut) above. Then, we can obtain the following derivation:

$$\dfrac{\dfrac{\begin{array}{c}\vdots\\ y\mathsf{R}w, w{:}B, \Gamma \Rightarrow \Delta', w{:}C, x{:}A\end{array} \quad \begin{array}{c}\vdots\\ x{:}A, \Pi \Rightarrow \Sigma\end{array}}{y\mathsf{R}w, w{:}B, \Gamma, \Pi \Rightarrow \Delta', w{:}C, \Sigma}(Cut)}{\Gamma, \Pi \Rightarrow \Delta', y{:}B\supset C, \Sigma}(R\supset)\dagger$$

where the application of cut is possible since the cut-height becomes smaller. The other cases, including (GRS) and (Mon), are similar to this case, though arguments for the rules without eigenvariable condition, such as (Mon), becomes simpler.

(iv) The cut labelled expression is principal in the left premise only: We divide our argument into cases, depending on the last applied rule of the *right* premise of (Cut), where we note that the cut labelled expression $x{:}A$ is not principal in the last rule because of our case (iv). But, the argument for this case is similar to (iii), so we omit the proof.

(v) The cut labelled formula is principal in both premises of cut: We have further three cases: $A \equiv B \vee C$, $B \& C$, or $B\supset C$ in the cut labelled expression $x{:}A$. Here we concentrate on the case of $x{:}B\supset C$. We have the following derivation:

$$\dfrac{\dfrac{\begin{matrix}\vdots\\ xRz, z{:}B, \Gamma \Rightarrow \Delta, z{:}C\end{matrix}}{\Gamma \Rightarrow \Delta, x{:}B\supset C}(R\supset)\dagger \quad \dfrac{\begin{matrix}\vdots\\ x{:}B\supset C, xRw, \Pi \Rightarrow \Sigma, w{:}B\end{matrix} \quad \begin{matrix}\vdots\\ w{:}C, x{:}B\supset C, xRw, \Pi \Rightarrow \Sigma\end{matrix}}{x{:}B\supset C, xRw, \Pi \Rightarrow \Sigma}(L\supset)}{xRw, \Gamma, \Pi \Rightarrow \Delta, \Sigma}(Cut),$$

where z is fresh in the lower sequent $\Gamma \Rightarrow \Delta, x{:}B\supset C$. From this derivation, we first construct the following derivation $\mathcal{D}_L$ with the help of hp-substitution with $[w/z]$:

$$\dfrac{\dfrac{\begin{matrix}\vdots\\ \Gamma \Rightarrow \Delta, x{:}B\supset C\end{matrix} \quad \begin{matrix}\vdots\\ x{:}B\supset C, xRw, \Pi \Rightarrow \Sigma, w{:}B\end{matrix}}{xRw, \Gamma, \Pi \Rightarrow \Delta, \Sigma, w{:}B}(Cut) \quad \begin{matrix}\vdots\\ w{:}B, xRw, \Gamma \Rightarrow \Delta, w{:}C\end{matrix}}{xRw, xRw, \Gamma, \Gamma, \Pi \Rightarrow \Delta, \Delta, \Sigma, w{:}C}(Cut),$$

where we note that the left application of cut is possible since the cut-height becomes smaller and that the final application of cut is possible because the weight of the cut labelled expression $w{:}B$ is smaller than that of $x{:}B\supset C$. Second, we also construct from our original derivation the following derivation $\mathcal{D}_R$.

$$\dfrac{\dfrac{\begin{matrix}\vdots\\ xRz, z{:}B, \Gamma \Rightarrow \Delta, z{:}C\end{matrix}}{\Gamma \Rightarrow \Delta, x{:}B\supset C}(R\supset) \quad \begin{matrix}\vdots\\ x{:}B\supset C, w{:}C, xRw, \Pi \Rightarrow \Sigma\end{matrix}}{w{:}C, xRw, \Gamma, \Pi \Rightarrow \Delta, \Sigma}(Cut),$$

where we note that the last application of cut is possible since the cut-height becomes smaller. Finally, we obtain the following derivation by $\mathcal{D}_L$ and $\mathcal{D}_R$.

$$\dfrac{\dfrac{\begin{matrix}\mathcal{D}_L\\ xRw, xRw, \Gamma, \Gamma, \Pi \Rightarrow \Delta, \Delta, \Sigma, w{:}C\end{matrix} \quad \begin{matrix}\mathcal{D}_R\\ w{:}C, xRw, \Gamma, \Pi \Rightarrow \Delta, \Sigma\end{matrix}}{xRw, xRw, xRw, \Gamma, \Gamma, \Gamma, \Pi, \Pi \Rightarrow \Delta, \Delta, \Delta, \Sigma, \Sigma}(Cut)}{\overline{xRw, \Gamma, \Pi \Rightarrow \Delta, \Sigma}},$$

where the double line means finitely many applications of the contraction and note that the application of cut is possible because the weight of the cut labelled expression $w{:}C$ is smaller than $x{:}B\supset C$. □

By Theorem 1, we can derive that labelled sequent calculi for *all* examples in Sects. 11.4.1, 11.4.2 and 11.4.3 admits the rule of cut.

11.6 Constructive Embedding from Extensions F into Modal Logics

This section establishes that **G3F(m)*** can be embedded into **G3K*** with some assumption. We first explain labelled sequent calculus for modal logic **K** developed in [21, 25].

Table 11.4 Labelled sequent calculus **G3K** (see [21])

(Axioms)	$\dfrac{}{x{:}P, \Gamma \Rightarrow \Delta, x{:}P}\,(Id)$ $\quad \dfrac{}{x\mathsf{R}y, \Gamma \Rightarrow \Delta, x\mathsf{R}y}\,(Rid)$ $\quad \dfrac{}{x{:}\bot, \Gamma \Rightarrow \Delta}\,(L\bot)$
(Logical rules)	$\dfrac{x{:}A, x{:}B, \Gamma \Rightarrow \Delta}{x{:}A\&B, \Gamma \Rightarrow \Delta}\,(L\&)$ $\quad \dfrac{\Gamma \Rightarrow \Delta, x{:}A \quad \Gamma \Rightarrow \Delta, x{:}B}{\Gamma \Rightarrow \Delta, x{:}A\&B}\,(R\&)$ $\dfrac{x{:}A, \Gamma \Rightarrow \Delta \quad x{:}B, \Gamma \Rightarrow \Delta}{x{:}A \vee B, \Gamma \Rightarrow \Delta}\,(L\vee)$ $\quad \dfrac{\Gamma \Rightarrow \Delta, x{:}A, x{:}B}{\Gamma \Rightarrow \Delta, x{:}A \vee B}\,(R\vee)$ $\dfrac{\Gamma \Rightarrow \Delta, x{:}A \quad x{:}B, \Gamma \Rightarrow \Delta}{x{:}A \supset B, \Gamma \Rightarrow \Delta}\,(L\supset)$ $\quad \dfrac{x{:}A, \Gamma \Rightarrow \Delta, x{:}B}{\Gamma \Rightarrow \Delta, x{:}A \supset B}\,(R\supset)$
(Modal Rules)	$\dfrac{y{:}A, x{:}\Box A, x\mathsf{R}y, \Gamma \Rightarrow \Delta}{x{:}\Box A, x\mathsf{R}y, \Gamma \Rightarrow \Delta}\,(L\Box)$ $\quad \dfrac{x\mathsf{R}y, \Gamma \Rightarrow \Delta, y{:}A}{\Gamma \Rightarrow \Delta, x{:}\Box A}\,(R\Box)^{a}$ $\dfrac{x\mathsf{R}y, y{:}A, \Gamma \Rightarrow \Delta}{x{:}\Diamond A, \Gamma \Rightarrow \Delta}\,(L\Diamond)^{a}$ $\quad \dfrac{x\mathsf{R}y, \Gamma \Rightarrow \Delta, y{:}A, x{:}\Diamond A}{x\mathsf{R}y, \Gamma \Rightarrow \Delta, x{:}\Diamond A}\,(R\Diamond)$

[a] y is fresh in the conclusion

11.6.1 *Labelled Sequent Calculus for* K

The modal syntax $\mathcal{ML}$ is an expansion of $\mathcal{L}$ with two modal operators $\Box$ and $\Diamond$, where we keep the same set Atom of atomic variables as $\mathcal{L}$. We also define $x{:}A$ and $x\mathsf{R}y$ similarly as before (note that we allow the expressions $x{:}\Box A$ and $x{:}\Diamond A$). Given finite multisets Γ and Δ of labelled modal formulas, we say that $\Gamma \Rightarrow \Delta$ is a *sequent* (here we allow the possibility that Δ may contain a relational atom). Table 11.4 provides a labelled sequent calculus **G3K** [21, 25] for modal logic **K**. Similarly to **G3F**(**m**)*, we may extend **G3K** with a *finite* set $*$ of geometric rule schemes as in [21] to write **G3K*** to mean the extension of **G3K** (for geometric rule schemes, recall Sect. 11.3.1). The notions of derivability, admissibility, etc., in **G3K*** are defined similarly to **G3F**(**m**)*. We note that, as we have done for **G3F**(**m**)* in the previous sections, it was shown in [21] that **G3K*** also enjoys height-preserving invertibility, height-preserving admissibility of substitution, weakening and contraction, and admissibility of cut.

11.6.2 *Embedding Theorem*

Now let us define our version of Gödel–Mckinsey–Tarski translation as follows:

Definition 9 (*Translation* $\boxplus$)

$$
\begin{aligned}
P^{\boxplus} &:= P\&\Box P,\\
\bot^{\boxplus} &:= \bot,\\
(A\&B)^{\boxplus} &:= A^{\boxplus}\&B^{\boxplus},\\
(A\vee B)^{\boxplus} &:= A^{\boxplus}\vee B^{\boxplus},\\
(A\supset B)^{\boxplus} &:= \Box(A^{\boxplus}\supset B^{\boxplus}),\\
(x:A)^{\boxplus} &:= x:A^{\boxplus},\\
(x\mathsf{R}y)^{\boxplus} &:= x\mathsf{R}y.
\end{aligned}
$$

For a finite multiset $\Gamma \equiv \varphi_1, \ldots, \varphi_n$ of labelled expressions, we define $\Gamma^{\boxplus} := \varphi_1^{\boxplus}, \ldots, \varphi_n^{\boxplus}$.

We note that the translation does not rewrite labels in labelled expressions.

Lemma 5 (i) **G3F*** $\vdash \Gamma \Rightarrow \Delta$ *implies* **G3K*** $\vdash \Gamma^{\boxplus} \Rightarrow \Delta^{\boxplus}$.
(ii) *Suppose that the following rule is admissible in* **G3K***.

$$\frac{x\mathsf{R}y, x{:}P\&\Box P, y{:}P\&\Box P, \Gamma \Rightarrow \Delta}{x\mathsf{R}y, x{:}P\&\Box P, \Gamma \Rightarrow \Delta}\ (TMon)$$

Then, **G3Fm*** $\vdash \Gamma \Rightarrow \Delta$ *implies* **G3K*** $\vdash \Gamma^{\boxplus} \Rightarrow \Delta^{\boxplus}$.

Proof First, we establish item (i) by induction on height n of derivation in **G3F***. Assume that there is a derivation of $\Gamma \Rightarrow \Delta$ in **G3F***. If the height of this derivation is 0, then $\Gamma \Rightarrow \Delta$ is an axiom or a zero-premise geometric rule scheme. If $\Gamma \Rightarrow \Delta$ is an axiom (that is, (Id) or $(L\bot)$), then the translation $\Gamma^{\boxplus} \Rightarrow \Delta^{\boxplus}$ is clearly derivable. If $\Gamma \Rightarrow \Delta$ is a zero-premise geometric rule scheme, then $\Gamma^{\boxplus} \Rightarrow \Delta^{\boxplus}$ is also a zero-premise geometric rule scheme which is of the form $\overline{S}^{\boxplus}, \Gamma'^{\boxplus} \Rightarrow \Delta^{\boxplus}$, since $\Gamma \Rightarrow \Delta$ is of the form $\overline{S}, \Gamma' \Rightarrow \Delta$ and $\overline{S}^{\boxplus} \equiv \overline{S}$. Let us consider the case where the height of the derivation is more than 0. Suppose that the last applied rule is $(R\supset)$, i.e., we have the following derivation:

$$\frac{\begin{array}{c}\vdots\\ x\mathsf{R}y, y{:}A, \Gamma \Rightarrow \Delta, y{:}B\end{array}}{\Gamma \Rightarrow \Delta, x{:}A\supset B}\ (R\supset).$$

By induction hypothesis, we straightforwardly obtain the following derivation in **G3K***:

$$\dfrac{\dfrac{\begin{array}{c}\vdots\\ x\mathsf{R}y, y{:}A^{\boxplus}, \Gamma^{\boxplus} \Rightarrow \Delta^{\boxplus}, y{:}B^{\boxplus}\end{array}}{x\mathsf{R}y, \Gamma^{\boxplus} \Rightarrow \Delta^{\boxplus}, y{:}A^{\boxplus}\supset B^{\boxplus}}\ (R\supset)}{\Gamma^{\boxplus} \Rightarrow \Delta^{\boxplus}, x{:}\Box(A^{\boxplus}\supset B^{\boxplus})}\ (R\Box),$$

whose end sequent is the result of the translation $\Gamma^{\boxplus} \Rightarrow \Delta^{\boxplus}, (x{:}A\supset B)^{\boxplus}$.

For the remaining other cases except (GRS), our argument is similar to the case just above. When the last applied rule is (GRS), it is straightforward to show that the translation is derivable in **G3K***, because our translation $(\cdot)^{\boxplus}$ does not rewrite any labels and $(x\mathsf{R}y)^{\boxplus} := x\mathsf{R}y$.

For item (ii), almost the same argument as in (i) works, but we comment on the case where the last applied rule is (Mon). That is,

$$\dfrac{\begin{array}{c}\vdots\\ x\mathsf{R}y, x{:}P, y{:}P, \Gamma \Rightarrow \Delta\end{array}}{x\mathsf{R}y, x{:}P, \Gamma \Rightarrow \Delta}\ (Mon).$$

Now we need to use the assumption of admissibility of $(TMon)$, which corresponds to the translation of the monotonicity rule (Mon) of atomic variables. Then, we apply induction hypothesis to the premise of (Mon) in the above derivation, and then we suffice to apply $(TMon)$ to obtain the following:

$$\dfrac{\begin{array}{c}\vdots\\ x\mathsf{R}y, x{:}P\&\Box P, y{:}P\&\Box P, \Gamma^{\boxplus} \Rightarrow \Delta^{\boxplus}\end{array}}{x\mathsf{R}y, x{:}P\&\Box P, \Gamma^{\boxplus} \Rightarrow \Delta^{\boxplus}}\ (TMon),$$

whose end sequent is $(x\mathsf{R}y)^{\boxplus}, (x{:}P)^{\boxplus}, \Gamma^{\boxplus} \Rightarrow \Delta^{\boxplus}$, as required. □

Lemma 6 (Main Lemma) *Let Γ, Δ be finite multisets of labelled expressions of the syntax $\mathcal{L}$,let Π, Σ be finite multisets of labelled atomic formulas of the syntax $\mathcal{L}$. Then,*

$$\mathbf{G3K}^* \vdash \Gamma^{\boxplus}, \Pi, \Box\Pi \Rightarrow \Sigma, \Delta^{\boxplus} \ \textit{implies}\ \mathbf{G3F}^* \vdash \Gamma, \Pi \Rightarrow \Sigma, \Delta.$$

Proof By induction on height n of the derivation of $\Gamma^{\boxplus}, \Pi, \Box\Pi \Rightarrow \Sigma, \Delta^{\boxplus}$ in **G3K***. If $n = 0$, $\Gamma^{\boxplus}, \Pi, \Box\Pi \Rightarrow \Sigma, \Delta^{\boxplus}$ is an axiom (there are just two cases: $(L\bot)$ and (Id)) or a zero-premise geometric rule scheme in **G3K***, so $\Gamma, \Pi \Rightarrow \Sigma, \Delta$ is also an axiom or a zero-premise geometric rule scheme in **G3F***.

If $n > 0$, we divide our argument into cases depending on the last rule of the derivation. Since Π and Σ are labelled atomic formulas for the syntax $\mathcal{L}$, the outermost logical connective of a labelled formula in the translations $\Gamma^{\boxplus}$ and $\Delta^{\boxplus}$ are never be the implication symbol $\supset$ nor the diamond $\Diamond$. So, the last applied logical rule must be other than the rules for $\supset$ and $\Diamond$. In what follows, we consider the following cases: (i) the last applied rule is one of $(L\vee)$, $(R\vee)$ and (GRS); (ii) the last applied rule is $(L\&)$ or $(R\&)$; (iii) the last applied rule is $(L\Box)$ or $(R\Box)$.

(i) The last applied rule is one of $(L\vee)$, $(R\vee)$ and (GRS): The straightforward application of induction hypothesis gives us the required derivation in **G3F***. For example, in the case of (GRS), the derivation ends with

$$\dfrac{\overline{T_1}[z_1/y_1], \overline{S}^{\boxplus}, \Gamma^{\boxplus}, \Pi, \Box\Pi \Rightarrow \Sigma, \Delta^{\boxplus} \quad \cdots \quad \overline{T_n}[z_n/y_n], \overline{S}^{\boxplus}, \Gamma^{\boxplus}, \Pi, \Box\Pi \Rightarrow \Sigma, \Delta^{\boxplus}}{\overline{S}^{\boxplus}, \Gamma^{\boxplus}, \Pi, \Box\Pi \Rightarrow \Sigma, \Delta^{\boxplus}} \ (GRS),$$

where we note that $\overline{S}^{\boxplus} \equiv \overline{S}$. Since $\overline{T_j}^{\boxplus} \equiv \overline{T_j}$, we can apply induction hypothesis to the premise to obtain the following derivation in **G3F*** by applying the same (GRS):

$$\dfrac{\overline{T_1}[z_1/y_1], \overline{S}, \Gamma, \Pi \Rightarrow \Sigma, \Delta \quad \cdots \quad \overline{T_n}[z_n/y_n], \overline{S}, \Gamma, \Pi \Rightarrow \Sigma, \Delta}{\overline{S}, \Gamma, \Pi \Rightarrow \Sigma, \Delta} \ (GRS)$$

(ii) The last applied rule is $(L\&)$ or $(R\&)$: We distinguish two further cases: (1) $P^{\boxplus} \equiv P\&\Box P$ is the principal formula, and (2) $(A\&B)^{\boxplus} \equiv A^{\boxplus}\&B^{\boxplus}$ is the principal formula. The latter case (2) is similar to the case (i). For the former case (1), we first suppose that the last applied rule is $(L\&)$, i.e., the derivation in **G3K*** is of the following form:

$$\dfrac{x{:}P, x{:}\Box P, \Gamma^{\boxplus}, \Pi, \Box\Pi \Rightarrow \Sigma, \Delta^{\boxplus}}{x{:}P\&\Box P, \Gamma^{\boxplus}, \Pi, \Box\Pi \Rightarrow \Sigma, \Delta^{\boxplus}} \ (L\&).$$

By induction hypothesis, we obtain, from the premise of $(L\&)$, the following derivation in the **G3F**(**m**)*:

$$x{:}P, \Gamma, \Pi \Rightarrow \Sigma, \Delta,$$

as required. Second for the case (1), we suppose that the last applied rule is $(R\&)$. Then, the last step of this derivation looks like:

$$\dfrac{\Gamma^{\boxplus}, \Pi, \Box\Pi \Rightarrow \Sigma, \Delta^{\boxplus}, x{:}P \quad \Gamma^{\boxplus}, \Pi, \Box\Pi \Rightarrow \Sigma, \Delta^{\boxplus}, x{:}\Box P}{\Gamma^{\boxplus}, \Pi, \Box\Pi \Rightarrow \Sigma, \Delta^{\boxplus}, x{:}P\&\Box P} \ (R\&).$$

Then, we apply induction hypothesis to the *left* premise to obtain the desired derivation:

$$\Gamma, \Pi \Rightarrow \Sigma, \Delta, x{:}P.$$

(iii) The last applied rule is $(L\Box)$ or $(R\Box)$: In this case, our strategy is: we first apply hp-invertibility to the implication in the premise of the derivation and second apply induction hypothesis. For example, let us consider the case of $(R\Box)$. The last step of the derivation is:

$$\dfrac{\begin{array}{c}\vdots\\ x\mathsf{R}y,\ \Gamma^{\boxplus},\ \Pi,\ \Box\Pi \Rightarrow \Sigma,\ \Delta^{\boxplus},\ y{:}A^{\boxplus}\supset B^{\boxplus}\end{array}}{\Gamma^{\boxplus},\ \Pi,\ \Box\Pi \Rightarrow \Sigma,\ \Delta^{\boxplus},\ x{:}\Box(A^{\boxplus}\supset B^{\boxplus})}\ (R\Box),$$

where y is fresh. We first apply hp-invertibility (of **G3K***) to the premise to obtain

$$x\mathsf{R}y,\ \Gamma^{\boxplus},\ \Pi,\ \Box\Pi,\ y{:}A^{\boxplus} \Rightarrow \Sigma,\ \Delta^{\boxplus},\ y{:}B^{\boxplus}$$

with preserving the height of the derivation. Second, now we can apply induction hypothesis to this sequent and then use the rule $(R\supset)$, i.e., :

$$\dfrac{\begin{array}{c}\vdots\\ x\mathsf{R}y,\ \Gamma,\ \Pi,\ y{:}A \Rightarrow \Sigma,\ \Delta,\ y{:}B\end{array}}{\Gamma,\ \Pi \Rightarrow \Sigma,\ \Delta,\ x{:}A\supset B}\ (R\supset).$$

□

Remark 1 This lemma is similar to the one given by Dyckhoff and Negri (see [8, Lemma 4]), but there is one important difference: we add a new assumption $\Box\Pi$ in **G3K**$^* \vdash \Gamma^{\boxplus}, \Pi, \Box\Pi \Rightarrow \Sigma, \Delta^{\boxplus}$, because of our modification of the translation sending an atomic variable P to $P\&\Box P$. In particular, we note that this modification plays a crucial role in the case (ii) in our proof of Lemma 6.

Example 1 In order to illustrate the idea of our proof of Lemma 6, let us consider the following derivation of $(x{:}P\supset P)^{\boxplus}$ in **G3K**:

$$\dfrac{\dfrac{\dfrac{\dfrac{\dfrac{}{x\mathsf{R}y,\ y{:}P,\ y{:}\Box P \Rightarrow y{:}P}\,(Id) \qquad \dfrac{\dfrac{\dfrac{}{y\mathsf{R}z,\ x\mathsf{R}y,\ y{:}P,\ y{:}\Box P,\ z{:}P \Rightarrow z{:}P}\,(Id)}{y\mathsf{R}z,\ x\mathsf{R}y,\ y{:}P,\ y{:}\Box P \Rightarrow z{:}P}\,(L\Box)}{x\mathsf{R}y,\ y{:}P,\ y{:}\Box P \Rightarrow y{:}\Box P}\,(R\Box)}{x\mathsf{R}y,\ y{:}P,\ y{:}\Box P \Rightarrow y{:}P\&\Box P}\,(R\&)}{x\mathsf{R}y,\ y{:}P\&\Box P \Rightarrow y{:}P\&\Box P}\,(L\&)}{x\mathsf{R}y \Rightarrow y{:}P\&\Box P\supset P\&\Box P}\,(R\supset)}{\Rightarrow x{:}\Box(P\&\Box P\supset P\&\Box P)}\,(R\Box).$$

From the left axiom (Id), i.e., the left premise of $(R\&)$ (we can disregard the right premise), we obtain the derivability of $x\mathsf{R}y,\ y{:}P \Rightarrow y{:}P$ in **G3F**. We also note that both the conclusion of $(R\&)$ and the conclusion of $(L\&)$ give us the derivability of the same sequent $x\mathsf{R}y,\ y{:}P \Rightarrow y{:}P$. Finally, we get from the next applications $(R\supset)$

and $(R\Box)$ the following derivation in **G3F**:

$$\dfrac{x\mathsf{R}y,\, y{:}P \Rightarrow y{:}P}{\Rightarrow x{:}P \supset P}\;(R\supset),$$

since we can apply Lemma 6 to both $x\mathsf{R}y,\, y{:}P,\, y{:}\Box P \Rightarrow y{:}P$ and $\Rightarrow x{:}\Box(P\&\Box P \supset P\&\Box P)$.

Theorem 2 (i) $\mathbf{G3F}^* \vdash \Gamma \Rightarrow \Delta$ iff $\mathbf{G3K}^* \vdash \Gamma^{\boxplus} \Rightarrow \Delta^{\boxplus}$.
(ii) *Suppose that the following rule is admissible in* $\mathbf{G3K}^*$*:*

$$\dfrac{x\mathsf{R}y,\, x{:}P\&\Box P,\, y{:}P\&\Box P,\, \Gamma \Rightarrow \Delta}{x\mathsf{R}y,\, x{:}P\&\Box P,\, \Gamma \Rightarrow \Delta}\;(TMon).$$

Then, $\mathbf{G3Fm}^* \vdash \Gamma \Rightarrow \Delta$ iff $\mathbf{G3K}^* \vdash \Gamma^{\boxplus} \Rightarrow \Delta^{\boxplus}$.

Proof It follows from each item of Lemma 5 that the left-to-right direction of the corresponding item holds. The right-to-left directions of both items are proved as special cases of Lemma 6 by putting $\Pi = \Sigma = \emptyset$, where we note that derivability in $\mathbf{G3F}^*$ implies derivability in $\mathbf{G3Fm}^*$. For the right-to-left direction of item (ii), we do not need to use admissibility of $(TMon)$. □

Theorem 2 uniformly captures embeddings from extensions of logic of strict implications into modal logics, as shown below. First of all, the following propositions give us a sufficient condition of applying Theorem 2(ii).

Proposition 2 *If* $x\mathsf{R}y,\, x{:}P,\, x{:}\Box P \Rightarrow y{:}\Box P$ *is derivable in* $\mathbf{G3K}^*$, *then*

$$\dfrac{x\mathsf{R}y,\, x{:}P\&\Box P,\, y{:}P\&\Box P,\, \Gamma \Rightarrow \Delta}{x\mathsf{R}y,\, x{:}P\&\Box P,\, \Gamma \Rightarrow \Delta}\;(TMon)$$

is admissible in $\mathbf{G3K}^*$.

Proof Assume that both $x\mathsf{R}y, x{:}P, x{:}\Box P \Rightarrow y{:}\Box P$ and $x\mathsf{R}y, x{:}P\&\Box P, y : P\&\Box P, \Gamma \Rightarrow \Delta$ are derivable in $\mathbf{G3K}^*$. It follows that $x\mathsf{R}y,\, x{:}P,\, x{:}\Box P,\, \Gamma \Rightarrow \Delta,\, y{:}\Box P$ by our assumption and admissibility of weakening. Then, we can derive our goal as follows:

$$\dfrac{\dfrac{\dfrac{\dfrac{\dfrac{}{x\mathsf{R}y, x{:}P, x{:}\Box P, y{:}P, \Gamma \Rightarrow \Delta, y{:}P}\;(Id)}{x\mathsf{R}y, x{:}P, x{:}\Box P, \Gamma \Rightarrow \Delta, y{:}P}\;(L\Box) \quad x\mathsf{R}y, x{:}P, x{:}\Box P, \Gamma \Rightarrow \Delta, y{:}\Box P}{x\mathsf{R}y, x{:}P, x{:}\Box P, \Gamma \Rightarrow \Delta, y{:}P\&\Box P}\;(R\&)}{x\mathsf{R}y, x{:}P\&\Box P, \Gamma \Rightarrow \Delta, y{:}P\&\Box P}\;(L\&) \quad y{:}P\&\Box P, x\mathsf{R}y, z{:}P\&\Box P, \Gamma \Rightarrow \Delta}{\dfrac{x\mathsf{R}y, x\mathsf{R}y, x{:}P\&\Box P, x{:}P\&\Box P, \Gamma, \Gamma \Rightarrow \Delta, \Delta}{x\mathsf{R}y, x{:}P\&\Box P, \Gamma \Rightarrow \Delta}}\;(Cut),$$

where the double line means finitely many applications of contraction. □

Proposition 3 *If a finite set* $*$ *of geometric rule schemes contains* $(Tran)$ *of Table* 11.3, *then* $x\mathsf{R}y,\, x{:}P,\, x{:}\Box P \Rightarrow y{:}\Box P$ *is derivable in* $\mathbf{G3K}^*$.

Proof

$$\dfrac{\dfrac{\dfrac{\dfrac{}{x\mathsf{R}y, y\mathsf{R}z, x\mathsf{R}z, x{:}P, x{:}\Box P, z{:}P \Rightarrow z{:}P}\ (Id)}{x\mathsf{R}y, y\mathsf{R}z, x\mathsf{R}z, x{:}P, x{:}\Box P \Rightarrow z{:}P}\ (L\Box)}{x\mathsf{R}y, y\mathsf{R}z, x{:}P, x{:}\Box P \Rightarrow z{:}P}\ (Tran)}{x\mathsf{R}y, x{:}P, x{:}\Box P \Rightarrow y{:}\Box P}\ (R\Box) \quad .$$

□

It follows from these propositions that a sequent calculus **G3Fm*** with (Mon) can be embedded into **G3K*** containing $(Tran)$ as a geometric rule scheme.

By Theorem 2(i), we obtain constructive embedding results for all examples of Sect. 11.4.3. By Theorem 2(ii) and Propositions 2 and 3, we can establish constructive embedding results for all examples of Sects. 11.4.1 and 11.4.2.

11.7 Soundness and Completeness of G3F(m)* for Kripke Semantics

This section establishes that **G3F**(**m**)* is sound and complete with respect to Kripke semantics.

11.7.1 Soundness

Recall that Var be the set of all labels. To establish the soundness of **G3F**(**m**)* for Kripke semantics, we need to lift Kripke semantics for $\mathcal{L}$-formulas up to the labelled expressions. Given $\mathfrak{M} = (S, R, V)$, an *assignment* is a function $f : \mathsf{Var} \to S$. Given a model $\mathfrak{M}$ and an assignment f, the *satisfaction relation* $\mathfrak{M}, f \models \varphi$ (read: φ holds in $\mathfrak{M}$ under f) for labelled expressions is defined by:

$$\mathfrak{M}, f \models x{:}A \text{ iff } \mathfrak{M}, f(x) \models A,$$
$$\mathfrak{M}, f \models x\mathsf{R}y \text{ iff } (f(x), f(y)) \in R.$$

A sequent $\Gamma \Rightarrow \Delta$ *holds in* $\mathfrak{M}$ *under* f if, whenever all of Δ hold in $\mathfrak{M}$ under f, $w{:}B$ holds in $\mathfrak{M}$ under f for some $w{:}B \in \Delta$. We say that $\Gamma \Rightarrow \Delta$ is *valid* in a model $\mathfrak{M}$ (notation: $\mathfrak{M} \models \Gamma \Rightarrow \Delta$) if $\mathfrak{M}, f \models \Gamma \Rightarrow \Delta$ for all assignments f. $\Gamma \Rightarrow \Delta$ is said to be *valid* in a class $\mathbb{M}$ of models (notation: $\mathbb{M} \models \Gamma \Rightarrow \Delta$) if $\mathfrak{M} \models \Gamma \Rightarrow \Delta$ for all models $\mathfrak{M} \in \mathbb{M}$. Let $*$ be a finite set of geometric rule schemes. We define $\mathbb{M}^*$ (or, $\mathbb{M}^*_{\mathbf{m}}$) as the class of all models (or, monotone models, respectively) whose underlying frames satisfy all corresponding geometric implications to $*$.

Theorem 3 (Soundness)

(i) *If* **G3F*** $\vdash \Gamma \Rightarrow \Delta$, *then* $\Gamma \Rightarrow \Delta$ *is valid in* $\mathbb{M}^*$.
(ii) *If* **G3Fm*** $\vdash \Gamma \Rightarrow \Delta$, *then* $\Gamma \Rightarrow \Delta$ *is valid in* $\mathbb{M}^*_{\mathbf{m}}$.

Proof It suffices to establish (ii) alone. Fix any model $\mathfrak{M} \in \mathbb{M}^*_{\mathbf{m}}$. By induction on height n of a derivation of $\Gamma \Rightarrow \Delta$ in **G3Fm***, we show that $\mathfrak{M} \models \Gamma \Rightarrow \Delta$. We only check the (seemingly unique nontrivial) case where the last applied rule is one of a finite set $*$ of geometric rule schemes. We divide our argument into two cases where the rule is zero-premise or not. First, we show that a zero-premise geometric rule scheme $\overline{S}, \Gamma' \Rightarrow \Delta$ is valid in $\mathbb{M}^*_{\mathbf{m}}$. Write $\mathfrak{M} = (S, R, V)$. By the assumption of $\mathfrak{M} \in \mathbb{M}^*_{\mathbf{m}}$, $\mathfrak{M}$ satisfies the corresponding geometric implication $\forall \boldsymbol{x}(S_1 \& \cdots \& S_m \supset \bot)$. Fix any assignment $f : \mathsf{Var} \to S$ and let $\boldsymbol{x} \equiv (x_1, \ldots, x_l)$. Since $\mathfrak{M}$ satisfies the corresponding geometric implication above, $\mathfrak{M}, f \not\models \overline{S}$ hence $\mathfrak{M}, f \not\models \overline{S}, \Gamma'$. This implies $\mathfrak{M}, f \models \overline{S}, \Gamma' \Rightarrow \Delta$.

Second, suppose that we have the following derivation:

$$\dfrac{\begin{matrix}\vdots \\ \overline{T_1}[z_1/y_1], \overline{S}, \Gamma' \Rightarrow \Delta\end{matrix} \quad \cdots \quad \begin{matrix}\vdots \\ \overline{T_n}[z_n/y_n], \overline{S}, \Gamma' \Rightarrow \Delta\end{matrix}}{\overline{S}, \Gamma' \Rightarrow \Delta}\ (GRS),$$

where $z_1, \ldots, z_n$ are fresh and $(GRS) \in *$. Fix any assignment f. Let σ be the corresponding geometric implication to (GRS). To show $\mathfrak{M}, f \models \overline{S}, \Gamma' \Rightarrow \Delta$, suppose that $\mathfrak{M}, f \models \overline{S}$ and $\mathfrak{M}, f \models \Gamma'$. Our goal is to show that $\mathfrak{M}, f \models w{:}C$ for some $w{:}C \in \Delta$. Since the underlying frame of $\mathfrak{M}$ satisfies the following geometric implication σ corresponding to (GRS):

$$\forall \boldsymbol{x}\,(S_1 \& \cdots \& S_m \supset \exists \boldsymbol{y} \bigvee_{1 \leqslant j \leqslant n} (T_{j1} \& \cdots \& T_{jn_j})),$$

$\mathfrak{M}, f \models \overline{S}$ implies that there exist $d_1, \ldots, d_n$ in the domain of $\mathfrak{M}$ such that all of $\overline{T_1}$, ..., $\overline{T_n}$ hold in $\mathfrak{M}$ under a variant of f such that we interpret all y_is by d_is, respectively. Define the following new assignment f' that assigns each of all variables expect z_is to the same value as f and sends $z_1, \ldots, z_n$ to $d_1, \ldots, d_n$, respectively. Then, it is clear that $\mathfrak{M}, f' \models \overline{T_1}[z_1/y_1]$, ..., $\mathfrak{M}, f' \models \overline{T_n}[z_n/y_n]$. Since $z_i s$ are fresh in $\overline{S}, \Gamma' \Rightarrow \Delta$, we also obtain from our assumption that $\mathfrak{M}, f' \models \overline{S}$ and $\mathfrak{M}, f' \models \Gamma'$. By induction hypothesis, $\mathfrak{M}, f' \models w{:}C$ for some $w{:}C \in \Delta$ hence $\mathfrak{M}, f \models w{:}C$ for some $w{:}C \in \Delta$, since $z_i s$ are fresh in $\overline{S}, \Gamma' \Rightarrow \Delta$. □

11.7.2 Completeness

In what follows in this subsection, we regard Γ and Δ as *possibly infinite* multisets of labelled expressions. We say that a possibly infinite sequent $\Gamma \Rightarrow \Delta$ is *derivable* in **G3F**(**m**)* if there are some finite $\Gamma' \subseteq \Gamma$ and some finite $\Delta' \subseteq \Delta$ such that **G3F**(**m**)$^* \vdash \Gamma' \Rightarrow \Delta'$ in the sense of Definition 4.

Definition 10 (*Saturation*) Let $\Gamma \Rightarrow \Delta$ be a possibly infinite sequent. We say that $\Gamma \Rightarrow \Delta$ is **G3F***-*saturated*, if it satisfies the following conditions:

(**unprov**) $\Gamma \Rightarrow \Delta$ is not derivable in **G3F***.
(**l&**) $x{:}A\&B \in \Gamma$ implies that $x{:}A, x{:}B \in \Gamma$.
(**r&**) $x{:}A\&B \in \Delta$ implies that $x{:}A \in \Delta$ or $x{:}B \in \Delta$.
(**l∨**) $x{:}A \vee B \in \Gamma$ implies that $x{:}A \in \Gamma$ or $x{:}B \in \Gamma$.
(**r∨**) $x{:}A \vee B \in \Delta$ implies that $x{:}A, x{:}B \in \Delta$.
(**l ⊃**) $x{:}A \supset B, x\mathsf{R}y \in \Gamma$ jointly imply that $y{:}A \in \Delta$ or $y{:}B \in \Gamma$.
(**r ⊃**) $x{:}A \supset B \in \Delta$ implies that $x\mathsf{R}y$, $y{:}A \in \Gamma$ and $y{:}B \in \Delta$ for some label y.
(**grs**) $S_1, \cdots, S_m \in \Gamma$ imply that $T_{j1}[z_j/y_j], \cdots, T_{jn_j}[z_j/y_j] \in \Gamma$ for some $j \in \{1, \cdots, n\}$ and some label z_j.

A possibly infinite sequent $\Gamma \Rightarrow \Delta$ is **G3Fm***-*saturated*, if it satisfies the above seven conditions except (**unprov**) as well as:

(**unprov′**) $\Gamma \Rightarrow \Delta$ is not derivable in **G3Fm***.
(**mon**) $x\mathsf{R}y, x{:}P \in \Gamma$ imply $y{:}P \in \Gamma$.

We note that (**grs**) is the corresponding condition to a nonzero-premise geometric rule scheme.

Lemma 7 (Saturation Lemma) *Let $\Gamma \Rightarrow \Delta$ be a finite sequent and suppose that* **G3F**(**m**)$^* \nvdash \Gamma \Rightarrow \Delta$. *Then, there exists a possibly infinite sequent $\Gamma^+ \Rightarrow \Delta^+$ such that $\Gamma \subseteq \Gamma^+$, $\Delta \subseteq \Delta^+$ and $\Gamma^+ \Rightarrow \Delta^+$ is* **G3F**(**m**)*-*saturated.*

Proof Fix an enumeration $(w_n)_{n\in\omega}$ of all labels Var. We inductively define a sequence $(\Gamma_n \Rightarrow \Delta_n)_{n\in\omega}$ of *finite* sequent $\Gamma_n \Rightarrow \Delta_n$ such that **G3F**(**m**)$^* \nvdash \Gamma_n \Rightarrow \Delta_n$. Let $(\varphi_n)_{n\in\omega}$ be an enumeration of all labelled formulas (i.e., except relational atoms) such that each φ_n occurs infinitely often. In what follows in this proof, we denote by $\{ GRS_i \mid 1 \leqslant i \leqslant N \}$ the finite set of all nonzero-premise geometric rule schemes in $*$ (recall that the original $*$ itself is finite).
(**Basis**) For $n = 0$, we define $\Gamma_0 := \Gamma$ and $\Delta_0 := \Delta$.
(**Inductive Step**) Suppose that we have defined $\Gamma_i \Rightarrow \Delta_i$ $(0 \leqslant i \leqslant n)$ such that **G3F**(**m**)$^* \nvdash \Gamma_i \Rightarrow \Delta_i$. Then we define $\Gamma_{n+1} \Rightarrow \Delta_{n+1}$ by the following procedure:

(Step 0) This step is for the calculus containing the rule (Mon), otherwise we can start from the next (**Step 1**). For all pairs $(x\mathsf{R}y, x{:}P) \in \Gamma_n \times \Gamma_n$, we add $y{:}P$ to Γ_n. That is, we define $\Gamma_n' := \Gamma_n \cup \{y{:}P \mid x\mathsf{R}y, x{:}P \in \Gamma_n \text{ for some } x\}$. Then, we still have **G3F**(**m**)$^* \nvdash \Gamma_n' \Rightarrow \Delta_n$ by (Mon). Then, we move to the next step.
(Step 1) This step is for the calculus having nonempty set $\{ GRS_i \mid 1 \leqslant i \leqslant N \}$ of nonzero-premise geometric rule schemes. We execute the following procedure for all nonzero-premise rules $\{ GRS_i \mid 1 \leqslant i \leqslant N \}$. If there is no such rules, we put $\Gamma_n'' := \Gamma_n'$ and go to (**Step 2**). Suppose that we have $(\Gamma_n')^{(i)} \Rightarrow \Delta_n$ $(1 \leqslant i < k)$ such

that each sequent is underivable in $\mathbf{G3F(m)}^*$. Now we deal with k-th geometric rule scheme (GRS_k). Let (GRS_k) have the following form:

$$\frac{\overline{T_1}[z_1/y_1], \overline{S}, \Gamma \Rightarrow \Delta \quad \cdots \quad \overline{T_n}[z_n/y_n], \overline{S}, \Gamma \Rightarrow \Delta.}{\overline{S}, \Gamma \Rightarrow \Delta} \; (GRS_k)$$

Let us consider all possible combinations of $\overline{S}$ in $(\Gamma_n')^{(k-1)}$ and let M be the number of such all combinations. We expand $(\Gamma_n')^{(k-1)} \Rightarrow \Delta_n$ into $(\Gamma_n')^{(k)} \Rightarrow \Delta_n$ as follows. Suppose that we have defined $(\Gamma_n')^{(k-1,i)} \Rightarrow \Delta_n (1 \leqslant i < M)$ such that $(\Gamma_n')^{(k-1,i)} \Rightarrow \Delta_n$ is unprovable in $\mathbf{G3F(m)}^*$ for all $1 \leqslant i < M$. Then, consider $(i+1)$-th combination of $\overline{S}$ in $(\Gamma_n')^{(k-1)}$. Let us write it as $\overline{S} \equiv S_1, \ldots, S_m$. By the above rule scheme and unprovability of $(\Gamma_n')^{(k-1,i)} \Rightarrow \Delta_n$, we can find some $j \in \{1, \ldots, n\}$ and some fresh z_j such that $\overline{T_j}[z_j/y_j], (\Gamma_n')^{(k-1,i)} \Rightarrow \Delta_n$ are unprovable. Then, we set up $(\Gamma_n')^{(k-1,i+1)} := \overline{T_j}[z_j/y_j], (\Gamma_n')^{(k-1,i)}$.
Finally, we define $(\Gamma_n')^{(k)} := (\Gamma_n')^{(k-1,M)}$. After when we check all rules in $\{GRS_i | 1 \leqslant i \leqslant N\}$, we put $\Gamma_n'' := (\Gamma_n')^{(N)}$ (where recall that N is the number of all nonzero-premise geometric rule schemes in $*$). Then, we move to the next step.

(Step 2) We execute the following procedure to define Γ_{n+1} and Δ_{n+1} in terms of the form of φ_n and then move back to **(Step 0)**.

(1) $\varphi_n \equiv x{:}A \& B$ and $\varphi_n \in \Gamma_n''$. Define $\Gamma_{n+1} := \Gamma_n'' \cup \{x{:}A, x{:}B\}$ and $\Delta_{n+1} := \Delta_n$. It is easy to verify $\mathbf{G3F(m)}^* \nvdash \Gamma_{n+1} \Rightarrow \Delta_{n+1}$ by $(L\&)$ and admissibility of contraction (Lemma 4).

(2) $\varphi_n \equiv x{:}A \& B$ and $\varphi_n \in \Delta_n$. Define $\Gamma_{n+1} := \Gamma_n''$ and Δ_{n+1} by:

$$\Delta_{n+1} := \begin{cases} \Delta_n \cup \{x{:}A\} & \text{if } \mathbf{G3F(m)}^* \nvdash \Gamma_n'' \Rightarrow \Delta_n \cup \{x{:}A\} \\ \Delta_n \cup \{x{:}B\} & \text{otherwise} \end{cases}$$

Since $\mathbf{G3F(m)}^* \nvdash \Gamma_n'' \Rightarrow \Delta_n$, we have $\mathbf{G3F(m)}^* \nvdash \Gamma_{n+1} \Rightarrow \Delta_{n+1}$ by $(R\&)$ and admissibility of contraction.

(3) $\varphi_n \equiv x{:}A \vee B$ and $\varphi_n \in \Gamma_n''$, it is similar to 2).

(4) $\varphi_n \equiv x{:}A \vee B$ and $\varphi_n \in \Delta_n$, it is similar to 1).

(5) $\varphi_n \equiv x{:}A \supset B$ and $\varphi_n \in \Gamma_n''$. Let $y_1, \cdots, y_k$ be all labels in Γ_n'' such that $x \mathsf{R} y_i \in \Gamma_n''$. Then, we expand $\Gamma_n'' \Rightarrow \Delta_n$ into $\Gamma_{n+1} \Rightarrow \Delta_{n+1}$ step by step by constructing $\nvdash (\Gamma_n'')^l \Rightarrow (\Delta_n)^l (1 \leqslant l \leqslant k)$. Suppose that we have defined $(\Gamma_n'')^i \Rightarrow (\Delta_n)^i$ for all $1 \leqslant i < l$ such that $\mathbf{G3F(m)}^* \nvdash (\Gamma_n'')^i \Rightarrow (\Delta_n)^i$. By $x{:}A \supset B \in (\Gamma_n'')^{l-1}$, $(L\supset)$ and admissibility of contraction, we define $(\Gamma_n'')^l \Rightarrow (\Delta_n)^l$ as

$$\begin{cases} (\Gamma_n'')^{l-1} \Rightarrow (\Delta_n)^{l-1} \cup \{y_l{:}B\} & \text{if } \mathbf{G3F(m)}^* \nvdash (\Gamma_n'')^{l-1} \Rightarrow (\Delta_n)^{l-1} \cup \{y_l{:}B\} \\ (\Gamma_n'')^{l-1} \cup \{y_l{:}A\} \Rightarrow (\Delta_n)^{l-1} & \text{if } \mathbf{G3F(m)}^* \nvdash (\Gamma_n'')^{l-1} \cup \{y_l{:}A\} \Rightarrow (\Delta_n)^{l-1}. \end{cases}$$

It is clear that $\mathbf{G3F(m)}^* \nvdash (\Gamma_n'')^l \Rightarrow (\Delta_n)^l$. Finally define: $\Gamma_{n+1} := (\Gamma_n'')^k$ and $\Delta_{n+1} := (\Delta_n)^k$.

(6) $\varphi_n \equiv x{:}A\supset B$ and $\varphi_n \in \Delta_n$. We choose a fresh labell y from Var not occurring in $\Gamma_n'' \Rightarrow \Delta_n$. Then, define $\Gamma_{n+1} := \Gamma_n'' \cup \{x\mathsf{R}y, y{:}A\}$ and $\Delta_{n+1} := \Delta_n \cup \{y{:}B\}$. It is easy to check that $\mathbf{G3F(m)}^* \nvdash \Gamma_{n+1} \Rightarrow \Delta_{n+1}$ by $\mathbf{G3F(m)}^* \nvdash \Gamma_n'' \Rightarrow \Delta_n$ and the rule of $(R\supset)$ and admissibility of contraction.
(7) Otherwise. Define $\Gamma_{n+1} := \Gamma_n''$ and $\Delta_{n+1} := \Delta_n$.

Finally, we define: $\Gamma^+ := \bigcup_{n\in\omega} \Gamma_n$ and $\Delta^+ := \bigcup_{n\in\omega} \Delta_n$. Clearly, $\Gamma \subseteq \Gamma^+$ and $\Delta \subseteq \Delta^+$. It is routine to check that $\Gamma^+ \Rightarrow \Delta^+$ is saturated. □

Definition 11 Let $\Gamma \Rightarrow \Delta$ be a saturated sequent. We define the derived model $\mathfrak{M} = (S, R, V)$ from $\Gamma \Rightarrow \Delta$ as follows:

- S is the set of labels occurring in $\Gamma \Rightarrow \Delta$.
- $(x, y) \in R$ iff $x\mathsf{R}y \in \Gamma$.
- $x \in V(P)$ iff $x{:}P \in \Gamma$.

Lemma 8 (Truth Lemma) *Let $\Gamma \Rightarrow \Delta$ be a saturated sequent and $\mathfrak{M} = (S, R, V)$ be the derived model from $\Gamma \Rightarrow \Delta$.*

(i) *$x{:}A \in \Gamma$ implies $\mathfrak{M}, x \models A$.*
(ii) *$x{:}A \in \Delta$ implies $\mathfrak{M}, x \not\models A$.*

Proof We prove (i) and (ii) by simultaneous induction on the number of the connectives of A. If $A \equiv P$ or $\bot$, then it is obvious. Otherwise, we only show the case where A is of the form $B\supset C$. For (i), assume $x{:}B\supset C \in \Gamma$, and assume $(x, y) \in R$ and $\mathfrak{M}, y \models B$. So, $x\mathsf{R}y \in \Gamma$. Then, by saturation, $y{:}B \in \Delta$ or $y{:}C \in \Gamma$, and then by induction hypothesis $\mathfrak{M}, y \not\models B$ or $\mathfrak{M}, y \models C$. But we already have $\mathfrak{M}, y \models B$. Therefore, we obtain $\mathfrak{M}, y \models C$.

For (ii), assume $x{:}B\supset C \in \Delta$. By saturation, $x\mathsf{R}y \in \Gamma$ and $y{:}B \in \Gamma$ and $y{:}C \in \Delta$ for some label y. By induction hypothesis, we obtain $\mathfrak{M}, y \models B$ and $\mathfrak{M}, y \not\models C$. By the definition of the derived Kripke model, we also obtain xRy. Therefore, $\mathfrak{M}, y \not\models x{:}B\supset C$, as required. □

Lemma 9 *Let $\Gamma \Rightarrow \Delta$ be a saturated sequent and $\mathfrak{M} = (S, R, V)$ be the derived model from $\Gamma \Rightarrow \Delta$. Then, the underlying valuation V of $\mathfrak{M}$ is monotone and the underlying frame (S, R) of $\mathfrak{M}$ satisfies all geometric implications corresponding to $*$.*

Proof By the condition (**mon**) of Definition 10, it is easy to see that the underlying valuation V of $\mathfrak{M}$ is monotone. Given any nonzero-premise geometric rule schemes (GRS), the condition (**grs**) of Definition 10 forces $\mathfrak{M}$ to satisfy the corresponding geometric implication to (GRS). So, let us focus on a zero-premise geometric rule scheme: $\overline{S}, \Pi \Rightarrow \Sigma$, where $\overline{S} := S_1, \ldots, S_m$. We show the corresponding first-order sentence $\forall \boldsymbol{x}(S_1 \& \cdots \& S_m \supset \bot)$ holds in $\mathfrak{M}$. Fix any list of labels $\boldsymbol{x}$ from W and suppose that $\mathfrak{M}, f \models \overline{S}$, where f sends each label x to itself. By the condition (**unprov**) (or (**unprov**$'$)) of Definition 10, $S_i \notin \Gamma$ for some $1 \leqslant i \leqslant m$. This means that $\mathfrak{M}, f \not\models \overline{S}$, as desired. □

Theorem 4 (i) *If* $\Gamma \Rightarrow \Delta$ *is valid in* $\mathbb{M}^*$, *then* **G3F**$^* \vdash \Gamma \Rightarrow \Delta$.
(ii) *If* $\Gamma \Rightarrow \Delta$ *is valid in* $\mathbb{M}^*_{\mathbf{m}}$, *then* **G3Fm**$^* \vdash \Gamma \Rightarrow \Delta$.

Proof We show (ii) alone. We show the contrapositive implication of (ii). Suppose **G3F**(**m**)$^* \nvdash \Gamma \Rightarrow \Delta$. By Lemma 7, we can find a possibly infinite saturated sequent $\Gamma^+ \Rightarrow \Delta^+$ such that $\Gamma \subseteq \Gamma^+$ and $\Delta \subseteq \Delta^+$. Let $\mathfrak{M} = (S, R, V)$ be the derived model from $\Gamma^+ \Rightarrow \Delta^+$. By Lemma 8, it is clear that $\mathfrak{M}, x \models C$ for all $x{:}C \in \Gamma$ and that $\mathfrak{M}, x \not\models C$ for all $x{:}C \in \Delta$. Define the derived assignment f as a function such that $f(x) = x$ for any $x \in S$. Then, by this assignment f, we obtain $\mathfrak{M} \not\models \Gamma \Rightarrow \Delta$. By Lemma 9, $\mathfrak{M} \in \mathbb{M}^*_{\mathbf{m}}$. Therefore, $\mathbb{M}^*_{\mathbf{m}} \not\models \Gamma \Rightarrow \Delta$, as required. □

By Theorem 4(i), we obtain completeness results for all examples of Sect. 11.4.3. By Theorem 4(ii), we can establish completeness results for all examples of Sects. 11.4.1 and 11.4.2.

11.8 Further Direction

There are several directions for further research of this work. Let us comment on four of these. The first direction is concerned with Visser's extension of basic propositional logic **BPL** by the Löb rule [32]: from $(\top \supset A) \supset A$ we may derive $\top \supset A$ or by the axiom $((\top \supset p) \supset p) \supset (\top \supset p)$ [30]. Visser [32] showed that the extension can be embedded into Gödel-Löb logic, i.e., modal logic **GL** extended by the axiom $\Box(\Box p \supset p) \supset \Box p$ via both the original Gödel–Mckinsey–Tarski translation $\Box$ and our translation $\boxplus$. It is natural to ask if we can provide a constructive embedding from **BPL** to **GL** via labelled sequent calculi. (We note that Negri [21] provided a cut-free and complete labelled sequent calculus for modal logic **GL**.)

Second, this paper did not consider the equality symbol between two labels. But it allows us to cover more frame properties such as *isolatedness* ($x\mathsf{R}y$ implies $x = y$, cf. [5]), *weak-transitivity* ($x\mathsf{R}y$ and $x\mathsf{R}z$ imply ($x = z$ or $x\mathsf{R}z$), cf. [18]), *connectedness* ($x\mathsf{R}y$ or $x = y$ or $y\mathsf{R}x$, cf. [32]). Note that these properties are still written in terms of a geometric implication extended with the equality symbol. The inclusion of the equality symbol as a new labelled atom will broaden the range of the correspondence between implicational logics (extensions of the logic **F** of strict implication) and modal logics. (For modal logic, Negri [21] dealt with on an extension of labelled formalism with equality between labels, cf. [24]).

Third, besides Gödel–Mckinsey–Tarski translation, there is another embedding, called Girard Translation (cf. [31]), from intuitionistic logic into modal logic **S4**. Is it possible to apply Dyckhoff and Negri's approach also to this embedding?

Finally, there is also a faithful translation from intuitionistic logic into Visser's basic propositional logic by Aghaei and Ardeshir [1], but its underlying semantic idea has not been clear so far. Can we apply Dyckhoff and Negri's approach to this translation to obtain the constructive embedding result via labelled sequent calculi?

Acknowledgments We would like to thank an anonymous reviewer for his/her invaluable comments. We also would like to thank Sara Negri for her sharing her draft [23] on a similar topic to our paper. We are grateful to Ryo Kashima for setting opportunities for the first author to give presentations on this topic at Tokyo Institute of Technology for giving helpful suggestions to us. The first author wishes to thank her supervisor Kengo Okamoto for a regular weekly discussion. The authors have presented material related to this paper at several occasions. We would like to thank the audiences of these events, including 2014 annual meetings of the Japan Association for Philosophy of Science in Japan, Trends in Logic XIII in Poland, the Second Taiwan Philosophical Logic Colloquium (TPLC 2014) in Taiwan, and the 49th MLG meeting at Kaga, Japan. The first author's visit to Taiwan for attending TPLC 2014 was supported by the grant from Tokyo Metropolitan University for graduate students. The work of the second author was partially supported by JSPS Core-to-Core Program (A. Advanced Research Networks) and JSPS KAKENHI, Grant-in-Aid for Young Scientists (B) 24700146 and 15K21025.

References

1. Aghaei, M., Ardeshir, M.: A bounded translation of intuitionistic propositional logic into basic propositional logic. Math. Log. Q. **46**, 199–206 (2000)
2. Ardeshir, M., Ruitenburg, W.: Basic propositional calculus I. Math. Log. Q. **44**, 317–343 (1998)
3. Cerrato, C.: Natural deduction based upon strict implication for normal modal logics. Notre Dome J. Form. Log. **35**, 471–495 (1994)
4. Chagrov, A., Zakharyaschev, N.: Modal Logic. Oxford University Press (1997)
5. Corsi, G.: Weak logics with strict implication. Math. Log. Q. **33**, 389–406 (1987)
6. Došen, K.: Modal translation in K and D. Diamond and Defaults, pp. 103–127 (1993)
7. Dragalin, A.: Mathmatical Intuitionism: Introduction to Proof Theory. American Mathematics Society (1988)
8. Dyckhoff, R., Negri, S.: Proof analysis in intermediate logics. Arch. Math. Log. **51**, 71–92 (2012)
9. Gödel, K.: Eine interpretation des intuitionistischen Aussagenkalküls. Ergebnisse Eines Mathematischen Kolloquiums **4**, 39–40 (1933)
10. Hughes, G., Cresswell, M.: A New Introduction to Modal Logic. Routledge, London (1996)
11. Ishigaki, R., Kashima, R.: Sequent calculi for some strict implication logics. Log. J. IGPL **16**(2), 155–174 (2008)
12. Ishii, K., Kashima, R., Kikuchi, K.: Sequent calculi for Visser's propositional logics. Notre Dame J. Form. Log. **42**(1), 1–22 (2001)
13. Kikuchi, K.: Relationships between basic propositional calculus and substructural logics. Bull. Sect. Log. **30**(1), 15–20 (2001)
14. Kikuchi, K., Sasaki, K.: A cut-free Gentzen formulation of basic propositional calculus. J. Log. Lang. Inf. **12**, 213–225 (2003)
15. Kleene, S.C.: Introduction to Metamathematics. North-Holland Public Co. (1952)
16. Lewis, C.I.: Implication and the algebra of logic. Mind **21**, 522–531 (1912)
17. Lewis, C.I.: A new algebra of strict implications and some consequents. J. Philos. Psychol. Sci. Methods **10**, 428–438 (1913)
18. Ma, M., Sano, K.: On extensions of basic propositional logic. In: Proceedings of the 13th Asian Logic Conference, pp. 170–200 (2015)
19. Mckinsey, J.C.C., Tarski, A.: Some theorems about the sentential calculi of Lewis and Heyting. J. Symbol. Log. **13**, 1–15 (1948)
20. Mints, G.: The Gödel-Tarski translations of intuitionistic propositional formulas. Correct Reason. 487–491 (2012)
21. Negri, S.: Proof analysis in modal logic. J. Philos. Log. **34**, 507–544 (2005)

22. Negri, S.: Proof analysis in non-classical logics. Logic Colloquium' 05,ASL Lecture Notes in Logic, vol. 28, pp. 107–128 (2008)
23. Negri, S.: The intensional side of algebraic-topological representation theorems. Submitted
24. Negri, S., Von Plato, J.: Structural Proof Theory. Cambridge University Press (2001)
25. Negri, S., Von Plato, J.: Proof Analysis. Cambridge University Press (2011)
26. Restall, G.: Subintuitionistic logics. Notre Dome J. Form. Log. **35**, 116–129 (1994)
27. Ruitenburg, W.: Constructive logic and the paradoxes. Modern Log. **1**, 271–301 (1991)
28. Sano, K., Ma, M.: Alternative semantics for Visser's propositional logics. In: Logic, Language, and Computation, volume 8984 of Lecture Notes in Computer Science, pp. 257–275 (2015)
29. Suzuki, Y., Ono, H.: Hilbert-style proof system for BPL. Technical Report IS-RR-97-0040F, Japan Advanced Institute of Science and Technology (1997)
30. Suzuki, Y., Wolter, F., Zakharyaschev, M.: Speaking about transitive frames in propositional languages. J. Log. Lang. Inf. **7**, 317–339 (1998)
31. Troelstra, A.S., Schwichtenberg, H.: Basic Proof Theory, 2nd edn. Cambridge University Press (2000)
32. Visser, A.: A propositional logic with explicit fixed points. Studia Logica **40**, 155–175 (1998)

Chapter 12
Common Knowledge and the Knowledge Account of Assertion

Syraya Chin-Mu Yang

Abstract In this chapter, I present the assertion account of common knowledge in the framework of a multi-agent system for the epistemic logic of knowledge and assertion: the propositional content of a formula φ is common knowledge to a group of agents G iff everyone in G knows that φ is true and that φ is asserted. Three current accounts of common knowledge, including the iterated account, the fixed-point account, and shared environment approach, will be examined. I argue that common knowledge arises from communication which results from overtly observable interactions among agents in a group. I then propose that assertion plays a substantial role in communication, and a fortiori, in the acquisition of common knowledge, given the knowledge account of assertion—one must assert φ only if one knows φ. I point out some semantic implications of the knowledge account of assertion in multi-agent systems, specifically, the transmission of individual knowledge to others, the transition of individual knowledge to common knowledge, and the luminosity of common knowledge. The assertion account of common knowledge is then proposed and justified by a class of Kripke models (referred to as TWC-models) appropriate for a multi-agent system of epistemic logic of common knowledge and assertion. The construction of TWC-models will be specified, and the related semantic rules will be given.

12.1 Introduction

The notion of common knowledge was first introduced into contemporary philosophy by Lewis [18] in his seminal study of convention. For Lewis, common knowledge should be presupposed as a prerequisite for a convention: in order for something to be a convention in a community, it must be common knowledge to the whole community. Aumann [1] further illustrated that common knowledge plays a significant role not only in game theory and economics of information but also in a variety of related fields whenever the process of exchanging information among a group of agents

S.C.-M. Yang (✉)
National Taiwan University, 1, Section 4, Roosevelt Road, Taipei 10617, Taiwan
e-mail: cmyang@ntu.edu.tw

S.C.-M. Yang et al. (eds.), *Structural Analysis of Non-Classical Logics*,
Logic in Asia: Studia Logica Library, DOI 10.1007/978-3-662-48357-2_12

such as Baysian statistical inference is involved. In the last few decades, several axiomatizations of epistemic logic based on certain characterization of common knowledge have been proposed, and the resulting systems have had a wide range of application in fields such as game theory, computer science, AI, and the theory of action, to mention a few (for the details, see Fagin et al. [9]; van Ditmarsch et al. [26]).

Roughly speaking, epistemic logic intends to theorize reasoning about epistemic states of agents, typically knowledge and beliefs. A system of epistemic logic at the propositional level can be constructed out of the classical propositional logic simply by (i) adding to the language in use some modal operators for *ascribing certain epistemic states*, such as knowledge, belief, or information, or whatever it could be, to agents, and then (ii) putting forth some suitable axioms to specify relations among these epistemic states. Applications of possible world semantics serve well as structural models for epistemic logic. Along this approach, a multi-agent system for the epistemic logic of common knowledge can be easily constructed. Since the ascription of knowledge to agents is purely externalistic, an axiomatization thus constructed and the notion of common knowledge thus characterized may shed a new light on the externalistic perspective of human knowledge.

In this paper, I shall only deal with multi-agent systems of epistemic logic at the propositional level. A fixed group G of agents with finitely many members, say n, and a language $\mathcal{L}_G$ defined by its BNF—$\varphi ::= p|\neg\varphi\,|\,\varphi\vee\psi\,|\mathcal{K}_i\,\varphi\,|\mathcal{E}_G\,\varphi\,|\mathcal{C}_G\,\varphi$ is assumed. Here each $\mathcal{K}_i\,\varphi(i = 1,\ldots, n)$ stands for 'The individual agent i knows φ', the modal operator '$\mathcal{E}_G$' for 'universal knowledge' so that '$\mathcal{E}_G\,\varphi$' means that 'Everyone in G knows φ', and '$\mathcal{C}_G$' for 'common knowledge' such that '$\mathcal{C}_G\,\varphi$' means that 'φ is common knowledge to all agents in G'. Hereafter, the indexical subscript '$_G$' in '$\mathcal{E}_G\,\varphi$' and '$\mathcal{C}_G\,\varphi$' will be omitted wherever there is no danger of confusion. Also, by 'a formula φ', I mean the propositional content of φ under the intended interpretation.

In the orthodox semantics for epistemic logic of knowledge with common knowledge, it is widely accepted to take the equivalence $\mathcal{E}\varphi =_{def} \mathcal{K}_1\,\varphi\wedge\mathcal{K}_2\,\varphi$' $\wedge\ldots\wedge\,\mathcal{K}_n\,\varphi$', or simply $\mathcal{E}\varphi =_{def} \bigwedge_{i\in G}\mathcal{K}_i\,\varphi$, as a definition of universal knowledge, and the notion of common knowledge can be characterized in terms of universal knowledge thus defined. At present, several accounts of common knowledge have been proposed. However, there are some intrinsic problems with the orthodox semantics. Some more appealing alternatives are called for.

In this paper, I propose a characterization of common knowledge in terms of the knowledge account of assertion in the framework of epistemic logic of knowledge and assertion: φ is common knowledge to a group of agents G iff everyone in G knows that φ is true and that φ is asserted, in symbols:

(CKA) $\quad \mathcal{C}\varphi \leftrightarrow \mathcal{E}(\varphi\wedge\mathcal{A}\,\varphi)$.

Here we need to add to the language in use an extra modal operator '$\mathcal{A}$' so that '$\mathcal{A}\,\varphi$' means that 'φ is an assertion', or 'φ is asserted by some agent i in G'.

I start with a survey of some notable characterizations of common knowledge in the framework of epistemic logic of knowledge, including the iterated account (which appeals to the equivalence '$\mathcal{C}\varphi \leftrightarrow \varphi \wedge \mathcal{E}\varphi \wedge \mathcal{E}\mathcal{E}\varphi \wedge \ldots \wedge \ldots$' as the required characterization), and the fixed-point account (which takes as an axiom schema '$\mathcal{C}\varphi \leftrightarrow \mathcal{E}(\varphi \wedge \mathcal{C}\varphi)$', known as the Fixed-Point Axiom). A brief description of the orthodox epistemic logic of knowledge and common knowledge will be given in due course.

Two main problems will be discussed. I first argue that the set of *accessibility relations* posited in the models involved for each agent is problematic; moreover, the posited group accessibility relation is ad hoc. Next, I point out that the current analysis by and large appeals to the iteration of universal knowledge, typically $\mathcal{E}\mathcal{E}\varphi$, the intended interpretation of which must be analyzed in terms of the intended interpretation of formulas of three prototypes—$\mathcal{K}_i\,\varphi$, $\mathcal{K}_i\mathcal{K}_i\,\varphi$, and $\mathcal{K}_i\mathcal{K}_j\,\varphi(i \neq j)$. Sticking to the orthodox semantics, for an agent i, $\mathcal{K}_i\,\varphi$ holds at a given state s simply because φ is true in all accessible states (with regard to a specified accessibility relation R_i for the agent i); $\mathcal{K}_i\mathcal{K}_i\,\varphi$ holds simply because $\mathcal{K}_i\,\varphi$ holds in all accessible states; and the same goes for $\mathcal{K}_i\mathcal{K}_j\,\varphi(i \neq j)$. It is striking that $\mathcal{K}_i\,\varphi$, $\mathcal{K}_i\mathcal{K}_i\,\varphi$, and $\mathcal{K}_i\mathcal{K}_j\,\varphi(i \neq j)$, under the intended interpretation represent three varieties of knowledge, as Davidson [5–7] rightly points out: $\mathcal{K}_i\,\varphi$ for 'factual knowledge', $\mathcal{K}_i\mathcal{K}_i\,\varphi$ for 'self-knowledge', and $\mathcal{K}_i\mathcal{K}_j\,\varphi(i \neq j)$ for 'knowledge of other minds'. The Davidsonian would insist that any semantics upon which a satisfactory characterization of common knowledge is proposed must be able to explain the differences in the acquisition of these three varieties of knowledge. But on the orthodox semantics, there is no difference among the way how we acquire factual knowledge, self-knowledge, and knowledge of other minds.

It can be shown that the two aforementioned problems have their roots in the acquisition of knowledge by virtue of ascribing something to agents. Accordingly, it shows no difference between ascribing self-knowledge and ascribing knowledge of other minds. But for human agents, the three varieties of knowledge should be acquired *via* different ways. The appeal to a uniform ascription becomes problematic. Things only get worse when we are concerned with multi-agent systems. As is well known, in a multi-agent system constructed by virtue of the ascription of knowledge to agents, a very substantial aspect of common knowledge has been entirely ignored, that is, communication and/or interaction among agents. It seems beyond reasonable doubt that common knowledge results from communication, and that the most common and effective way of communication is *via* some overtly observable interactions among agents in the group. Some noticeable characteristics of common knowledge based on such a communication-oriented approach, e.g., luminosity, cumulativeness, and transmission, will be noted.

It is somewhat interesting to notice that although the iterated account and the fixed-point account failed, they do suggest a promising approach by indicating some sort of modality, say $\mathcal{X}$, weaker than $\mathcal{C}\varphi$ but stronger than $\mathcal{E}\ldots\mathcal{E}\varphi$, for any finite number of iterations of $\mathcal{E}$, namely $\mathcal{C}\varphi \rightarrow \mathcal{X}\varphi$ and $\mathcal{X}\varphi \rightarrow \mathcal{E}^n\,\varphi$ (for any n). In searching for such a desired modality, I further examine the shared environment approach and argue that common knowledge can be attained only *via* a certain type of communication-oriented speech act. Following this line of thought and a lesson learnt from the shared environment approach, it can be suggested that $\mathcal{X}\varphi$ should signify some epistemic modality which is embedded in some sort of outwardly observable, or perceptible, speech act of human agents in a certain shared situation so that the required luminosity, cumulativeness, and the transmission of knowledge among a group of agents can be guaranteed. I then suggest that assertion plays a substantial role in communication, and a fortiori, in the acquisition of common knowledge. In particular, if we stick to the knowledge account of assertion—one must assert φ only if one knows φ, the epistemic modality embedded in assertion should be the best candidate for $\mathcal{X}\varphi$. This consideration will lead to a desired characterization of common knowledge: φ is common knowledge to a group of agents G iff everyone in G knows that φ is true and that φ is an assertion, as (CKA) so formulated.

Finally, I show that (CKA) can be justified in a class of models, referred to as TWC-models, for the logic of common knowledge with the knowledge account of assertion. The construction of a TWC-model will be described. It can be shown that in TWC-models, only a single accessibility relation is posited; neither a set of accessibility relations for every agent, nor the alleged group accessibility relation is required. And the aforementioned three varieties of knowledge involved in the analysis of universal knowledge can be illuminated by virtue of some basic presuppositions of the knowledge account of assertion, so that the difference in the ways of the acquisition of these three forms of knowledge can be explained.

12.2 Common Knowledge in the Framework of Orthodox Epistemic Logic of Knowledge

We have noted that in multi-agent systems it is straightforward to define universal knowledge by the equivalence $\mathcal{E}\varphi =_{def} \mathcal{K}_1 \varphi \wedge \mathcal{K}_2 \varphi' \wedge \ldots \wedge \mathcal{K}_n \varphi'$, and then to characterize common knowledge in terms of universal knowledge. Intutively, the notion of common knowledge and that of universal knowledge have a very close kinship in that φ is common knowledge to a group G of agents only if φ is shared by all agents in G, that is, everyone in G knows φ. However, it was soon realized that if common knowledge is to serve as a prerequisite for some desired actions based on a series of interaction of agents in a given group, such as in the cases like the well-known Muddy Children Puzzle and Coordinated Attack, the acquisition of universal knowledge may not be sufficient to guarantee the success of the desired actions. In some cases, it is required that not only everyone knows φ but also everyone knows that everyone knows φ. Still, in some other cases, the fact that everyone knows that everyone knows that everyone knows φ is not good enough. Some theorists have shown that in some special cases, when limited to a finite number of iterations of universal knowledge, the desired actions can never be guaranteed. It is then tempting to put forth a more general formulation of common knowledge in terms of an infinitary conjunction of iterated universal knowledge. That is to say, the notion of common knowledge can be conceptually analyzed in terms of the conjunction that everyone in G knows φ, and everyone knows that everyone knows φ, and everyone knows that everyone knows that everyone knows φ, and so on ad infinitum. In symbols,

$$(\mathrm{C}_{\mathrm{iter}}) \quad \mathcal{C}\varphi =_{def} \varphi \wedge \mathcal{E}\varphi \wedge \mathcal{E}\mathcal{E}\varphi \wedge \ldots \textit{ad infinitum}.$$

or, more simply $\mathcal{C}\varphi \leftrightarrow \bigwedge_{k\in\mathbb{N}} \mathcal{E}^k \varphi$.[1]

Historically, this approach, known as the iterated account, can be traced back to Aumann [1] where he took ($\mathrm{C}_{\mathrm{iter}}$) as an informal formulation of common knowledge and showed that this formulation is equivalent to a formal definition of common knowledge based on the framework of Baysian-theoretic approach to probability.

It was soon realized that there is an intrinsic difficulty with the iterated account in the framework of multi-agent systems for human agents, due to the *finiteness* constraint that the standard propositional/first-order logic imposes on the length of formulas of the language in use—a well-formed formula should be *finitary*. Of course, from a logical point of view, a formal language containing formulas of infinite length can be allowed, if a certain nonclassical logic is adopted.[2] But, the intended interpretation of a formula of infinite length would be beyond the cognitive capability of human agents. The meaning of ($\mathrm{C}_{\mathrm{iter}}$) as a whole thus becomes problematic, let alone taken as an explicit definition of some other concept. Moreover, one can find no formula of the language in use being logically equivalent to ($\mathrm{C}_{\mathrm{iter}}$). Consequently, there is no room for a legitimate axiomatization of common knowledge to human agents based on this account. An alternative account is called for.

A more appealing approach is to take the modal operator $\mathcal{C}$ as primitive and put forth some appropriate axioms. Interestingly, one may find that the formulation ($\mathrm{C}_{\mathrm{iter}}$) paves the way for an appealing axiom. Although in different cases different numbers of iterated universal knowledge may be required, and although in some cases even any feasible finite number of iterations of universal knowledge is not sufficient to guarantee the success of a desired action, there is no need to appeal to the conjunction of infinitely many conjuncts of iterated universal knowledge. As a matter of fact, it is striking that if the modal operator $\mathcal{E}$ can be treated as some sort of increasing function, then every agent in G will get more and more information by virtue of a recursive application of $\mathcal{E}$. Eventually, to a certain extent or at a certain point, the accumulative information will be sufficient enough for all agents to be aware of the fact that not only everyone knows φ, but also φ itself is a common knowledge. Accordingly, we may have the following (definition-like) equation:

$$(\mathrm{FP}) \quad \mathcal{C}\varphi \leftrightarrow \mathcal{E}(\varphi \wedge \mathcal{C}\varphi).$$

[1] Sometimes, the notation '$\mathcal{E}^{n+1}\varphi$' can be introduced as an abbreviation of '$\mathcal{E}\mathcal{E}^n\varphi$'; by convention, '$\mathcal{E}^0\varphi$' is just '$\varphi$'.

[2] Several logic systems of knowledge and common knowledge based on this equivalence have been proposed, e.g., Halpern and Moses [13], Mertens and Zamir [21], Fagin et al. [9]. In particular, Baltag et al. [2] construct an epistemic logic containing infinitary operators used in the standard modeling of common knowledge. It is worth mentioning that Lismont and Mongin ([20]: 129, footnote 1) briefly note that some logicians prefer to take certain infinitary logic as the required underlying system for a desired logic of common knowledge, such as Kaneko and Nagashima's works in 1991 and 1993, and a paper of Heifetz in 1994.

This is in general referred to as the fixed-point axiom, which states that φ is common knowledge if and only if everyone knows both that φ holds and that it is common knowledge as well. Note that the definiens part in (FP), namely '$\mathcal{E}(\varphi \wedge \mathcal{C}\varphi)$' indirectly captures the basic iterative intuition of common knowledge, as the occurrence of $\mathcal{C}\varphi$ in the definiens displays the desired cumulative sequence of inferences of the form $\mathcal{C}\varphi \rightarrow \mathcal{E}^k \varphi$, for all $k > 1$.

A closer examination shows that (FP) is merely an application of Tarski's [25] well-known fixed-point theorem, which states that an increasing function f on the domain of a complete lattice $\langle A, \leq \rangle$, say $f \colon A \rightarrow A$, will have at least one fixed point, namely an element x in A such that $f(x) = x$. Here, we may take $f(x) = \mathcal{E}(\varphi \wedge x)$ as an increasing function operating on the set of formulas of the language in use. (FP) can then be construed as saying that the iteration of the modality $\mathcal{E}$ will eventually lead to a fixed point, i.e., $\mathcal{C}\varphi$. The legitimacy of (FP) can be thus justified.

Some might argue that (FP), thus formulated, runs into circularity as $\mathcal{C}\varphi$ is contained as a component of the proposed definiens. We have been taught that a circular definition is problematic and unacceptable. Interestingly, in the last few decades, there has been a growing inclination to accept circular definitions for some fundamental concepts, if only they are well behaved and informative. Noticeably, Gupta and Belnap [12], in defence of a revision theory of truth, argue that circular definitions can be meaningful and useful as well. They put forth a general theory of circular definition which is both philosophically illuminating and logically elegant.

The involvement of circularity in the formulation of (FP) may not be a threat to the legitimacy of (FP). From a philosophical point of view, (FP) substantially indicates the complete *transparency* (or luminosity, in Williamson's [28] term), an ultimately intrinsic property, of common knowledge (to all agents) in that for a formula φ to be qualified as common knowledge everyone must know that it is common knowledge, in symbols $\mathcal{C}\varphi \rightarrow \mathcal{E}\mathcal{C}\varphi$. Intuitively, this also suggests a significant role that common knowledge plays in the transmission of knowledge: it is impossible for an agent i to know that φ is common knowledge without accepting that any other agent knows that it is common knowledge as well. The transmission of (individual) knowledge (of an agent) to some others can be then guaranteed by the transition of (individual) knowledge (of some agents) to common knowledge.

So far, several axiomatizations of epistemic logic of knoweldge and common knowledge based on the fixed-point account can be found in Halpern and Moses [13], Lismont and Mongin [20], Milgrom [22], Monderer and Samet [23], and some others. In particular, Halpern and Moses ([13]: 571–572) present a logic of knowledge with common knowledge by adding a greatest fixed-point operator and illustrating how common knowledge and its variants can be formally defined as greatest fixed points (for the details, see Halpern and Moses [13]: Appendix A, pp. 580–583).

In spite of the seemingly acceptable justification of the legitimacy of (FP) from mathematical and philosophical viewpoints, some misgivings remain, insofar as a multi-agent system of epistemic logic for human agents in ordinary discourse, rather than agents of some other sort, is concerned. To this we turn our attention next.

12.3 Two Main Problems with the Orthodox Semantic

Let us start with a brief description of the orthodox semantics for epistemic logic of knowledge and common knowledge.

First, I take as the starting point, the basic language $\mathcal{L}_G$ as defined above, i.e., $\varphi ::= p|\neg\varphi\,|\,\varphi\vee\psi\,|\,\mathcal{K}_i\,\varphi\,|\,\mathcal{E}\varphi\,|\,\mathcal{C}\varphi$, and a required frame $\mathfrak{F}$ of the form $\langle S, \{R_i\}_{i\in n}\rangle$, where S, a set of (epistemic) states, and each R_i, a binary (accessibility) relation on S, i.e., $R_i \subseteq S \times S$. A Kripke model $\mathcal{M}$ *on the frame* $\mathcal{F}$ is a triple $\langle S, \{R_i\}_{i\in n}, V^P\rangle$ where P is any choice of a countable set of proposition letters, and $V^P : P \rightarrow 2^S$ is a valuation function, assigning to each $p \in P$ a set $V^P(p) \subseteq S$ of states in which p is true. The semantic rules for propositional connectives are standard and the semantic rule for the knowledge operators $\mathcal{K}_i$'s is given by the clause that $\mathcal{K}_i\,\varphi$ is true at a state s iff φ is true at *all* states t such that R_ist holds, in symbols

$(\mathcal{K}^S)$ $\quad \mathcal{M}, s \models \mathcal{K}_i\,\varphi$ iff $\forall t \in S, R_ist \rightarrow \mathcal{M}, t \models \varphi$.

For simplicity, let us assume that the frame in use is based on ***S5***-models, wherein all the accessibility relations R_i are equivalence relations.[3] That is, we would have a class of Kripke models of the form $\mathcal{M} = \langle S, \{\sim_i\}_{i\in G}, V\rangle$, where associated with each $i \in G$, there is an equivalence relation $\sim_i$ on S. The semantic rule for the universal knowledge operator $\mathcal{E}$ is straightforward:

$(\mathcal{E}^S)$ $\quad \mathcal{M}, s \models \mathcal{E}\varphi$ iff $\forall i \in G, \mathcal{M}, s \models \mathcal{K}_i\,\varphi$

The semantics for the common knowledge operator, then, is given by taking the reflexive and transitive closure R_G of the union of R_i ranging over agents i in G, and stipulating that

$(\mathcal{C}^S)$ $\quad \mathcal{M}, s \models \mathcal{C}\varphi$ iff $\forall t \in S, R_Gst \rightarrow \mathcal{M}, t \models \varphi$.

where $R_G := (\bigcup_{i\in G} \sim_i)^*$ which is the reflexive transitive closure of $\bigcup_{i\in G} \sim_i$.[4]

At the moment, the semantics thus constructed is widely accepted for multi-agent systems of epistemic logic of common knowledge in general, and both the iterated account and fixed-point account of common knowledge work well on this framework. However, insofar as a multi-human-agent system is concerned, there are some misgivings over the orthodox semantics. Here we will focus on two main problems. The first has something to do with the legitimacy of the posited accessibility relations, while the second comes from a Davidsonian challenge.

---

[3]It is noteworthy that the characterization of common knowledge based on ***S5***-models would validate negative introspection—$\neg\mathcal{C}\varphi \rightarrow \mathcal{C}\neg\mathcal{C}\varphi$. However, as far as multi-human-agent systems of epistemic logic is concerned, it seems rather problematic to claim that to a group of agents G, that φ is not common knowledge is common knowledge, provided that φ is not common knowledge.

[4]For the details of the construction of a logic system of knowledge $(\boldsymbol{S5})_{\mathcal{C}}$, by taking '$\mathcal{C}$' as primitive, and (FP) as an axiom schema, see Fagin et al. [9]; van Ditmarsch et al. [26].

12.3.1 The Legitimacy of the Posit Accessibility Relations in Kripke's Models

Inheriting from the standard semantics for a mono-agent system of epistemic logic, a set of binary relations $\{R_i\}_{i \in G}$ is posited in the required frame so that, associated with each agent i in G, there is an accessibility relation R_i in a given model to identify the so-called epistemic possibilities of the agent. Recall that Hintikka [17] posited an epistemic notion of accessibility relation in a Kripke model to specify a designated class of epistemic possibilities (for the agent) out of the universe of possible states. Intuitively, any ascription of a certain epistemic attitude to agents in a model, typically knowing, requires a partition of the whole collection, the universe, of epistemic possibilities (or scenarios, in Hintikka's term) into two parts: those which are compatible with the given epistemic possibility under investigation and those which are not. It is in this sense that an epistemic logic of knowledge offers us a way of systematically specifying the set of epistemic states compatible with what an agent knows. The very epistemic concept can be then characterized by the algebraic properties of the posited accessibility relation. However, it is questionable exactly what it is that counts as a legitimate partitioning of states. Apparently, Hintikka appealed to agents' logically possible experience. As Hendricks and Symons ([16]: 143) construe, 'the logical possible experiences' mean experiences 'pertaining to possibilities of error that any account of knowledge must exclude'. Accordingly, the primary concern of the posited accessibility relation is, in Hendricks and Symons' [16] words, 'to limit the set of citable possible worlds carrying potential error.'

But, as Hendricks and Symons ([16]: 142) rightly remark, 'if my only criterion for partitioning is logical consistency, then I will find scenarios that are compatible with my model that undermine the very possibility of knowledge ...How can I be sure that my inclusion or exclusion of scenarios is legitimate?' We would be in no position to offer an objective response to this question. If so, the objectivity of the agent's knowledge characterized in terms of the posited accessibility relations would become problematic. In particular, it is natural to assume that for distinct agents, say i and j, the associated accessibility relations R_i and R_j should be different. Accordingly, the states involved in the truth conditions of $\mathcal{K}_i\ \varphi$ and $\mathcal{K}_j\ \varphi$ in the same epistemic state may be different. The truth of $\mathcal{K}_i\ \varphi$ will be determined by the set of states that are possible from the agent i's epistemic viewpoint, while the truth of $\mathcal{K}_j\ \varphi$, j's viewpoint.

Things will only get worse, if we consider the legitimacy of the alleged group accessibility relation R_G. In particular, if we stick to the epistemic notion of accessibility relation, it is difficult to interpret exactly what R_G is supposed to mean. Recall that $R_G$is supposed to identify a set of states so that what counts as common knowledge in a given state s can be determined by what holds in every state of this set . Nonetheless, in speaking of positing an accessibility relation to specify the set of states that *every* agent can access simultaneously, we should bear in mind that 'every' is a quantifier ranging over the set of all agents, rather than a singular term used to designate a possibly unspecified individual agent. The very group of agents here can hardly be treated as an individual agent whatsoever. Just like it would be a bit awkward to claim what happens to a so-called average man, it would be a bit awkward to say that such and such a set of epistemic states constitutes as a partition of all epistemic possibilities for the very group of agents. The best we can say about the alleged group accessibility relation for the very group of agents is that it is so posited in order to classify the set of states t which are compatible with *what is common knowledge* among G in s. But, construed in this way, the posited group accessibility relation is not only ad hoc, but also circular.

It is noteworthy that in the required Kripke models for the orthodox epistemic logic of knowledge and belief, two accessibility relations are posited with different constraints—one for the knowledge operator $\mathcal{K}$ and the other for the belief operator $\mathcal{B}$. Elsewhere [29], I have argued that this is misleading, and suggested that if we accept Williamson's knowledge-first epistemology wherein belief can be characterized in terms of knowledge, no accessibility relation for the belief operator is required. There, a class of models, referred to as TW-models, are constructed, and the sole accessibility relation is posited to specify the so-called nearby cases—the cases similar to the one where the agent is actually in. It is then appealing to construct a class of models for a multi-agent system with only a sole accessibility relation. As Hendricks and Symons ([16]: 153) insightfully point out,

> Epistemic-logical principles or axioms building up modal systems are relative to an agent who may or may not validate these principles. Indices on accessibility relations will not suffice for epistemological and cognitive pertinence simply because there is nothing particularly epistemic about being indices. The agents are inactive, hence indifference.

If we can have a class of models with a sole accessibility relation, the aforementioned problems can then be dissolved. We need not posit a set of distinct accessibility relations for each agent. Nor would we need posit the alleged R_G.

Following this line of thought, semantically we should be able to characterize both universal knowledge and common knowledge in a framework with only a single accessibility relation. If this can be done, we may have a more promising analysis of common knowledge by virtue of some weaker epistemic modality so that we can get rid of the uneasy dilemma between the commitment to circularity involved in (FP) and the acceptance of a formulation with infinite length suggested by ($\mathrm{C_{iter}}$). Interestingly, (FP) and ($\mathrm{C_{iter}}$) together suggest an appealing middle course. On the one hand, to avoid the involvement of circularity embedded in (FP), all that we need for a satisfactory characterization of common knowledge $\mathcal{C}\varphi$ is to find some kind of modality, say $\mathcal{X}\varphi$ such that $\mathcal{X}\varphi$ is weaker than $\mathcal{C}\varphi$ itself, that is, $\mathcal{C}\varphi \rightarrow \mathcal{X}\varphi$. On the other hand, to be free from any formulation of infinitary length, the desired modality $\mathcal{X}\varphi$ must be stronger than the conjunction of finite iteration of universal knowledge. That is, $\mathcal{X}\varphi \rightarrow \mathcal{E}^n\varphi$ holds for any arbitrary finite $n \in \mathbb{N}$. In short, we are searching for some modality $\mathcal{X}\varphi$ such that $\mathcal{C}\varphi \rightarrow \mathcal{X}\varphi$ and $\mathcal{X}\varphi \rightarrow \mathcal{E}^n\varphi$ hold for any arbitrary n. If such a modality $\mathcal{X}\varphi$ can be constructed in the desired framework, we would be able to show that both $\mathcal{C}\varphi \rightarrow \mathcal{X}\varphi$ and $\mathcal{X}\varphi \rightarrow \varphi \wedge \mathcal{E}\varphi \wedge \ldots \wedge \mathcal{E}^n\varphi \wedge \ldots$ hold. Now, if we accept ($\mathrm{C_{iter}}$) as the pre-theoretic account of common knowledge in that $\mathcal{C}\varphi \leftrightarrow (\mathrm{C_{iter}})$ holds, we would then have $\mathcal{C}\varphi \leftrightarrow \mathcal{X}\varphi \leftrightarrow (\varphi \wedge \mathcal{E}\varphi \wedge \ldots \wedge \mathcal{E}^n\varphi \wedge \ldots)$. This equivalence would show that the proposed modality $\mathcal{X}\varphi$ would (i) indirectly capture the basic idea of iterated approach within the desired system, and also (ii) captures the idea of the transparency of common knowledge without being committed to circularity. We can then have a characterization of common knowledge, namely $C\varphi \leftrightarrow X\varphi$.

Still, there is a second problem with the orthodox semantics, to which I now turn.

12.3.2 A Davidsonian Challenge

The current analysis of common knowledge appeals to universal knowledge, especially a sequence of iterated universal knowledge, the typical example being a formula of the form $\mathcal{E}\mathcal{E}\varphi$—'Everyone knows that everyone knows φ.' Clearly the truth condition of $\mathcal{E}\mathcal{E}\varphi$ in a state is based on the intended interpretation of three more basic formulas, viz. $\mathcal{K}_i\,\varphi$, $\mathcal{K}_i\mathcal{K}_i\,\varphi$ and $\mathcal{K}_i\mathcal{K}_j\,\varphi$ (for any i and j in G and $i \neq j$). Naturally, one will find that the orthodox semantics treats the truth conditions of these three formulas indifferently. More specifically, in the orthodox semantics, for an agent i, $\mathcal{K}_i\,\varphi$ holds simply because φ is true in all accessible states (with regard to R_i); the same semantic rule goes to $\mathcal{K}_i\mathcal{K}_i\,\varphi$ and $\mathcal{K}_i\mathcal{K}_j\,\varphi - \mathcal{K}_i\mathcal{K}_i\,\varphi$ holds in a state simply because $\mathcal{K}_i\,\varphi$ holds in all accessible states (with regard to R_i) and $\mathcal{K}_i\mathcal{K}_j\,\varphi$ holds in a state simply because $\mathcal{K}_j\,\varphi$ holds in all accessible states (with regard to R_i). It looks as if there is no difference in the ways how the agent i knows φ, $\mathcal{K}_i\,\varphi$, and $\mathcal{K}_j\,\varphi$, respectively. After all, knowledge acquisition in the orthodox framework of epistemic logic is merely a matter of *ascribing* knowledge to agents (by system designers or programmers).

However, from an epistemological point of view, for any human agents i and j, under the intended interpretation, $\mathcal{K}_i\,\varphi$ stands for *factual knowledge* (i.e., knowledge of the external world), $\mathcal{K}_i\mathcal{K}_i\,\varphi$ for *self-knowledge*, and $\mathcal{K}_i\mathcal{K}_j\,\varphi$, for *knowledge of other minds*. Davidson in a series of papers in the 1980 s [8] argued that they are three varieties of knowledge of human agents. Davidson insists that 'each of the three varieties of knowledge is indispensable' and that they are 'mutually irreducible.' Now, if we accept the indispensability and irreducibility of these three forms of knowledge, the standard analysis of common knowledge in the framework of orthodox epistemic logic would be unacceptable. There are significantly intrinsic differences in the ways that human agents acquire knowledge of these three distinct types. Epistemologically, the three forms of knowledge substantially represent distinct intrinsic properties and nature of human knowledge in different aspects. The Davidsonian would insist that any semantics upon which a satisfactory characterization of common knowledge is proposed must be able to explain the differences in the acquisition of these three varieties of knowledge.

It is not my intention here to discuss the pros and cons of the aforementioned Davidsonian challenge. Rather, I want to focus on the pursuit of the aforementioned modality $X\varphi$ and to see if such a desired modality can be characterized in a framework of epistemic logic wherein the proposed semantics can explain the difference in the ways of the acquisition of the three varieties of knowledge.

Although Davidson emphasizes the irreducibility and indispensability of three forms of knowledge, he maintains that they must be mutually dependent. Davidson argues that 'knowledge of other minds is possible only if one has knowledge of the world'; also, 'we are not in a position to attribute thoughts to others unless we know what we think.' He also notes that being in a position to attribute thoughts to others is prerequisite to having knowledge of other minds. This indicates the dependency of knowledge of other minds on self-knowledge. In view of the specified mutual dependency, there must be something in common to the acquisition of these three sorts of knowledge if common knowledge is to be characterized in terms of these three types of knowledge. It seems to me that this should play a key role in any satisfactory account of common knowledge. Interestingly, Davidson has already pointed out a substantial concept which plays a key role in multi-human-agent systems but the orthodox semantic treatment has completely ignored, namely communication. According to Davidson, a given agent possesses knowledge of other minds only if intersubject communication is possible: 'there is no propositional thought without communication.' Communication is also crucial to self-knowledge. Although Davidson accepts the first-person authority, he insists that even when we know φ, we may not be in a position to know that we know φ, unless we can communicate with others so that they can know what we know. Moreover, communication mainly hinges upon overt behaviors of agents. In particular, knowledge of other minds can be acquired *via* observations of one's behaviors, specifically, one's speech acts. This line of thought will pave the way for a communication-oriented approach to common knowledge.

12.4 Toward a Communication-Oriented Approach to Common Knowledge

We have remarked that a Davidsonian challenge shows that common knowledge arises from communication which lies in the interactions among agents in a fixed group. We have also noted that the current accounts of common knowledge do not address the interaction among agents. Barwise [3] rightly points out that although it is widely accepted (e.g., Aumann [1]; Halpern and Moses [13]) that the fixed-point account is equivalent to the iterated approach, to prove the equivalence of these two approaches, some assumptions are required. He then argues that these assumptions are simply false because the transparency of common knowledge cannot be illuminated explicitly. To overcome this difficulty, Barwise adopts the so-called shared-environment approach due originally to Clark and Marshall [4]. As Barwise ([3]: 379) notes:

> [C]ommon knowledge per se, the notion captured by the fixed-point analysis, is not actually all that useful. It is a necessary but not a sufficient condition for action. What suffices in order for common knowledge to be useful is that it arises in some fairly straightforward shared situation. The reason this is useful is that such shared situations provide a basis for perceivable situated action; action that then produces further shared situations. That is, what makes a shared environment work is not just that it gives rise to common knowledge, but also that it provides a stage for maintaining common knowledge through the maintenance of a shared environment.

Roughly speaking, on this account, two agents i and j have common knowledge of φ just in case there is a situation s such that

$$s \models \varphi$$
$$s \models \mathcal{K}_i\, \varphi$$
$$s \models \mathcal{K}_j\, \varphi$$

Here, '$s \models \alpha$' means that α is a fact obtaining in the situation s. The underlying thought is to identify common knowledge with perception, or awareness, of a certain situation, 'part of which includes the fact in question, but another part of which includes the very awareness of the situation by all agents' (Barwise [3]: 368).

One can see a great merit of this approach, that is, the shared environment should be able to guarantee the transition of knowledge from individual knowledge to common knowledge. Barwise ([3]: 369) argues that, although the fixed-point approach gives the best conceptual analysis of the pre-theoretic notion of common knowledge, the shared environment plays a role in our understanding of common knowledge. In particular, it sheds a new light on our understanding of how common knowledge usually arises and is maintained over an extended interaction.

Surely, in some cases we may have common knowledge based upon a certain shared environment/situation. But if the acquisition of common knowledge has to, and can only, appeal to a shared environment/situation, it would be extremely difficult in practice to acquire a large amount of common knowledge. After all, it may happen that in some situation it would be rather difficult for all agents to be simultaneously aware of what happens in the given shared environment. Be that as it may, this approach offers no explanation for the transmission of knowledge. Barwise simply assumes that φ is common knowledge to a fixed group of agents when everyone observes in a shared state s that φ is true in s and that everyone knows φ in s. This may sufficiently explain the transition of individual knowledge to common knowledge but no explanation of how the agent i knows φ, given that the agent j knows φ. It would be too far-fetched to claim that, for a formula φ to be common knowledge to a group, everyone knows φ automatically. In ordinary discourse, it happens more often that some form of transmission of knowledge from a few agents to others is required.

Clearly the ignorance of transmission of knowledge in the shared environment approach is due to the lack of communication. Ever since early 1990s, a large number of theorists of epistemic logic have echoed Davidson's appeal to communication, maintaining that communication plays a substantial role in the acquisition of common knowledge. For example, Halpern and Moses ([13]: 551) note that 'when communication is not guaranteed, it is impossible to attain common knowledge.' A similar view can be found in a series of works of Fagin et al. [9, 10]. They further argue that 'even when communication is guaranteed, common knowledge may still not be attained when there is no bound on the time it takes for message to be delivered.' ([10]: 90) The main reason is that at this point, the transmission of knowledge among individual agents and the transition of individual knowledge to common knowledge should be guaranteed by some simultaneous changes of agents' epistemic states. As Fagin et al. ([10]: 91, 98) rightly remark, when a not commonly known statement is transited to a piece of common knowledge, a simultaneous change in all relevant agents' knowledge (states) must involve. In other words, in the absence of certain events that are guaranteed to hold simultaneously, common knowledge is not attained.

Following this line of thought, an important question arises: How is it possible for an agent in a fixed group to make sure that her individual knowledge can be transmitted to others simultaneously via communication, and a fortiori, be transited to common knowledge simultaneously via communication? The most promising approach, as I see it, is to appeal to some sort of observable speech acts by virtue of which the agent's knowledge can be delivered. More importantly, the proposed speech acts must signify some kind of epistemic modality which can be characterized in the framework of Kripke models for multi-agent systems of epistemic logic. Now, the problem is: What kind of speech act can do the job?

In what follows I propose that the required simultaneity in communication can be guaranteed by a kind of overtly observable speech act, known as 'assertion,' provided that the knowledge account of assertion is well grounded, or assumed.

12.5 The Appeal to the Knowledge Account of Assertion

Historically, the appeal to assertion for communication can be traced back to Frege. As is well known, Frege took it for granted that there are thoughts, which enjoy a mode of being in the so-called third realm and can be grasped by a human agent. Having grasped a thought, the agent can further make a judgement to see whether the very thought holds or not. For Frege, making a judgement is 'inwardly to recognize something as true,' which is essentially an inner metal activity. Now, if the agent intends to express a true judgment, the given judgement must be manifested outwardly by uttering a (declarative) sentence. Frege entitled this kind of speech act as assertion. Accordingly, assertion aims at the manifestation of true judgement. An assertion can be treated as an outward sign of judgement—a kind of overt speech act, observable by others. Consequently, the propositional content (the thought) of an assertion can be transmitted from the asserter to the hearer, who thereby grasps the propositional content (the thought) of the assertion. Furthermore, if we take assertion as a specific way of expressing knowledge, making an assertion would have the function of 'sharing knowledge' with other agents in a group of agents. It is in this sense that assertion plays a substantial role in a theory of communication

If assertion can be furthermore treated as a kind of (epistemic) modality to be signified by an extra modal operator, say $\mathcal{A}$, so that the truth condition of a formula $\mathcal{A}\varphi$ can be specified in the framework of the epistemic logic of knowledge and assertion, we may have a characterization of common knowledge in terms of assertion.

In the last few decades, several versions of the logic of assertion thus described have been proposed (See Rescher [24]; Gullvåg [11]). The required Kripke models can be constructed by putting forth a specified accessibility relation $R^{\mathcal{A}}$ (preferably equivalent relation so that **S5** models are accepted), and then stipulate that

$$(\mathcal{A}^S) \quad \mathcal{M}, s \models \mathcal{A}\varphi \quad \text{iff} \quad \forall t \in S,\ R^{\mathcal{A}}st \rightarrow \mathcal{M}, t \models \varphi$$

Unfortunately, there is no explicit connection between knowledge and assertion displayed in such a framework. In fact, neither Frege's original conception of assertion, nor any semantic treatment of the logic of assertion has appealed to knowledge, let alone to common knowledge. This is partly because of the lack of a satisfactory philosophical account of assertion. Some philosophers insist that whatever an agent asserts must be true—the so-called truth norm of assertion; some others maintain that an agent can only assert justified beliefs—the justified belief norm of assertion, or the norm of warranted belief. Both can easily find some substantial support in recent literatures.

In order to be treated as an (epistemic) modality, assertion must 'bear some epistemic import' in that when an assertion is made the agent holds a certain epistemic attitude to the propositional content of the given assertion. In particular, if we intend to take assertion as an ideal guarantee for the transmission of knowledge, the propositional contents of assertions must be knowledge. Recently, a third account of the norm of assertion, known as the knowledge account, has been proposed, which states that for a given proposition p, one asserts p only if one knows p, in symbols $\mathcal{A}\varphi \rightarrow \mathcal{K}\varphi$. [27, 28] Now, if we can stipulate a certain semantic treatment for $\mathcal{A}\varphi$ in the framework of a multi-agent system for a logic of knowledge such that $\mathcal{A}\varphi$ can be characterized in terms of $\mathcal{K}\varphi$, we would be able to characterize common knowledge in terms of assertion.

In a previous work [30], I have constructed a class of models, referred to as TWA-models, which is appropriate for a logic of knowledge and assertion, and satisfies the knowledge account of assertion.[5] In this paper, we shall show that a class of models, taken as extensions of TWA-models, can be constructed to serve as the required models for the logic of knowledge and common knowledge, wherein the notion

[5]In fact, Yang [29] presented a class of TW-models for an epistemic logic of knowledge and belief which satisfy the main theses of Timothy Williamson's *knowledge-first epistemology*, proposed in his *Knowledge and its Limits*, which can be summarized in what follows:

- Knowing is a state of mind
- Knowing is factive
- The broadness of knowing(Externalist approach)
- The primeness of knowing (Knowledge first!)
- Take knowledge as central to our understanding of belief.
- Cognitive-homeless thesis
- The knowledge account of assertion—Assert *p* only if one knows that *p*
- The knowledge account of evidence—One's knowledge is just one's evidence.

Note that TWA-models are essentially extensions of TW-models and can be used to justify the knowledge account of assertion. A justification of the knowledge account of evidence needs some other kind of models, which will be proposed somewhere else.

of common knowledge can be characterized in terms of the knowledge account of assertion

For the sake of self-containedness, I will give a brief description of TWA-models for a mono-agent system of the epistemic logic of knowledge and assertion without detailed explanation in what follows. Let us fix a language for an epistemic logic with modal operators '$\mathcal{K}$' (for knowledge) and '$\mathcal{A}$' (for assertion); the set $\mathcal{L}_{\mathcal{A}}$ of formulas of the language in use can be defined as $\varphi ::= p | \neg \varphi | \varphi \rightarrow \psi | \mathcal{K}\varphi | \mathcal{A}\varphi$. A TWA-model is a tuple of the form $\mathcal{M} = \langle S, R, \delta, \lambda, V^P \rangle$, where

S, a nonempty set of states;
$R \subseteq S \times S$, a partial ordering with reflexivity to serve as the required accessibility relation on S;
δ: $S \rightarrow \wp(\mathcal{L})$ such that for any $s \in S$, $\delta(s) \subseteq \{\varphi | \mathcal{M}, s \models \varphi, \ \varphi \in \mathcal{L}\}$;
λ: $S \rightarrow \wp(\mathcal{L})$ such that for any $s \in S$, $\lambda(s) \subseteq \delta(s)$;
$V^P : P \rightarrow 2^S$ is a valuation which assigns to each $p \in P$, a set $V^P(p) \subseteq S$ of states in which p is true.

Note that when a state s is in $V^P(p)$, we say that V^P assigns p a truth value 'True,' or more straightforwardly, V^P makes p true in s.

Here, the introduction of δ, referred to as the *ipk*-function, is to capture Williamson's original notion of '*being in a position to know a proposition in a state*'. For Williamson, the fact that a sentence is true in all nearby cases (i.e., all accessible states, or all possible epistemic states) would not be sufficient for an agent to know it. It may happen that some propositions appear to be true in all nearby cases but, in the very state, the agent is not in a position to know them. The agent would thereby not be able to know them. For a more convincing reason, a formula $\varphi \in \delta(s)$ will be interpreted as saying that *the agent is actually in a position to know φ in a state s* (See Yang [29]: 326–329, for a detailed explanation).

The semantic rules for atomic formulae, negation, and material implication are standard. And the semantic rule for $\mathcal{K}\varphi$ can be given:

($\mathcal{K}^S$) $\mathcal{M}, s \models \mathcal{K}\varphi$ iff $\forall t \in S(Rst \rightarrow \mathcal{M}, t \models \varphi) \wedge \varphi \in \delta(s)$.

The second condition in ($\mathcal{K}^S$), namely '$\varphi \in \delta(s)$,' indicates the requirement that to know φ, the agent must *be actually in a position to know* φ in the given state.

The function λ here is introduced in order to indicate explicitly that assertion is a kind of intentional speech act in that in making an assertion, the agent must be doing so with intention. Accordingly, a formula $\varphi \in \lambda(s)$ is to mean that *the agent has the intention of asserting φ in s*, or *the agent intends to assert φ*.[6] The condition '$\lambda(s) \subseteq \delta(s)$' shows that when the agent has the intention of asserting φ, she must be actually in a position to know what she intends to assert. After all, assertion is a kind of intentional speech act, and if we accept the knowledge account of assertion, it would be unacceptable to claim that someone would intend to assert something that she does not know. Moreover, in view of the assertoric force of the

[6] As Davidson ([8]: 90) rightly remarked, there are no such conventions governing the formation of intentions. So I can only put forth a primitive function here.

knowledge account, the agent must *know* that she *knows whatever she intends to assert*. Accordingly, we have the following semantic rule for the modal operator $\mathcal{A}$ for assertion:

$$(\mathcal{A}^S) \quad \mathcal{M}, s \models \mathcal{A}\varphi \text{ iff } \forall t \in S(Rst \rightarrow \mathcal{M}, t \models \mathcal{K}\varphi) \wedge \varphi \in \lambda(s) \wedge \mathcal{K}\varphi \in \delta(s).$$

The first condition, $\forall t \in S(Rst \rightarrow \mathcal{M}, t \models \mathcal{K}\varphi)$, simply sticks to the knowledge account of assertion: 'One asserts p only if one knows p,' which can be characterized in terms of the semantic stipulation 'only if $\mathcal{K}\varphi$ is true in all nearby cases.' The second condition, $\varphi \in \lambda(s)$, indicates that to assert φ, the agent must *have the intention of asserting* φ in the given state, apart from the given fact that the agent *knows* φ *in all nearby cases*. The third condition merely suggests that the agent must *be actually in a position to know* that she *knows whatever she intends to assert*. This will be able to validate $\mathcal{A}\varphi \rightarrow \mathcal{K}\mathcal{K}\varphi$ in TWA-models, though she may not know what she is doing, namely, $\mathcal{K}\mathcal{A}\varphi$ may not hold.[7] The semantic rule $(\mathcal{A}^S)$ for $\mathcal{A}\varphi$ is then sufficient to characterize the concept of assertion in terms of knowledge.

Now, let us take a closer examination to see how to characterize common knowledge in terms of the knowledge account of assertion in the framework of the epistemic logic of knowledge and assertion.

We have already noted that to attain common knowledge in a group of agents, communication must be guaranteed and that communication aims at sharing knowledge. One can see clearly that on the basis of the knowledge account of assertion, assertion, when made by some agent in a group, aims at sharing knowledge: the agent intends to share whatever she knows with the others by virtue of making an assertion. It follows that assertion can guarantee communication. Along this line of thought, it is appealing to claim that common knowledge arises from assertion, given that communication is essential to the acquisition of common knowledge. The notion of common knowledge can thereby be characterized in terms of the knowledge account of assertion. The remainder of this paper is then devoted to the formulation and justification of the desired characterization in the framework of a multi-agent system of the epistemic logic of knowledge and assertion.

However, before we go into the details, it is noteworthy to specify some intrinsically epistemic features of assertion, taken as a kind of speech act performed by some agent in a community—presumably, a multi-human-agent system in character. Intuitively we simply take these features as fundamental assumptions and treat them as the guidelines for the construction of the desired models

For the sake of convenience, let us assume that a fixed finite set of agents G is given and that a language in use $\mathcal{L}_{AG}$ is defined as $\varphi ::= p | \neg\varphi | \varphi \rightarrow \psi | \mathcal{K}_i\varphi | \mathcal{A}_i\varphi$ (for all $i \in G$). From an epistemic point of view, these intrinsic features of assertion can be formulated as the following assumptions.

Assumption 1 (KAA) $\mathcal{A}_i\varphi \rightarrow \mathcal{K}_i\varphi$ *(The knowledge account of assertion)*

---

[7] Davidson ([8]: 91) notes that 'It is a mistake to suppose that if an agent is doing something intentionally, he must know that he is doing it.' This indicates that $\mathcal{A}\varphi \rightarrow \mathcal{K}\mathcal{A}\varphi$ may not hold. But it seems beyond reasonable doubt to claim that the agent must know that she knows what she asserts, otherwise, it would be difficult to show how she could do this intentionally.

This is merely a constitutional formulation of the knowledge account of assertion in multi-agent systems: Everyone must know whatever she asserts. We may take this as a basic assumption.

It is worth noting that the knowledge account of assertion, as its original version shows, is a normative rule in character. Recall Williamson's formulation ([27]: 494, [28]: 243):

(The knowledge rule) One must: assert p only if one knows p.

The 'must' here is used in a normative sense. In practice, it happens occasionally that someone might violate it, so does Williamson admit ([27]: 511). Bearing this normative sense in mind, the assertion account of common knowledge shows ideally that assertion normally produces common knowledge. The present work intends to take this as an assumption for the construction of the desired models.[8]

Assumption 2 (LKA) $\mathcal{A}_i\,\varphi \rightarrow \mathcal{K}_i\mathcal{K}_i\,\varphi$ *(The luminosity of self-knowledge over assertion:)*

We have already noted that, although the well-known KK principle (i.e., $\mathcal{K}_i\,\varphi \rightarrow \mathcal{K}_i\mathcal{K}_i\,\varphi$) fails to hold in knowledge-first epistemology, the luminosity of self-knowledge over assertion holds: when the agent asserts φ, she must already know that she knows φ.

Assumption 3 (PC) $\mathcal{A}_i\,\varphi \rightarrow \mathcal{K}_i\mathcal{K}_j\,\varphi\ (i \neq j)$ *(Principle of Charity)*

When an agent asserts something, she knows that all others (hearers) must know what she asserts, if the very assertion guarantees the success of intended communication. One can easily find that this is merely an application of the well-known Principle of Charity, typically in Davidson's program of radical interpretation. Clearly, this assumption highlights the Davidsonian way of acquiring knowledge of other minds.

Assumption 4 (TK) $\mathcal{A}_i\,\varphi \rightarrow \mathcal{K}_j\,\varphi$, for all $j \in G$ and $i \neq j$ *(Transmission of knowledge).*

This is merely a logical consequence of Assumption 3. As knowing is factive, so $\mathcal{K}_i\mathcal{K}_j\,\varphi \rightarrow \mathcal{K}_j\,\varphi$; also given $\mathcal{A}_i\,\varphi \rightarrow \mathcal{K}_i\mathcal{K}_j\,\varphi$, $\mathcal{A}_i\,\varphi \rightarrow \mathcal{K}_j\,\varphi$ follows. Accordingly, once an assertion has been made by an agent, ideally all others (the hearers) must know whatever the agent asserts. This to a certain extent justifies the claim that assertion aims at sharing knowledge.

Assumption 5 (OA) $\mathcal{A}_i\,\varphi \rightarrow \mathcal{K}_j\mathcal{A}_i\,\varphi$, for all $j \in G$ and $i \neq j$ *(Observability of assertion).*

---

[8] I am indebted to an anonymous referee for reminding me of making this remark to show explicitly the implication of the normative character of the knowledge rule of assertion, and its impact on the acquisition of common knowledge. Bearing this in mind, misgivings over $\mathcal{A}_i\,\varphi \rightarrow \mathcal{K}_i\,\varphi$ could be put aside.

Since assertion is a kind of overtly observable speech act, when an agent makes an assertion, ideally all others know immediately and spontaneously that she makes an assertion. It is then beyond reasonable doubt to maintain that Assumption 5 together with Assumption 3 indicates that assertion guarantees successful communication.

At this stage, one can see clearly that the knowledge account of assertion in a multi-human-agent system can explain the difference of the acquisition of three varieties of knowledge. First, Assumption 1 (i.e., $\mathcal{A}_i\,\varphi \rightarrow \mathcal{K}_i\,\varphi$) and Assumption 4 (i.e., $\mathcal{A}_i\,\varphi \rightarrow \mathcal{K}_j\,\varphi$) show that everyone in G can acquire the propositional content of φ, typically a piece of factual knowledge, via an assertion made by some agent. For convenience, we may introduce an extra modal operator '$\mathcal{E}$' to signify the *universal knowledge* of φ—'Everyone knows φ' by '$\mathcal{E}\varphi$.' Thus, $\mathcal{A}_i\,\varphi \rightarrow \mathcal{E}\varphi$ holds. Furthermore, Assumption 2 (i.e., $\mathcal{A}_i\,\varphi \rightarrow \mathcal{K}_i\mathcal{K}_i\,\varphi$) shows that self-knowledge can be guaranteed by assertion. Finally, the acquisition of knowledge of other minds can be justified by Assumption 3 (i.e., $\mathcal{A}_i\,\varphi \rightarrow \mathcal{K}_i\mathcal{K}_j\,\varphi$) and Assumption 5 (i.e., $\mathcal{A}_i\,\varphi \rightarrow \mathcal{K}_j\mathcal{A}_i\,\varphi$); hence $\mathcal{A}_i\,\varphi \rightarrow \mathcal{K}_j\mathcal{K}_i\,\varphi$.

Some remarks should be made. So far, one may notice that in speaking of $\mathcal{A}_i\,\varphi \rightarrow \mathcal{E}\varphi$, it does not matter who the speaker is: no matter who asserts φ, $\mathcal{E}\varphi$ always holds. An assertion always renders a universal knowledge. To cope with this fact, we may stipulate that the formula '$\mathcal{A}\varphi$' means that someone asserts φ, or more briefly, 'φ is an assertion to the group G,' or 'φ is asserted knowledge to the group G.' Accordingly, we would have $\mathcal{A}\varphi \rightarrow \mathcal{E}\varphi$.

Following the aforementioned assumptions, we can easily get $\mathcal{A}_i\,\varphi \rightarrow \mathcal{E}\mathcal{E}\varphi$ as well, apart from $\mathcal{A}_i\,\varphi \rightarrow \mathcal{E}\varphi$. Again it would be arbitrary whoever makes the assertion, we then have $\mathcal{A}\varphi \rightarrow \mathcal{E}\mathcal{E}\varphi$. It would be then tempting to generalize this result to the extent that given an assertion of φ, if $\mathcal{A}\varphi \rightarrow \mathcal{E}^k\,\varphi$ holds, so would $\mathcal{A}\varphi \rightarrow \mathcal{E}^{k+1}\,\varphi$. Now, recall that we are searching for some kind of epistemic modality $X\varphi$ such that $\mathcal{C}\varphi \rightarrow X\varphi$ and $X\varphi \rightarrow \mathcal{E}^n\,\varphi$ hold for any arbitrary finite n. Of course at this stage, we need to introduce into the language an extra modal operator '$\mathcal{C}$' for 'common knowledge.' Now, if the above generalization can be justified, we can show, by a simple application of induction, that $\mathcal{A}\varphi \rightarrow \mathcal{E}^n\,\varphi$ hold for any arbitrary finite $n \in \mathbb{N}$. We then would have $\mathcal{A}\varphi \rightarrow \varphi \wedge \mathcal{E}\varphi \wedge \mathcal{E}\mathcal{E}\varphi \wedge \ldots \mathcal{E}^n\,\varphi \wedge \ldots$ *ad infinitum*. If it can be shown at the same time that $\mathcal{C}\varphi \rightarrow \mathcal{A}\varphi$ holds, $\mathcal{C}\varphi \leftrightarrow \mathcal{A}\varphi$ follows straightforwardly. This would serve as the required characterization of common knowledge.

Nonetheless, a justification of the aforementioned generalization, i.e., from $\mathcal{A}\varphi \rightarrow \mathcal{E}^n\,\varphi$ to $\mathcal{A}\varphi \rightarrow \mathcal{E}^{n+1}\,\varphi$ is tantamount to the acceptance of an application of the Axiom 4 in modal logic to universal knowledge (that is, $\mathcal{E}\varphi \rightarrow \mathcal{E}\mathcal{E}\varphi$). Since we have shown that Axiom 4 fails to hold in knowledge-first epistemology, it would not hold in the logic of knowledge and the knowledge account of assertion. So we cannot derive $\mathcal{A}\varphi \rightarrow \mathcal{E}^{k+1}\,\varphi$ from $\mathcal{A}\varphi \rightarrow \mathcal{E}^k\,\varphi$, although we do have $\mathcal{A}\varphi \rightarrow \mathcal{E}\varphi$ and $\mathcal{A}\varphi \rightarrow \mathcal{E}\mathcal{E}\varphi$. A seemingly promising attempt perhaps is to put forth a more general assumption of the luminosity of assertion such that $\mathcal{A}\varphi \rightarrow \mathcal{E}\mathcal{A}\varphi$ holds. If so, then given both $\mathcal{A}\varphi \rightarrow \mathcal{E}\mathcal{A}\varphi$ and $\mathcal{A}\varphi \rightarrow \mathcal{E}^k\,\varphi$, $\mathcal{A}\varphi \rightarrow \mathcal{E}^{k+1}\,\varphi$ would follow straightforwardly (just a routine deduction in propositional modal logic). Intuitively, this seems appealing because we have already had Assumption 5, i.e., $\mathcal{A}_i\,\varphi \rightarrow \mathcal{K}_j\mathcal{A}_i\,\varphi$. Nonetheless, we are in no position to claim that $\mathcal{A}_i\,\varphi \rightarrow \mathcal{K}_i\mathcal{A}_i\,\varphi$ holds as well, though assertion is a kind of intentional action—one might not know that one is making an assertion. In other words, while complete transparency, or luminosity, holds for common knowledge simultaneously and immediately, assertion would not. Be that as it may, we would have $\mathcal{C}\varphi \leftrightarrow \mathcal{A}\varphi$ as the desired characterization of common knowledge. But this would not be acceptable simply because this would give rise to the collapse of common knowledge to assertion: whatever is asserted becomes common knowledge, and vice versa.

Interestingly, one may find that our discussion so far suggests a much more appealing way out. The problem of deriving $\mathcal{A}\varphi \rightarrow \mathcal{E}^{k+1}\varphi$ from $\mathcal{A}\varphi \rightarrow \mathcal{E}^{k}\varphi$ lies in the consideration that one may not know that one makes an assertion; hence $\mathcal{A}\varphi \rightarrow \mathcal{E}\mathcal{A}\varphi$ fails to hold. However, one can easily find that although in a multi-human-agent system assertion per se cannot be transparent, the transparency of universal knowledge of assertion appears to be beyond reasonable doubt. That is, whenever someone asserts φ, and if everyone knows the fact that someone asserts φ, then everyone knows that everyone knows this fact, in symbols $(\mathcal{A}\varphi \wedge \mathcal{E}\mathcal{A}\varphi) \rightarrow \mathcal{E}\mathcal{E}\mathcal{A}\varphi$. This is substantially a weakened form of $\mathcal{A}\varphi \rightarrow \mathcal{E}\mathcal{A}\varphi$ resulting from adding the information that everyone already knows that someone makes an assertion of φ to the antecedent. Since $\mathcal{A}\varphi$ is already implied by $\mathcal{E}\mathcal{A}\varphi$, we may just formulate this as '$\mathcal{E}\mathcal{A}\varphi \rightarrow \mathcal{E}\mathcal{E}\mathcal{A}\varphi$'. Let us take this as a extra basic assumption, referred to as the Luminosity of Universal Knowledge of Assertion in multi-agent systems:

Assumption 6 (LUKA) $\mathcal{E}\mathcal{A}\varphi \rightarrow \mathcal{E}\mathcal{E}\mathcal{A}\varphi$ *(Luminosity of universal knowledge of assertion)*

The Assumption 6 has an important consequence in that it improves all agents' epistemic state of φ from $\mathcal{E}^{k}\varphi$ to $\mathcal{E}^{k+1}\varphi$, given that φ is asserted. It thus paves a way to get the desired result—given that $\mathcal{E}\mathcal{A}\varphi \rightarrow \mathcal{E}^{k}\varphi$ for any arbitrary $k \in \mathbb{N}$, $\mathcal{E}\mathcal{A}\varphi \rightarrow \mathcal{E}^{k+1}\varphi$ holds as well. Hence, $\mathcal{E}\mathcal{A}\varphi \rightarrow \mathcal{E}^{n}\varphi$ holds for any arbitrary $n \in \mathbb{N}$. We can thereby have:

(*) $\mathcal{E}\mathcal{A}\varphi \rightarrow \varphi \wedge \mathcal{E}\varphi \wedge \mathcal{E}\mathcal{E}\varphi \wedge \ldots \mathcal{E}^{n}\varphi \wedge \ldots$ *ad infinitum*.

Now, as we may treat $\mathcal{C}\varphi \leftrightarrow \varphi \wedge \mathcal{E}\varphi \wedge \mathcal{E}\mathcal{E}\varphi \wedge \ldots \mathcal{E}^{n}\varphi \wedge \ldots$ *ad infinitum* as a pre-theoretic characterization of common knowledge, it is appealing to take $\mathcal{E}\mathcal{A}\varphi$ as the kind of epistemic modality $\mathcal{X}\varphi$, provided we can further show that $\mathcal{C}\varphi \rightarrow \mathcal{E}\mathcal{A}\varphi$. That is to say, both $\mathcal{C}\varphi \rightarrow \mathcal{E}\mathcal{A}\varphi$ and $\mathcal{E}\mathcal{A}\varphi \rightarrow \mathcal{E}^{n}\varphi$ hold for any arbitrary $n \in \mathbb{N}$. We can then take the equivalency $\mathcal{C}\varphi \leftrightarrow \mathcal{E}\mathcal{A}\varphi$, as the desired characterization of common knowledge. However, sometimes we may want to notify explicitly that when everyone knows that someone asserts φ, everyone knows φ automatically and spontaneously. To formulate this explicitly, we shall write '$\mathcal{E}\varphi \wedge \mathcal{E}\mathcal{A}\varphi$' instead of '$\mathcal{E}\mathcal{A}\varphi$.' For simplicity, we may write as '$\mathcal{E}(\varphi \wedge \mathcal{A}\varphi)$' instead. Thus, we can take $\mathcal{E}(\varphi \wedge \mathcal{A}\varphi)$ as the required epistemic modality $\mathcal{X}\varphi$ such that not only $\mathcal{C}\varphi \rightarrow \mathcal{E}(\varphi \wedge \mathcal{A}\varphi)$ holds but also $\mathcal{E}(\varphi \wedge \mathcal{A}\varphi) \rightarrow \mathcal{E}^{n}\varphi$ holds for any arbitrary $n \in \mathbb{N}$. The required characterization of common knowledge can be then formulated by the following equivalence:

(CKA) $\mathcal{C}\varphi \leftrightarrow \mathcal{E}(\varphi \wedge \mathcal{A}\varphi)$.

In words, the propositional content of a formula φ is common knowledge to a group of agents G if and only if everyone knows φ and also everyone knows that someone asserts φ.

What remains is to show that (CKA) can be explicitly justified in the framework of a multi-agent system of the epistemic logic of common knowledge with the knowledge account of assertion. Of course, the required models in such a framework, referred to as TWC-models, will be substantially extensions of TWA-models for a multi-agent system.

12.6 TWC-Models and the Assertion Account of Common Knowledge

12.6.1 TWC-models

First, the language in use $\mathcal{L}_C$ is defined as $\varphi ::= p | \neg \varphi | \varphi \rightarrow \psi | \mathcal{K}_i \varphi | \mathcal{A}_i \varphi | \mathcal{E}\varphi | \mathcal{A}\varphi | \mathcal{C}\varphi$ (for all $i \in G$). As usual, other logical connectives can be introduced in the standard way. A TWC-model for a multi-agent system can be obtained from a TWA-model described above by replacing the functions δ and λ in a TWA-model with a pair of functions δ_i and λ_i, for each individual agent $i$in G. That is, a TWC-model is a tuple of the form

$$\mathcal{M} = \langle S, R, \{\delta_i\}_{i \in n}, \{\lambda_i\}_{i \in n}, V^P \rangle,$$

S, a nonempty set of states;
$R \subseteq S \times S$, a partial ordering with reflexivity and transitivity as the required accessibility relation on S;
$V^P : P \rightarrow 2^S$, a valuation, assigning to each $p \in P$, a set $V^P(p) \subseteq S$ of states in which p is true. When a state s is in $V^P(p)$, we say that V^P assigns p a truth value 'True,' or more straightforwardly, V^Pmakes p true in s;
$\delta_i : S \rightarrow \wp(\mathcal{L}_C)$ with some more conditions to be specified later;
$\lambda_i : S \rightarrow \wp(\mathcal{L}_C)$ with some more conditions to be specified later.

This completes our construction of TWC-models. Clearly, in TWC-models, a sole accessibility relation is posited in order to specify the so-called 'nearby cases' in a more metaphysical sense, while the set of all epistemic possibilities for an agent i in a given state s is to be identified by virtue of the function δ_i in that $\varphi \in \delta_i(s)$ indicates that the agent *is actually in a position to know* φ in s. And $\varphi \in \lambda_i(s)$ indicates that the agent i has the intention of asserting φ in s. Thus, we need neither assume the existence of a set of accessibility relations, nor would we need a group accessibility relation R_G.

We then put forth some extra conditions on $\{\delta_i\}_{i \in n}$ and $\{\lambda_i\}_{i \in n}$ so that all basic assumptions of the knowledge account of assertion in multi-agent systems, i.e., Assumption 1–6, can be validated.

Condition 1 (S-KAA) If $\varphi \in \lambda_i(s)$, then $\varphi \in \delta_i(s)$ (The knowledge account of assertion).

One has the intention of asserting φ only if one is actually in a position to know φ. Clearly, this condition is sufficient to validate Assumption 1, i.e., $\mathcal{A}_i\, \varphi \rightarrow \mathcal{K}_i\, \varphi$.

Condition 2 (S-LKA) If $\varphi \in \lambda_i(s)$, then $\mathcal{K}_i\, \varphi \in \delta_i(s)$ (The luminosity of self-knowledge over assertion).

One has the intention of asserting φ only if one is actually in a position to know that one knows φ. This is to validate Assumption 2, i.e., $\mathcal{A}_i\, \varphi \rightarrow \mathcal{K}_i \mathcal{K}_i\, \varphi$.

Condition 3 (S-PC) If $\varphi \in \lambda_i(s)$, then for all $j \in G$ and $i \neq j$, $\mathcal{K}_j\, \varphi \in \delta_i(s)$ (The Principle of Charity).

When one has the intention of asserting φ, not only must one be actually in a position to know φ, more importantly, one must assume that the others are also actually in a position to know φ. Otherwise, one would not make such an assertion. This is a prerequisite for success of communication by assertion. And so this would validate Assumption 3 (The Principle of Charity), i.e., $\mathcal{A}_i\, \varphi \rightarrow \mathcal{K}_i \mathcal{K}_j\, \varphi$, for all $j \in G$ and $i \neq j$.

Condition 4 (S-TK) If $\varphi \in \lambda_i(s)$, then for all $j \in G$ and $i \neq j$, $\varphi \in \delta_j(s)$ (Transmission of knowledge).

One has the intention of asserting φ, only if one takes it for granted that all others are actually in a position to know φ. Hence, once the very assertion of φ is performed (i.e., $\mathcal{A}_i\, \varphi$ holds), $\mathcal{K}_j\, \varphi$ holds simultaneously. This condition thereby guarantees the transmission of knowledge from an agent to others. We may then have: $\mathcal{A}_i\, \varphi \rightarrow \mathcal{K}_j\, \varphi$

Condition 5 (S-OA) If $\varphi \in \lambda_i(s)$, then for all $j \in G$ and $i \neq j$, $\mathcal{A}_i\, \varphi \in \delta_j(s)$ (Observability of assertion).

One has the intention of asserting φ, only if all other agents are actually in a position to know that one asserts φ. This is simply due to the basic assumption that assertion is a kind of overtly observable speech act, and hence ideally guarantees communication in a group of agents. Accordingly, Assumption 5 (i.e., $\mathcal{A}_i\, \varphi \rightarrow \mathcal{K}_j \mathcal{A}_i\, \varphi$) is validated in TWC-models.

Condition 6 (S-LUKA) If $\varphi \in \lambda_i(s)$, then, if for all $l \in G$, $\mathcal{A}_i\, \varphi \in \delta_l(s)$, then $\mathcal{E}\mathcal{A}_i\, \varphi \in \delta_l(s)$ (Luminosity of universal knowledge of assertion).

This condition will validate Assumption 6, i.e., $\mathcal{E}\mathcal{A}\, \varphi \rightarrow \mathcal{E}\mathcal{E}\mathcal{A}\, \varphi$.

Having specified these conditions for the construction of TWC-models, let us now turn our attention to the details of semantics.

12.6.2 The Semantics

Based on TWC-models, the semantic rules will be stipulated as follows:

(p)	$\mathcal{M},s \models p$	iff	V makes p true at s.
(N_{eg})	$\mathcal{M},s \models \neg\varphi$	iff	It is not the case that $\mathcal{M},s \models \varphi$
(Imp)	$\mathcal{M},s \models \varphi \rightarrow \psi$	iff	either it is not the case that $\mathcal{M},s \models \varphi$ or it is the case that $\mathcal{M}, s \models \psi$.
$(\mathcal{K}_i)$	$\mathcal{M},s \models \mathcal{K}_i\,\varphi$	iff	$\forall t \in S(Rst \rightarrow \mathcal{M},t \models \varphi) \wedge \varphi \in \delta_i(s)$.
$(\mathcal{E})$	$\mathcal{M},s \models \mathcal{E}\varphi$	iff	$\forall i \in G \rightarrow \mathcal{M},s \models \mathcal{K}_i\,\varphi$
$(\mathcal{A}_i)$	$\mathcal{M},s \models \mathcal{A}_i\,\varphi$	iff	$\forall t \in S(Rst \rightarrow \mathcal{M},t \models \mathcal{K}_i\,\varphi) \wedge \varphi \in \lambda_i(s) \wedge \mathcal{K}_i\,\varphi \in \delta_i(s)$.
$(\mathcal{A})$	$\mathcal{M},s \models \mathcal{A}\varphi$	iff	$\exists i \in G \wedge \mathcal{M},s \models \mathcal{A}_i\,\varphi$
$(\mathcal{C})$	$\mathcal{M},s \models \mathcal{C}\varphi$	iff	$\exists i \in G\ (\forall t \in S(Rst \rightarrow \mathcal{M},t \models \mathcal{A}_i\,\varphi) \wedge \forall l \in G \rightarrow (\varphi \in \delta_l(s) \wedge \mathcal{A}_i\,\varphi \in \delta_l(s))$.

12.6.3 Basic Assumptions Validated and Some Other Results

All basic assumptions Assumption 1–6 can be validated in TWC-models by checking the semantic rules and the aforementioned conditions. We then have the following theorem:

Theorem 1 *The following statements of implication hold in the class of TWC-models:*

1	$\models \mathcal{A}_i\,\varphi \rightarrow \mathcal{K}_i\,\varphi$	*(Assumption 1 by S-KAA)*
2	$\models \mathcal{A}_i\,\varphi \rightarrow \mathcal{K}_i\mathcal{K}_i\,\varphi$	*(Assumption 2 by S-LKA)*
3	$\models \mathcal{A}_i\,\varphi \rightarrow \mathcal{K}_j\,\varphi,\ \forall j \in G \wedge i \neq j$	*(Assumption 6 by S-TK).*
4	$\models \mathcal{A}_i\,\varphi \rightarrow \mathcal{E}\varphi$	*(from 1 and 3)*
5	$\models \mathcal{A}_i\,\varphi \rightarrow \mathcal{K}_i\mathcal{K}_j\,\varphi \forall j \in G \wedge i \neq j$	*(Assumption 3 by S-PC).*
6	$\models \mathcal{A}_i\,\varphi \rightarrow \mathcal{K}_i\mathcal{E}\varphi$	*(from 2 and 5)*
7	$\models \mathcal{A}_i\,\varphi \rightarrow \mathcal{K}_j\mathcal{A}_i\,\varphi$	*(Assumption 5 by S-OA)*
8	$\models \mathcal{A}_i\,\varphi \rightarrow \mathcal{K}_j\mathcal{E}\varphi$	*(from 7 and 4)*
9	$\models \mathcal{A}_i\,\varphi \rightarrow \mathcal{E}\mathcal{E}\varphi$	*(from 6 and 8)*
10	$\models \mathcal{E}\mathcal{A}\varphi \rightarrow \mathcal{E}\mathcal{E}\mathcal{A}\varphi$	*(Assumption 6 by S-LUKA)*
11.	$\models \mathcal{K}_i\,\varphi \rightarrow \varphi$	*(Factivity of knowledge)*

12.6.4 A Justification of (CKA)

Having shown that all basic assumptions are valid in the constructed TWC-models, (CKA) $\mathcal{C}\varphi \leftrightarrow \mathcal{E}(\varphi \wedge \mathcal{A}\varphi)$ can be justified easily. But justification of two lemmas should be helpful:

Lemma 2 $\models \mathcal{C}\varphi \rightarrow \mathcal{E}(\varphi \wedge \mathcal{A}\varphi)$

Lemma 3 $\models \mathcal{E}(\varphi \wedge \mathcal{A}\varphi) \rightarrow \mathcal{C}\varphi$

A justification of Lemma 2, $\mathcal{C}\varphi \rightarrow \mathcal{E}(\varphi \wedge \mathcal{A}\varphi)$, is quite straightforward. In fact, the desired result immediately follows from the semantic rules for $\mathcal{C}\varphi$, $\mathcal{E}\varphi$, $\mathcal{A}_i\,\varphi$ and $\mathcal{A}\varphi$. Here, the Distributive Law—$\mathcal{E}(\varphi \wedge \mathcal{A}\varphi) \leftrightarrow (\mathcal{E}\varphi \wedge \mathcal{E}\mathcal{A}\varphi)$—is required.

A justification of Lemma 3, $\mathcal{E}(\varphi \wedge \mathcal{A}\varphi) \rightarrow \mathcal{C}\varphi$, is a bit more complicated. We take it for granted that the pre-theoretic equivalence of common knowledge with (C_{iter}) holds, that is,

(C1) $\mathcal{C}\varphi \leftrightarrow \varphi \wedge \mathcal{E}\varphi \wedge \mathcal{E}\mathcal{E}\varphi \wedge \ldots \mathcal{E}^n\,\varphi \wedge \ldots$ *ad infinitum*

So to justify Lemma 3, all that is required at the core is to show that

(C2) $\mathcal{E}(\varphi \wedge \mathcal{A}\varphi) \rightarrow \varphi \wedge \mathcal{E}\varphi \wedge \mathcal{E}\mathcal{E}\varphi \wedge \ldots \mathcal{E}^n\,\varphi \wedge \ldots$ *ad infinitum*

Intuitively, (C2) can be reformulated as

(C2*) $\mathcal{E}(\varphi \wedge \mathcal{A}\varphi) \rightarrow \mathcal{E}^0\,\varphi \wedge \mathcal{E}^1\,\varphi \wedge \mathcal{E}^2\,\varphi \wedge \ldots \wedge \mathcal{E}^n\,\varphi \wedge \mathcal{E}^{n+1}\,\varphi \wedge \ldots$ *ad infinitum*

which can be justified by showing that all of the following implications hold:

(C2-1) $\mathcal{E}(\varphi \wedge \mathcal{A}\varphi) \rightarrow \mathcal{E}^0\,\varphi\ (=\varphi)$
(C2-2) $\mathcal{E}(\varphi \wedge \mathcal{A}\varphi) \rightarrow \mathcal{E}^1\,\varphi\ (=\mathcal{E}\varphi)$
(C2-3) $\mathcal{E}(\varphi \wedge \mathcal{A}\varphi) \rightarrow \mathcal{E}^2\,\varphi\ (= \mathcal{E}\mathcal{E}\varphi)$
⋮ ⋮
(C2-*n*) $\mathcal{E}(\varphi \wedge \mathcal{A}\varphi) \rightarrow \mathcal{E}^n\,\varphi$
(C2-*n*+1) $\mathcal{E}(\varphi \wedge \mathcal{A}\varphi) \rightarrow \mathcal{E}^{n+1}\,\varphi$
⋮ ⋮

Obviously, (C2-1), (C2-2), and (C2-3) can be proved easily from (1), (4), (9), and (11). To justify the cases when $n \geq 2$, (C2*) suggests that this can be justified by induction on the number of the iterated $\mathcal{E}$. Since the basic step has been done, all that is required is to show the inductive step holds as well, i.e., to show that given $\mathcal{E}(\varphi \wedge \mathcal{A}\varphi) \rightarrow \mathcal{E}^n\,\varphi$, $\mathcal{E}(\varphi \wedge \mathcal{A}\varphi) \rightarrow \mathcal{E}^{n+1}\,\varphi$ holds. Since $\mathcal{E}\varphi \rightarrow \mathcal{E}\mathcal{E}\varphi$ would not hold in general, and so when $n \geq 2$, we cannot get $\mathcal{E}\varphi \rightarrow \mathcal{E}^{n+1}\,\varphi$ directly from $\mathcal{E}\varphi \rightarrow \mathcal{E}^n\,\varphi$. Instead, we have to show that

(+) If $\mathcal{E}\mathcal{A}\varphi \rightarrow \mathcal{E}^n\,\varphi$ then $\mathcal{E}\mathcal{A}\varphi \rightarrow \mathcal{E}^{n+1}\,\varphi$.

Clearly, by Assumption 6, hence (10), in any state s, if both $\mathcal{A}\varphi$ and $\mathcal{E}\mathcal{A}\varphi$ hold, then $\mathcal{E}\mathcal{E}\mathcal{A}\varphi$ holds as well because $\mathcal{E}\mathcal{A}\varphi \rightarrow \mathcal{E}\mathcal{E}\mathcal{A}\varphi$. Now, given, as the hypothesis of induction, $\mathcal{E}\mathcal{A}\varphi \rightarrow \mathcal{E}^n\,\varphi$, we do have $\mathcal{E}\mathcal{A}\varphi \rightarrow \mathcal{E}\mathcal{E}^n\,\varphi$. Hence, $\mathcal{E}\mathcal{A}\varphi \rightarrow \mathcal{E}^{n+1}\,\varphi$.

This completes the induction of the desired justification; hence the justification of (C2). Now, an equivalency follows immediately from Lemmas 2 and 3, that is

Theorem 4 *(CKA)* $\mathcal{C}\varphi \leftrightarrow \mathcal{E}(\varphi \wedge \mathcal{A}\varphi)$.

This is precisely the desired characterization of common knowledge in terms of the knowledge account of assertion. We may call this the assertion account of common knowledge, for short.

Corollary 5 $\mathcal{C}\varphi \rightarrow \mathcal{E}\mathcal{C}\varphi$

This can be easily proved by showing that $\mathcal{E}(\varphi \wedge \mathcal{A}\varphi) \rightarrow \mathcal{E}\mathcal{E}(\varphi \wedge \mathcal{A}\varphi)$. This corollary shows that the complete transparency, or the luminosity, of common knowledge, an important intrinsic property of common knowledge, can be also justified on the assertion account of common knowledge.

12.7 Concluding Remarks

(CKA) as the proposed characterization of common knowledge has several merits. First of all, we have to posit nothing but one single accessibility relation to specify all nearby cases, or accessible states, for all agents in the fixed group. Second, as the required framework shows clearly that communication is guaranteed by assertion, the proposed characteristics of the knowledge account of assertion in multi-agent systems will clarify the relations among individual knowledge, universal knowledge, and common knowledge. Typically, it illuminates the transition of individual knowledge to common knowledge, as well as the transmission of individual knowledge among agents in a given group. The proposed basic assumptions also explain the difference in the ways of attaining the three varieties of knowledge. Finally, the luminosity of common knowledge can be justified easily as well. Note that it is an easy routine to justify $\mathcal{C}\varphi \rightarrow \mathcal{E}\mathcal{C}\varphi$ by $\mathcal{E}(\varphi \wedge \mathcal{A}\varphi) \rightarrow \mathcal{E}\mathcal{E}(\varphi \wedge \mathcal{A}\varphi)$. Interestingly, the so-called 'complete transparency,' or luminosity, of universal knowledge of assertion can pave a way to justify the complete transparency of common knowledge. We can then conclude that for human agents in ordinary discourse, we do have common knowledge if the knowledge account of assertion, taken as a kind of overtly observable speech act, is granted.

Some may notice that the role that the knowledge account of assertion plays in the framework we proposed here has a very close kinship with that of a public announcement in multi-agent systems. It would be very helpful to give a detailed discussion on public announcement and then put forth some substantial comparison and contrast between the proposed assertion account of common knowledge and some logics of public announcement. However, it strikes me that we could have more substantial comparisons, only when an axiomatization, based on the proposed semantics, of the epistemic logic of common knowledge with the knowledge account of assertion is available. We may then be able to address the comparison and contrast among these logics. But for the sake of the limitation of the space, we can only present such an axiomatization somewhere else. The interested reader is referred to van Ditmarsch et al. ([26], especially Chap. 4) for further information and discussion about the logic of public announcement and its semantics.

One final remark has something to do with the methodology I adopt. In a sense, my approach follows a guideline, which is something like what Hendricks and Symons ([16]: 150) suggested: 'Find a plausible epistemological story to tell about the semantics of epistemic logic.' Some might think that the appeal to the knowledge account of assertion to characterize the very notion of common knowledge would complicate the construction of the required models. But to be compared with the commitment to circularity or to an infinite conjunction of the formulation involved in the orthodox treatment, I believe that our approach is less costly, but more satisfactory nonetheless.

Acknowledgments An earlier version of this chapter was presented at the Second Taiwan Philosophical Colloquium (TPLC-2014). I thank the participants and audience for valuable criticism and remarks. I am grateful to anonymous referees for their helpful comments and suggestions. This paper is one of a series of papers working on the construction of models to justify Williamson's main theses in his knowledge-first epistemology. I am deeply indebted to Robert Goldblatt, Kok Yong Lee, Joshua Sack, and Tomoyuki Yamada for their discussion and comments on this paper and previous work. This paper is partly supported by a grant from the Ministry of Science and Technology (MOST), Taiwan, No.: MOST 103-2410-H-002-174-MY3

References

1. Aumann, R.J.: Agreeing to disagree. Annuals Stat. **4**, 1236–1239 (1976)
2. Baltag, A., Moss, L.S., Solecki, S.: The logic of public announcements, common knowledge, and private suspicions. In: TARK'98: Proceedings of the 7th Conference on Theoretical Aspects of Rationality and Knowledge, pp. 43–56. Morgan Kaufmann Publishers, San Francisco (1998)
3. Barwise, J.: Three views of common knowledge. In: TARK'88: Proceedings of the 2nd Conference on Theoretical Aspects of Reasoning about Knowledge, pp. 365-379. Morgan Kaufmann Publishers, San Francisco (1988)
4. Clark, H.H., Marshall, C.R.: Definite reference and mutual knowledge. In: Joshi, A.K., Webber, B.L., Sag, I.A. (Eds.) Elements of Discourse Understanding, pp. 10–63. Cambridge University Press, Cambridge (1981)
5. Davidson, D.: First person authority. In: Subjective, Intersubjective, Objective: Philosophical Essays Vol. 3, pp. 3-14. Clarendon Press, Oxford (1984/2001)
6. Davidson, D.: Knowing one's own mind. In: Subjective, Intersubjective, Objective: Philosophical Essays Vol. 3, pp. 15-38. Clarendon Press, Oxford (1987/2001)
7. Davidson, D.: Three varieties of knowledge. In: Subjective, Intersubjective, Objective: Philosophical Essays Vol. 3, pp. 205-220. Clarendon Press, Oxford (1988/2001)
8. Davidson, D.: Subjective, intersubjective, objective—Philosophical essays, vol. 3. Clarendon Press, Oxford (2001)
9. Fagin, R., Halpern, J.Y., Moses, Y., Vardi, M.Y.: Reasoning about knowledge. MIT Press, Cambridge, Mass. (1995)
10. Fagin, R., Halpern, J.Y., Moses, Y., Vardi, M.Y.: Common knowledge revisited. In: Hendricks, V.F., Jørgensen, K.F., Pedersen, S.A. (Eds.) Knowledge Contributors, pp. 87-104. Kluwer, Dordrecht (2003)
11. Gullvåg, I.: The logic of assertion. Theoria **44**, 75–116 (1978)
12. Gupta, A., Belnap, N.D.: The revision theory of truth. MIT Press, Cambridge, Mass. (1993)
13. Halpern, J.Y., Moses, Y.: Knowledge and common knowledge in a distributed environment. J. Assoc. Comput. Mach. **37**, 549–587 (1990)
14. Halpern, J.Y., Moses, Y.: A guide to completeness and complexity for modal logics of knowledge and belief. Artif. Intell. **54**, 319–379 (1992)
15. Hendricks, V.F., Jørgensen, K.F., Pedersen, S.A. (Eds.): Knowledge Contributors. Kluwer, Dordrecht (2003)
16. Hendricks, V.F., Symons, J.: Where's the bridge? Epistemology and epistemic logic. Philos. Stud. **128**, 137–167 (2006)

17. Hintikka, J.: Knowledge and belief: An introduction to the logic of the two notions. Cornell University Press, Ithaca (1962)
18. Lewis, D.: Convention: A philosophical study. Blackwell, Oxford (1969)
19. Lismont, L., Mongin, P.: On the logic of common belief and common knowledge. Theor. Decis. **37**, 75–106 (1994)
20. Lismont, L., Mongin, P.: Belief closure: A semantics of common knowledge for modal propositional logic. Math. Soc. Sci. **30**, 127–153 (1995)
21. Mertens, J.-F., Zamir, S.: Formulation of Bayesian analysis for games with incomplete information. Int. J. Game Theor. **14**, 1–29 (1985)
22. Milgrom, P.: An axiomatic characterization of common knowledge. Econometrica **49**, 219–222 (1981)
23. Monderer, D., Samet, D.: Approximating common knowledge with common beliefs. Game Econ. Behav. **1**, 170–190 (1989)
24. Rescher, N.: Topics in philosophical logic. Kluwer D. Reidel, Dordrecht (1968)
25. Tarski, A.: A lattice-theoretical fixpoint theorem and its applications. Pacific J. Math. **5**, 285–309 (1955)
26. van Ditmarsch, H., van der Hoek, W., Kooi, B.: Dynamic epistemic logic. Springer, Amsterdam (2008)
27. Williamson, T.: Knowing and asserting. Philos. Rev. **105**, 489–523 (1996)
28. Williamson, T.: Knowledge and its limits. Oxford University Press, Oxford (2000)
29. Yang, S.C.-M.: TW-models for logic of knowledge-cum-belief. In: Downey, R., Brendle, J., Goldblatt, R., Kim, B. (Eds.) The Proceedings of 12th Asian Logic Conference, pp. 314–337. World Scientific Publishing Co., Singapore (2013)
30. Yang, S.C.-M.: A defence of the knowledge account of assertion: From a model- theoretical perspective. In: Hung, T.-W. (Ed.) Communicative Action, pp. 33–50. Springer, Berlin (2014)

MIX
Papier aus verantwortungsvollen Quellen
Paper from responsible sources
FSC® C105338

If you have any concerns about our products,
you can contact us on
ProductSafety@springernature.com

In case Publisher is established outside the EU,
the EU authorized representative is:
Springer Nature Customer Service Center GmbH
Europaplatz 3, 69115 Heidelberg, Germany

Printed by Libri Plureos GmbH
in Hamburg, Germany